Molecular Virology: A Clinical Approach

Molecular Virology: A Clinical Approach

Editor: Orlando Johnston

MURPHY & MOORE
www.murphy-moorepublishing.com

www.murphy-moorepublishing.com

MURPHY & MOORE

Cataloging-in-Publication Data

Molecular virology : a clinical approach / edited by Orlando Johnston.
p. cm.
Includes bibliographical references and index.
ISBN 978-1-63987-376-0
1. Molecular virology. 2. Molecular microbiology. 3. Virology. I. Johnston, Orlando.
QR389 .M65 2022
579.2--dc23

Murphy & Moore Publishing,
1 Rockefeller Plaza,
New York City, NY 10020, USA

ISBN 978-1-63987-376-0 (Hardback)

Contents

Preface

The branch of virology which studies viruses at the molecular scale is called molecular virology. Viruses have a considerable level of biological diversity, and are able to parasitize and infect all kinds of life forms. On the basis of their replication strategy, viruses can be categorized into various classes such as single stranded DNA, double stranded DNA, single stranded RNA-positive sense and single stranded RNA-negative sense. Viruses can cause different types of infections in the hosts. A few such types of infections are acute infection, abortive infection, persistent infection and chronic infection. The study of the replication cycle of viruses is also undertaken within this field. This book contains some path-breaking studies in the field of molecular virology. It attempts to understand the multiple branches that fall under this discipline and how such concepts have clinical applications. Scientists and students actively engaged in this field will find this book full of crucial and unexplored concepts.

The researches compiled throughout the book are authentic and of high quality, combining several disciplines and from very diverse regions from around the world. Drawing on the contributions of many researchers from diverse countries, the book's objective is to provide the readers with the latest achievements in the area of research. This book will surely be a source of knowledge to all interested and researching the field.

In the end, I would like to express my deep sense of gratitude to all the authors for meeting the set deadlines in completing and submitting their research chapters. I would also like to thank the publisher for the support offered to us throughout the course of the book. Finally, I extend my sincere thanks to my family for being a constant source of inspiration and encouragement.

Editor

Section 1

Retrovirus - Molecular Biology and Pathogenesis

1

Molecular Biology and Pathogenesis of Retroviruses

Shailendra K. Saxena and Sai V. Chitti

Abstract

Retroviruses consist of a varied family of enveloped RNA viruses with positive-sense RNAs that replicate in a host cell through the process of reverse transcription. Retroviruses belong to the Retroviridae family that typically carries their genetic material in the form of ribonucleic acid, while the genetic material of their hosts is in the form of deoxyribonucleic acid. Infections with a number of retroviruses can lead to serious conditions, such as AIDS, a range of malignancies, neurological diseases, and added clinical conditions. In addition, some can even become integrated as DNA in the germ line and passed as endogenous viruses from generation to generation. Surprisingly, retroviruses do not appear to straightforwardly activate host innate defenses. On the other hand, attention in these viruses extends beyond their disease causing capabilities. For example, studies on the retroviruses led to the discovery of oncogenes, understanding of mechanisms that regulate eukaryotic gene expression, and these are proving to be valuable research tools in molecular biology and have been used successfully in gene therapy. The central goals of retrovirology today are the treatment and the prevention of human and non-human diseases and to use this virus in research.

Keywords: retrovirus, Retroviridae, reverse transcriptase, replication, immune responses, ART

1. Introduction

During the past few decades retrovirus has done an adequate amount of harm to the human life and became a big threat globally. These are the group of viruses that belong to the family Retroviridae and that typically carry their genetic material in the form of ribonucleic acid (RNA), while the genetic material of their hosts is in the form of deoxyribonucleic acid (DNA). Retroviruses are named for an enzyme known as reverse transcriptase (RT), which was discovered independently in 1971 by American virologists Howard Temin and David Baltimore for which they have received Nobel Prize in physiology and medicine in the year

1975. Retroviridae is a family of enveloped, obligate parasites with single-stranded positive-sense RNA (ssRNA) that replicate in a host cell through the process of reverse transcription. The activity of RT makes it feasible for genetic material from a retrovirus to become permanently integrated into the DNA genome (provirus) of an infected cell.

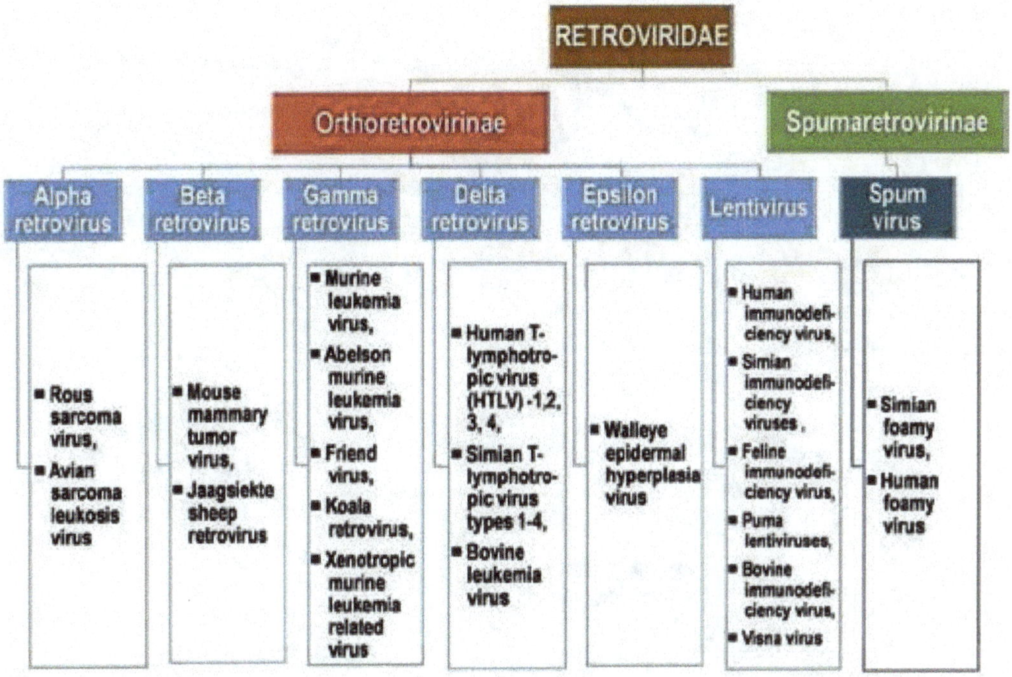

Figure 1. Classification of *Retroviridae* family of viruses.

Retroviridae is subdivided into *Orthoretrovirinae* and *Spumaretrovirinae* (Figure 1). Under *Orthoretrovirinae* the various genus are *Alpharetrovirus* (Rous sarcoma virus, avian sarcoma leukosis virus), *Betaretrovirus* (Mouse mammary tumor virus, Jaagsiekte sheep retrovirus), *Gammaretrovirus* (murine leukemia virus, Abelson murine leukemia virus, Friend virus, koala retrovirus, xenotropic murine leukemia-related virus), *Deltaretrovirus* (Human T-lymphotropic virus (HTLV) types 1–4, simian T-lymphotropic virus types 1–4, Bovine leukemia virus), *Epsilonretrovirus* (Walleye epidermal hyperplasia virus), and *Lentivirus* (human immunodeficiency virus (HIV), simian immunodeficiency viruses (SIV), feline immunodeficiency virus, puma lentiviruses, bovine immunodeficiency virus, caprine arthritis encephalitis virus, visna virus) are present, whereas under *Spumaretrovirinae* only one genus is present *spumavirus* (simian foamy virus, human foamy virus) [1]. The retroviruses host range include human, murine, feline (cat), avian (birds), and bovine (pig), and it is dependent upon the viral envelope, glycoproteins and structural proteins, involved in integration. Infections with a number of retroviruses can lead to serious conditions, such as AIDS, a range of malignancies, neurological diseases, and added clinical conditions [2]. In addition, some retroviruses can even become integrated as DNA in the germ line and passed as endogenous viruses from generation to generation. Using retrovirus in research has built up the need to advance the investigation in detail regarding the viral particles and genomes, their modes of replication, integration, and host immune evasion. The basic replication of retroviruses includes that (Figure 2) the ssRNA become double-stranded DNA (dsDNA) and gets into the host genetic material and employs host machinery for the synthesis of new virions.

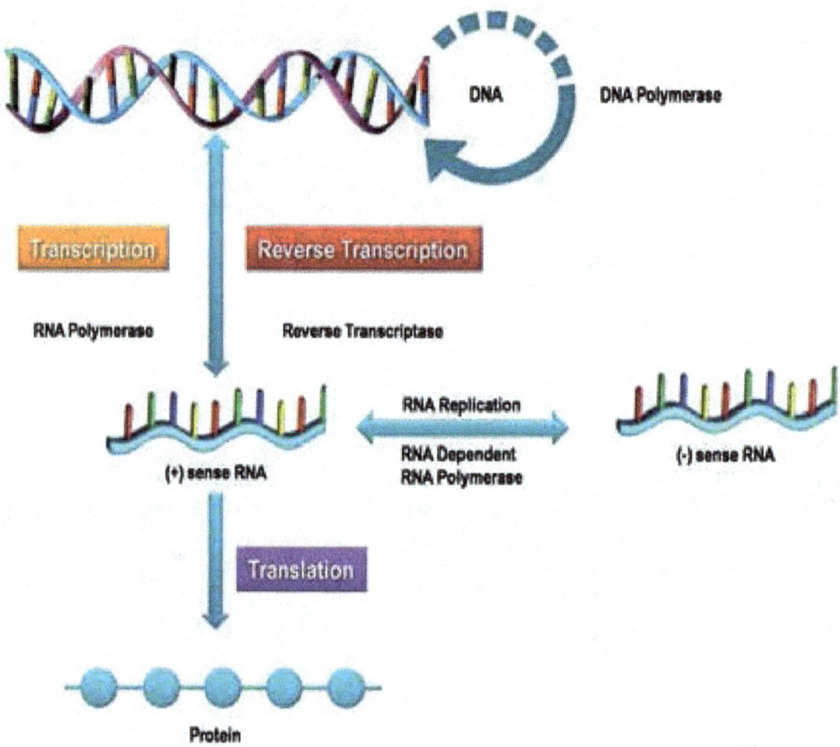

Figure 2. The reverse of Cricks' central dogma that occurs in retroviruses. RNA genome is converted by reverse transcriptase into double-stranded DNA, followed by integration into the host genome, transcription and translation of viral proteins occurs along with the host.

Interesting Points about Retroviruses

- Retroviruses contain RNA as genetic material but have DNA-dependent steps in their replication.
- Replicates via reverse transcription because of the presence of reverse transcriptase enzyme.
- Integrase transfers the viral DNA into the cell nucleus and viral dsDNA is covalently and randomly integrated into the cell's genome.
- Retroviruses that can transform host cells at high rates contain gene sequences such as viral oncogenes and proto-oncogenes.
- Human retroviruses can cause immune deficiencies, cancer, and neurological diseases.

2. Retrovirus structure, genome, and proteins

The typical retrovirus structure is enveloped, spherical to pleomorphic in shape, and they have diameter of 80–100 nm. The different genuses of retrovirus virions (Figure 3) have diverse morphology, but they have their same virion component, which includes the outer envelope coat, two copies of the genetic material, and the viral proteins. Envelope consists of lipids that are obtained from the host plasma membrane during budding process and the glycoprotein such as gp120 and gp41 in case of HIV [3]. The retroviral envelope serves three separate functions that includes the outer lipid bilayer protects from the extracellular environment, it also aids in the entry and way out of host cells through endosomal membrane trafficking, and the facility to straightforwardly enter cells by fusing with their membranes.

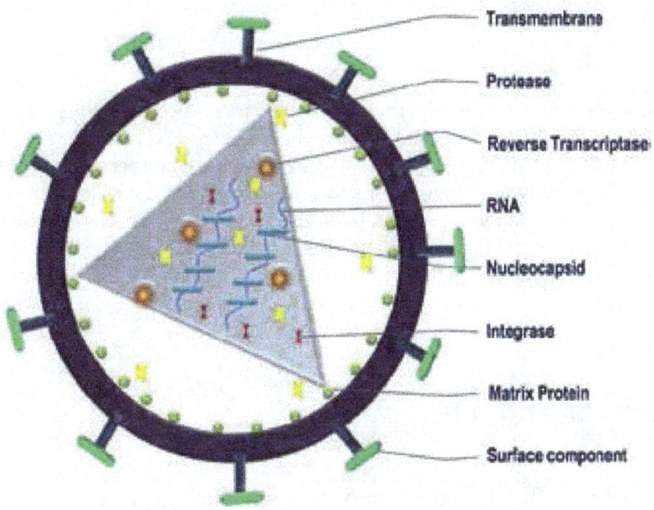

Figure 3. Schematic cross section through a retroviral particle: Showing retrovirus components.

The genome of retrovirus is monopartite, linear, dimeric, ssRNA (+) of about 8–10 kb, with a 5′-cap and a 3′poly-A tail (Figure 4). The group-specific gene (*gag*), *pol*, *pro*, envelope (*env*) genes are flanked between the R regions. The 5′-long terminal repeats (LTRs) consist of U3 (unique sequence), R primer binding site (PBS), and U5 regions. The 3′ end consists of a polypurine tract (PPT), U3, and R regions. The R region is a short repeated sequence at each end of the genome used during the reverse transcription to ensure correct end-to-end transfer in the growing chain. U5, on the other hand, is a short exceptional arrangement in the middle of R and PBS [4]. PBS consists of 18 bases corresponding to 3′ end of tRNA primer. L region is an untranslated leader region that gives the sign for packaging of the genome RNA. The retroviral protein includes gag, protease, pol, and env proteins. Gag is the primary retroviral structural protein responsible for orchestrating the majority of steps in viral assembly. Most of these assembly steps occur through interactions with three gag subdomains—matrix (MA), capsid (CA), and nucleocapsid (NC). The gag subdomains are structurally discrete but have functionally overlapping roles in the viral assembly process [5, 6].

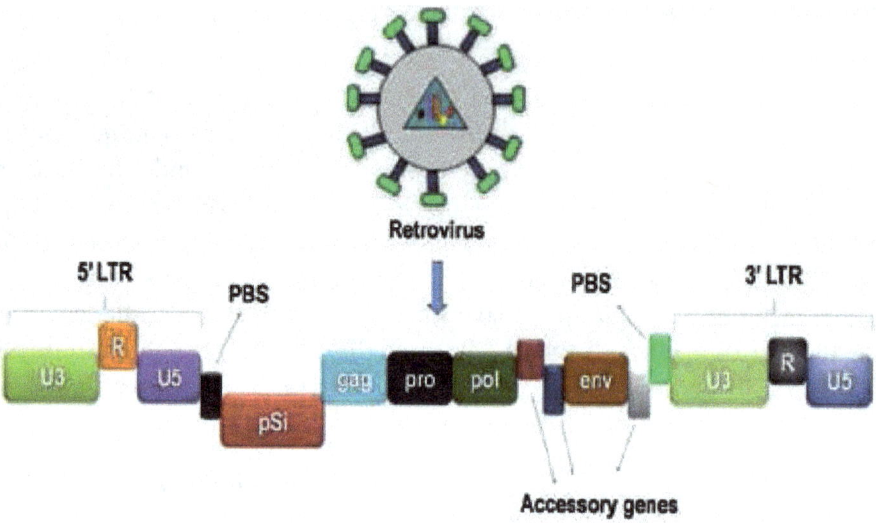

Figure 4. Retrovirus structure and genome: The three significant protein coding genes include group-specific gene (gag), pol, envelope (env) genes, which are flanked between the R regions and codes for capsid, reverse transcriptase, integrase, protease, and envelope proteins, respectively. Retroviruses are characterized by the 5′ and 3′ long terminal repeat (LTR) sequences, which are thought to control and gene expression. The LTRs each contain two unique non-protein-coding sequences, called U5 at the 5′ end and U3 at the 3′ end, which encodes controlling elements. PBS consists of 18 bases corresponding to 3' end of tRNA primer. The R region is a short repeated sequence at each end of the genome used during the reverse transcription to ensure correct end-to-end transfer in the growing chain.

3. Genetic variations and retroviruses

Retroviruses, similar to all RNA viruses, show a high mutation rate. The real component for creating genetic variation within retroviral populations is because of the polymerization error during DNA synthesis by RT, which does not have a proofreading activity [7]. The reason for this high genetic variation is because of viral mutation rate, recombination rate, rate of replication, size of the viral population, and selective forces [8, 9]. Genetic variation has been documented extensively in populations of HIV type 1. This genetic adaptability has significant consequences for the evolution of HIV-1 and other retroviruses and their impact on human health [8]. Because of the genetic variations, retroviruses are expanding its host range; for example, HIV-1 can switch from using the CCR5 co-receptor to using the CXCR4 co-receptor [10]. This difference in retroviral populations provides a mechanism for retroviruses to escape host immune responses and expand resistance to all known antiretroviral drugs [7, 11].

4. Replication of retroviruses

Replication is a multistep process; each step is crucial for the virus entry and multiplies itself in the host cell. The study of retroviruses particle assembly, budding, and release has been especially rich in terms of the exchange of concepts and techniques with related areas of cell biology [12]. There are seven steps in the replication cycle of the retrovirus (Figure 5). The initial step is attachment, in which the retrovirus utilizes one of its glycoproteins to attach to one or more particular cell-surface receptors on the host cell. Some retroviruses likewise utilize an optional receptor, referred to as the co-receptor. The second and third steps are penetration and uncoating, individually. Retroviruses infiltrate the host cell by direct fusion of the virion envelope with the plasma membrane of the host. The fourth step is replication, which happens after the retrovirus undergoes partial uncoating thereby releasing its genome and three essential enzymes (RT, integrase, and pol gene coding enzymes). At this stage, the RNA genome is converted by RT into double-stranded DNA, followed by integration into the host genome, transcription and translation of viral proteins along with the host. The fifth step is assembly, in which retrovirus capsids are assembled in an immature form. The sixth step is budding, in which the immature viral particle acquires the host plasma membrane, and the final step is maturation and release, in which the gag and pol proteins of the retrovirus are cleaved by the retroviral protease, thus forming the mature and infectious form of the virus [13]. The retrovirus replication is well studied in case of HIV virus. HIV replicates million of time per day, destroying the host immune cells and eventually causing disease progression. During HIV replication the virus recognizes host cell such as CD4+ T- lymphocyte. Entry of HIV into the cells requires certain substances on the cell surface such as CD4 receptor and co-receptors such as CCR5 and CXCR4 [14]. These receptors interact with protein complexes that are embedded in the viral envelope. The viral proteins consist of extracellular gp120 and trans-membrane gp41 proteins. When HIV approaches the target cell, the gp120 binds with the cell surface receptor, this process is termed as attachment. Following co-receptor binding results in a conformational change in gp120, this allows gp41 to unfold and extend its hydrophobic terminal into the cell membrane. gp41 then folds back on itself, this causes the virus to move close toward the cell and facilitates the fusion of their membranes. The viral nucleocapsid then enters the cell and releasing two viral RNA strands and three essential replication enzymes; integrase, protease, and RT. HIV RT is a heterodimer composed of two subunits (p66 and p51). At first, RT begins the reverse transcription of viral RNA; it consists of two catalytic domains —ribonuclease H active site and polymerase active site. In the polymerase active site, single-stranded viral RNA is transcribed into an RNA-DNA double helix. These RNA-DNA hybrids are cleaved into individual stands by ribonuclease H. The polymerase then completes the remaining strand into DNA double helix (dsDNA). After the formation of dsDNA, integrase moves into action, it cleaves each dinucleotide from 3′ end of the DNA creating two sticky ends. Integrase then transfers the viral DNA into the cell nucleus and viral dsDNA is covalently and randomly integrated into the cell's genome [15]. The host cell genome now contains the genetic information of HIV virus. Activation of the host cell induces the transcription of proviral DNA by Pol II produces viral spliced and unspliced messenger RNAs. This messenger RNA now migrates into the cytoplasm, where building blocks for a new virus were synthesized. Some of the building blocks have to be processed by the viral protease where longer proteins are cleaved into small core proteins. The processing of viral proteins is crucial to create an infectious virus. Translation of unspliced viral RNAs produces env, gag, and gag-pol polyproteins. The two viral RNA strands with three enzymes come together and core proteins assemble around them forming a capsid, which is an immature virus particle. Capsid leaves the host cell by acquiring new envelope of host and viral proteins (mature virus) such as gp120 and gp41; this process is known as budding. Recent reports suggest that during this process of budding, clathrin is recruited into the HIV particle with high specificity [16]. These matured virus become ready to infect the other cells [15]. A critical aspect of viral replication is the assembly of virus particles, which are subsequently released as progeny virus. While a great deal of attention has been focused on better understanding this phase of the viral lifecycle, many aspects of the molecular details remain poorly understood.

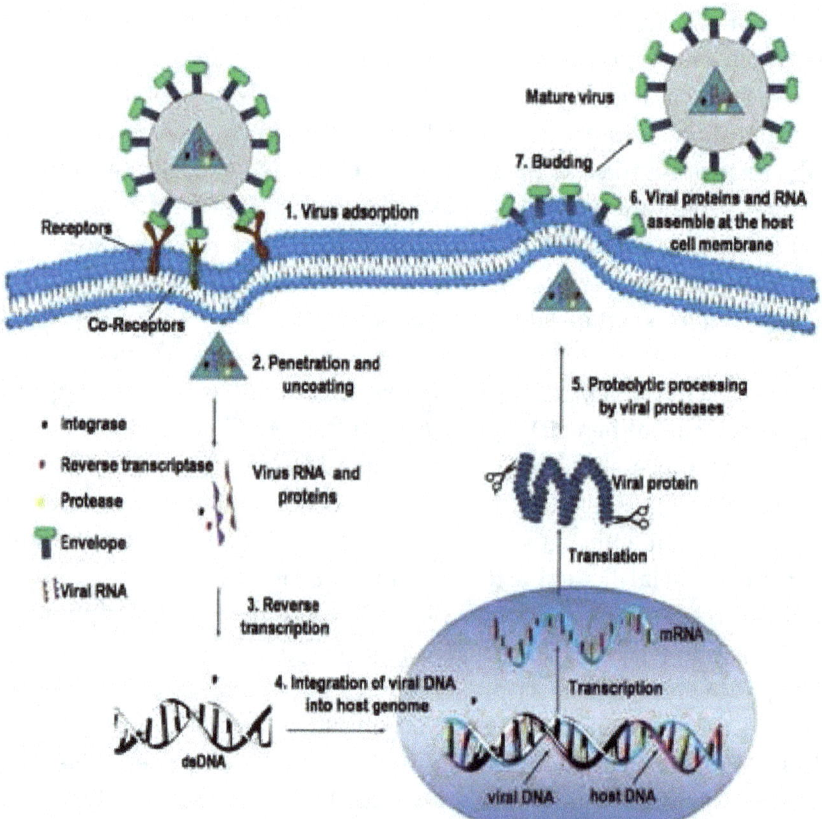

Figure 5. Replication of retroviruses: There are seven steps in the replication cycle of the retrovirus. The initial step is attachment, in which the retrovirus utilizes one of its glycoproteins to attach to one or more particular cell-surface receptors on the host cell. Some retroviruses likewise utilize an optional receptor, referred to as the co-receptor. The second step is penetration and uncoating, individually. Retroviruses infiltrate the host cell by direct fusion of the virion envelope with the plasma membrane of the host. The third step is replication, which happens after the retrovirus undergoes partial uncoating thereby releasing its genome and three essential enzymes. At this stage, the RNA genome is converted by reverse transcriptase into double-stranded DNA (dsDNA). The fourth step is integration, in which retrovirus dsDNA integrates into the host genome followed by transcription and translation of the viral proteins occurs. The fifth step includes proteolytic processing of viral proteins. The sixth step includes assembly of viral proteins and RNA. The seventh step is budding, in which the immature viral particle acquires the host plasma membrane and final step is maturation and release, in which the gag and pol proteins of the retrovirus are cleaved by the retroviral protease, thus forming the mature and infectious form of the virus.

A provirus can be transmitted through the germ line from parents to offspring as an endogenous retrovirus [17, 18]. Human endogenous retroviruses (HERVs) account for about 8% of the human genome [19]. Exogenous retroviruses seem to have arisen from endogenous retrotransposons by acquisition of a cellular envelope gene [20]. The existence of HERVs has been identified for many years, but their abundance in the genome was not predicted by earlier studies [21]. Retroviral genome gets into human genome and by de novo insertion followed by activation of downstream proto-oncogenes, or by gene disruption [22]. Retrovirus integration does not occur in resting (Go phase) cells yet rather requires that cells be in the S phase (DNA synthesis) of their mitotic cycle. Since the mitotic phase is induced by a binary recognition event, integration and virus reproduction perhaps require that the invaded target T-cell interact with the appropriate B-cell-processed antigen complex. Once the viral cDNA integrates, transcription to mRNA proceeds at some rate that depends on the details of the infecting virus and the invaded cell [23]. It is also reported that enhancer and promoter elements in retroviral LTRs can influence the transcription of next genes that can result in transcriptional activation or gene silencing and which may result in abnormal expression of tissue-specific proteins [24]. The human genome contains many endogenous retroviral sequences, and these have been suggested to play important roles in a number of physiological and pathological processes. Researchers also found that ERVs also take part in the body's immune defense against regular bacterial and viral pathogens. HERVs are classified into three broad classes (I, II, III). Analysis of the draft human genome has so far found only three HERV proviruses with complete open reading frames for gag, pol, and env, which are considered as essential viral genes, and at least one of these HERVs is mutated at a critical residue in the reverse transcriptase domain of pol [25]. HERVs have frequently been reported as etiological cofactors in chronic diseases such as cancer, autoimmunity, and neurological disease.

5. Mode of transmission

Most of the retroviruses transmission occurs through cell to cell, mother to fetus transmission, and through biological fluids. Cell-to-cell transmission of retroviruses is much more efficient as compared with cell-free conditions, and as retroviruses reach through the tight cell-cell interface, they are out of reach of the immune system [26]. Retrovirus employs various mechanisms of immune evasion, however, and can destroy the immune system or subvert it to enable successful transmission [27].

6. Immune system and retroviruses

The human immune system needs to manage with various pathogens, ranging from RNA viruses to 30-foot-long tapeworms [28]. Although we have gained much understanding of innate immune recognition of many microbial pathogens, currently we have very little knowledge about innate immune responses against retroviral infections [29]. The immune system retroviruses (ISRV) are defined as a retrovirus (HIV) whose target is T4-positive T-helper cells of the immune system that requires stimulation by antigens to reproduce. T-cells are part of the response mechanism that defends the body against attacking agents and are stimulated to reproduce by such agents [23]. Antiviral responses are characteristically marked by stimulation of type I interferon through various mechanisms that recognize viral nucleic acids. These responses restrain the viral replication by various mechanisms and activate the adaptive immune responses with the help of antigen presenting cells and also aids in developing memory and viral clearance. But, quite number of reports suggests that viruses also have developed a variety of means for circumventing innate immune responses, ensuring their survival and transmission. Surprisingly retroviruses do not appear to straightforwardly activate host innate defenses. It was observed that generalized immune activation and increased amount of cytokines and immunoglobulins along with the progressive loss of CD4+ T-cells was reported during HIV-1 infection [30].

Generally, viral infection triggers innate immune sensors to produce type I interferon. However, this is not the case with retroviruses; the reason is still not known. The recent reports suggest various molecules it includes: TREX1, which is a cytosolic exonuclease that degrades DNA [31] derived from HIV or endogenous retroelements, thereby preventing the accumulation of cytosolic DNA, which would otherwise trigger innate immunity. In a study on Trex1(-/-) mouse cells and human CD4(+) T-cells and macrophages in which TREX1 was inhibited by RNA-mediated interference, cytosolic HIV-DNA accumulated and HIV infection induced type I interferon that inhibited HIV replication and spreading [32]. The recent study on innate immune sensors during retroviral infection has identified the enzyme cyclic guanosine monophosphate–adenosine monophosphate (cGAMP) synthase (cGAS) which triggers the cytosolic DNA and activates the production of type I interferons and other cytokines. The mechanism in which these sensors act by includes firstly viral DNA binds and activates cGAS, which catalyzes the synthesis of a cGAMP isomer from adenosine triphosphate (ATP) and guanosine triphosphate (GTP). This cGAMP isomer that is termed has 2′3′-cGAMP contains both 2′-5′ and 3′-5′ phosphodiester linkages. cGAMP then binds and activates the endoplasmic reticulum protein stimulator of IFN genes (STING) and functions as a second messenger. STING activates the NF- κB, interferon regulatory factor 3 (IRF3) to induce interferons and other cytokines through the activation of protein kinases IκB kinase (IKK) and TANK-binding kinase 1 (TBK1) [32–35]. In our view, there is currently insufficient understanding about how the retroviral infection, in general, is sensed by the innate immune system. If the innate immune part is traced, that will aid in understanding the activation of adaptive immunity and development of antiviral against retroviruses.

7. Antiretroviral therapy

There are about 34 million HIV-1–infected people in the world [36], and this number plainly says that there is an urgent unmet need for investigation of antiretroviral therapy (ART), and management of this worldwide risk is highly desired [37]. ART is treatment of people infected with the retroviruses using anti-retroviral drugs. The goal of antiretroviral therapy is to reduce the amount of virus in infected individual body (viral load) to a level that can no longer be detected with ongoing treatment blood test. Considerable advances in ART have been made since the introduction of zidovudine (3'-azido-3'-deoxythymidine—AZT) in 1987 [38]. The regular treatment consists of a grouping of at least three drugs called as highly active antiretroviral therapy (HAART) that hold back the viral replication within the host cell and thereby reduces the viral load.

Six classes of antiretroviral agents (Table 1) currently exist (specifically towards HIV); they include the following:

1. Nucleoside reverse transcriptase inhibitors (NRTIs) such as abacavir, emtricitabine, and tenofovir [39] takes action by interfering with the HIV replication cycle through competitive inhibition of reverse transcriptase enzyme a key enzyme in replication and thereby terminates the DNA formation. These can also terminate the DNA formation by incorporating into the proviral DNA; the reason is that NRTIs are structurally similar to the DNA nucleoside bases.

2. Non-nucleoside reverse transcriptase inhibitors (NNRTIs) includes niverapine, efavirenz, and etravirine, which acts by non-competitive binding of NNRTIs at the hydrophobic pocket of p66 subunit of the enzyme results in a conformational change and alters the active site and limits RT activity. The limitation of these drugs is that it has a low genetic barrier, i.e., a single mutation in RT genome induces a high-level of phenotypic resistance and prevents its use [40–42].

3. Protease inhibitors (PIs) include indinavir, atazanavir, darunavir, and tipranavir. This retrovirus protease is a 99-amino-acid, aspartic acid protein, which plays an important role in the maturation of virus particles late in the viral life cycle. During or immediately after viral budding from an infected cell, proteases systematically cleaves individual proteins from the gag and gag-pol polypeptide precursors into functional subunits for viral capsid formation. Protease inhibitors act as competitive inhibitors that directly bind to protease and put off the subsequent cleavage of polypeptides. It has recently been suggested that PIs can directly inhibit lymphocyte apoptosis and this effect may contribute to an immunologic benefit independently of an antiviral effect [43–45].

4. Integrase inhibitors (INSTIs) such as dolutegravir and raltegravir are used in combination with a protease inhibitor and target the strand transfer step of retroviral DNA integration. These are approved by FDA in 2007. Integration is essential for viral replication and is thus an attractive target for novel chemotherapy [46]. The integrase enzyme is responsible for transfer of virus-encoded DNA to the host cell chromosome, a necessary event in retrovirus replication [15] INSTIs active against a wide range, including both CCR5 coreceptor and CXCR4 coreceptor–using strains [47].

5. Fusion inhibitors (FIs) include enfuvirtide—act extracellularly to prevent the fusion. It is a peptide based on the gp41 sequence that specifically prevents membrane fusion by competitively binds to gp41 and preventing the conformational change of gp41 required to complete the final step in the fusion process [48].

6. Chemokine receptor antagonist. This small molecule such as maraviroc [49] selectively and reversibly binds the CCR5 coreceptor, blocking the V3 loop interaction and inhibiting fusion of the cellular membranes. Using these inhibitors individually or in combination the virus replication process is slowed and the retroviruses find it more difficult to overcome this combined attack. ART has the potential both to reduce mortality and morbidity rates among infected people, and to improve their quality of life [50].

	Antiretroviral agents	Examples	Mode of action
1.	Nucleoside reverse transcriptase inhibitors (NRTIs)	Abacavir, Emtricitabine	Competitive inhibition of reverse transcriptase enzyme a key enzyme in replication and there by terminates the DNA formation.
2.	Non-nucleoside reverse transcriptase inhibitors (NNRTIs)	Niverapine, Efavirenz, Etravirine	Non-competitive binding of NNRTIs at the hydrophobic pocket of p66 subunit of the reverse transcriptase enzyme results in a conformational change and alters the active site and limits enzyme activity.
3.	Protease inhibitors (PIs)	Indinavir, Atazanavir, Darunavir, Tipranavir	Act as competitive inhibitors that directly bind to protease and put off the subsequent cleavage of polypeptides which is an important step in viral maturation.
4.	Integrase inhibitors (INSTIs)	Dolutegravir, Raltegravir	Target the strand transfer step of retroviral DNA integration.
5.	Fusion inhibitors (FIs)	Enfuvirtide	Specifically prevents membrane fusion by competitively binding to gp41 and preventing the conformational change of gp41 required to complete the final step in the fusion process.
6.	Chemokine receptor antagonist	Maraviroc	It binds selectively and reversibly binds the CCR5 coreceptor, blocking the V3 loop interaction and inhibiting fusion of the cellular membranes.

Table 1. Classes of antiretroviral agents and their mode of action

8. Conclusions

The central goals of retrovirology nowadays are the treatment and the prevention of human and non-human diseases and to use this virus in research. Recent studies have shown that retroviruses can be used in a number of ways such as model for biological research, for understanding of genes, molecular and cell biology studies. On the other hand, attention in these viruses extends beyond their disease causing capabilities, discovery of oncogenes, understanding of mechanisms that regulate eukaryotic gene expression was possible because of the study on retroviruses. The complete understanding of retrovirus could help the researchers and clinicians to use them in various fields of biology and medicine for the development of new methodologies and techniques. Ongoing investigation on application of retroviruses in gene therapy and anti-cancer agents makes these type a widely studying group. The way retroviruses enter and target the specific cells and integrate itself into the host genome was very fascinating to the scientists globally, and these can be used as models to develop new vectors that could be employed in research. Collaborative international project needs to be taken up to understand the complete life cycle of retroviruses. The reason is that this not only aids in developing antiviral, but also gives us idea where gained knowledge could be applied in other fields such as engineering and material sciences and to develop new technologies.

Acknowledgements

Authors are grateful to the Director, CCMB and Council of Scientific and Industrial Research (CSIR-CCMB), India, for the encouragement and support for this work. SK Saxena is also supported by US National Institute of Health Grants: R37DA025576 and R01MH085259.

Author details

Shailendra K. Saxena* and Sai V. Chitti

*Address all correspondence to: shailen@ccmb.res.in; shailen1@gmail.com

CSIR–Centre for Cellular and Molecular Biology (CCMB), Hyderabad, India

References

[1] Zhang W, Cao S, Martin JL, Mueller JD, Mansky LM. Morphology and ultrastructure of retrovirus particles. AIMS Biophys. 2015; 2(3): 343-369.

[2] Varmus H. Retroviruses. Science. 1988; 240(4858): 1427-1435.

[3] Wyatt R, Sodroski J. The HIV-1 envelope glycoproteins: Fusogens, antigens, and immunogens. Science. 1998; 280(5371): 1884-1888.

[4] Coffin JM. Structure and classification of retroviruses. In: Levy, JA. The *Retroviridae* 1 (1st ed.). New York: Plenum; 1992. p. 20. ISBN 0-306-44074-1.

[5] Rabson AB, Graves BJ. Synthesis and processing of viral RNA. In: Coffin JM, Hughes SH, Varmus HE, editors. Retroviruses. Cold Spring Harbor (NY): Cold Spring Harbor Laboratory Press; 1997.

[6] Bharat TA, Davey NE, Ulbrich P, Riches JD, de Marco A, Rumlova M, Sachse C,Ruml T, Briggs JA. Structure of the immature retroviral capsid at 8 Å resolution by cryo-electron microscopy. Nature. 2012; 487(7407): 385-389.

[7] Freed EO. HIV-1 assembly, release and maturation. Nat Rev Microbiol. 2015; 13(8): 484-496.

[8] Svarovskaia ES, Cheslock SR, Zhang WH, Hu WS, Pathak VK. Retroviral mutation rates and reverse transcriptase fidelity. Front Biosci. 2003; 8: 117-134.

[9] Mansky LM, Temin HM. Lower in vivo mutation rate of human immunodeficiency virus type 1 than that predicted from the fidelity of purified reverse transcriptase. J Virol. 1995; 69(8): 5087-5094.

[10] Anderson JA, Teufel RJ 2nd, Yin PD, Hu WS. Correlated template-switching events during minus-strand DNA synthesis: A mechanism for high negative interference during retroviral recombination. J Virol. 1998; 72(2): 1186-1194.

[11] Connor RI, Sheridan KE, Ceradini D, Choe S, Landau NR. Change in coreceptor use correlates with disease progression in HIV-1–infected individuals. J Exp Med. 1997; 185(4): 621-628.

[12] Jekle A, Schramm B, Jayakumar P, Trautner V, Schols D, De Clercq E, Mills J, Crowe SM, Goldsmith MA. Coreceptor phenotype of natural human immunodeficiency virus with nef deleted evolves in vivo, leading to increased virulence. J Virol. 2002; 76(14): 6966-6973.

[13] Bieniasz PD. The cell biology of HIV-1 virion genesis. Cell Host Microbe. 2009; 5(6): 550-558.

[14] Pedersen FS, Pyrz M, Duch, M. Retroviral replication. In: eLS. 2011.

[15] Saxena SK, Gupta A, Bhagyashree K, Saxena R, Arora N, Banerjee AK, Tripathi AK, Chandrasekar MJ, Gandhi N, Nair MP. Targeting strategies for human immunodeficiency virus: A combinatorial approach. Mini Rev Med Chem. 2012; 12(3): 236-254.

[16] Pommier Y, Johnson AA, Marchand C. Integrase inhibitors to treat HIV/AIDS. Nat Rev Drug Discov. 2005; 4(3): 236-248.

[17] Zhang F, Zang T, Wilson SJ, Johnson MC, Bieniasz PD. Clathrin facilitates the morphogenesis of retrovirus particles. PLoS Pathog. 2011; 7(6).

[18] Temin HM. Reverse transcription in the eukaryotic genome: Retroviruses, pararetroviruses, retrotransposons, and retrotranscripts. Mol Biol Evol. 1985; 2(6): 455-468.

[19] Black SG, Arnaud F, Palmarini M, Spencer TE. Endogenous retroviruses in trophoblast differentiation and placental development. Am J Reprod Immunol. 2010; 64(4): 255-264.

[20] Gim JA, Han K, Kim HS. Identification and expression analysis of human endogenous retrovirus Y (HERV-Y) in various human tissues. Arch Virol. 2015; 160(9): 2161-2168.

[21] Malik HS, Henikoff S, Eickbush TH. Poised for contagion: evolutionary origins of the infectious abilities of invertebrate retroviruses. Genome Res. 2000; 10(9): 1307-1318.

[22] Boeke JD, Stoye JP. Retrotransposons, endogenous retroviruses, and the evolution of retroelements. In: Coffin JM, Hughes SH, Varmus HE, editors. Retroviruses. Cold Spring Harbor (NY): Cold Spring Harbor Laboratory Press: 1997, pp. 343-435.

[23] Löwer R. The pathogenic potential of endogenous retroviruses: Facts and fantasies. Trends Microbiol. 1999; 7(9): 350-356.

[24] Cooper LN. Theory of an immune system retrovirus. Proc Natl Acad Sci U S A. 1986; 83(23): 9159-9163.

[25] Griffiths DJ. Endogenous retroviruses in the human genome sequence. Genome Biol. 2001; 2(6).

[26] Mayer J, Sauter M, Rácz A, Scherer D, Mueller-Lantzsch N, Meese E. An almost-intact human endogenous retrovirus K on human chromosome 7. Nat Genet. 1999; 21(3): 257-258.

[27] Jin J, Sherer NM, Heidecker G, Derse D, Mothes W. Assembly of the murine leukemia virus is directed towards sites of cell-cell contact. PLoS Biol. 2009; 7(7).

[28] Kane M, Case LK, Kopaskie K, Kozlova A, MacDearmid C, Chervonsky AV, Golovkina T. Successful transmission of a retrovirus depends on the commensal microbiota. Science. 2011; 334(6053): 245-249.

[29] Medzhitov R, Littman D. HIV immunology needs a new direction. Nature. 2008; 455(7213): 591.

[30] Gao D, Wu J, Wu YT, Du F, Aroh C, Yan N, Sun L, Chen ZJ. Cyclic GMP-AMP synthase is an innate immune sensor of HIV and other retroviruses. Science. 2013; 341(6148): 903-906.

[31] Manel N, Littman DR. Hiding in plain sight: How HIV evades innate immune responses. Cell. 2011; 147(2): 271-274.

[32] Stetson DB, Ko JS, Heidmann T, Medzhitov R. Trex1 prevents cell-intrinsic initiation of autoimmunity. Cell. 2008; 134(4): 587-598.

[33] Yan N, Regalado-Magdos AD, Stiggelbout B, Lee-Kirsch MA, Lieberman J. The cytosolic exonuclease TREX1 inhibits the innate immune response to human immunodeficiency virus type 1. Nat Immunol. 2010; 11(11): 1005-1013.

[34] Sun L, Wu J, Du F, Chen X, Chen ZJ. Cyclic GMP-AMP synthase is a cytosolic DNA sensor that activates the type I interferon pathway. Science. 2013; 339(6121): 786-7891.

[35] Wu J, Sun L, Chen X, Du F, Shi H, Chen C, Chen ZJ. Cyclic GMP-AMP is an endogenous second messenger in innate immune signaling by cytosolic DNA. Science. 2013; 339(6121): 826-830.

[36] Tanaka Y, Chen ZJ. STING specifies IRF3 phosphorylation by TBK1 in the cytosolic DNA signaling pathway. Sci Signal. 2012; 5(214).

[37] Saxena SK, Tiwari S, Nair MP. A global perspective on HIV/AIDS. Science. 2012; 337(6096): 798.

[38] Singh G, Pai RS, Mustafa S. Nanostructured delivery systems: Augmenting the delivery of antiretroviral drugs for better management of HIV/AIDS. Crit Rev Ther Drug Carrier Syst. 2015; 32(6): 503-533.

[39] Cox SW, Apéria K, Albert J, Wahren B. Comparison of the sensitivities of primary isolates of HIV type 2 and HIV type 1 to antiviral drugs and drug combinations. AIDS Res Hum Retroviruses. 1994; 10(12): 1725-1729.

[40] Spano JP, Poizot-Martin I, Costagliola D, Boué F, Rosmorduc O, Lavolé A, Choquet S, Heudel PE, Leblond V, Gabarre J, Valantin MA, Solas C, Guihot A, Carcelain G, Autran B, Katlama C, Quéro L. Non-AIDS-related malignancies: Expert consensus review and practical applications from the multidisciplinary CANCERVIH Working Group. Ann Oncol. 2015.

[41] De Clercq E. Non-nucleoside reverse transcriptase inhibitors (NNRTIs): Past, present, and future. Chem Biodivers. 2004; 1(1): 44-64.

[42] Joly V, Yeni P. Non-nucleoside reverse transcriptase inhibitors. Ann Med Interne. 2000; 151(4): 260-267.

[43] De Béthune MP. Non-nucleoside reverse transcriptase inhibitors (NNRTIs), their discovery, development, and use in the treatment of HIV-1 infection: A review of the last 20 years (1989-2009). Antiviral Res. 2010; 85(1): 75-90.

[44] Weichold FF, Bryant JL, Pati S, Barabitskaya O, Gallo RC, Reitz MS Jr. HIV-1 protease inhibitor ritonavir modulates susceptibility to apoptosis of uninfected T cells. J Hum Virol. 1999; 2(5): 261-269.

[45] Sloand EM, Kumar PN, Kim S, Chaudhuri A, Weichold FF, Young NS. Human immunodeficiency virus type 1 protease inhibitor modulates activation of peripheral blood CD4(+) T cells and decreases their susceptibility to apoptosis in vitro and in vivo. Blood. 1999; 94(3): 1021-1027.

[46] Chavan S, Kodoth S, Pahwa R, Pahwa S. The HIV protease inhibitor Indinavir inhibits cell-cycle progression in vitro in lymphocytes of HIV-infected and uninfected individuals. Blood. 2001; 98(2): 383-389.

[47] Hazuda DJ, Felock P, Witmer M, Wolfe A, Stillmock K, Grobler JA, Espeseth A, Gabryelski L, Schleif W, Blau C, Miller MD. Inhibitors of strand transfer that prevent integration and inhibit HIV-1 replication in cells. Science. 2000; 287(5453): 646-650.

[48] Hicks C, Gulick RM. Raltegravir: The first HIV type 1 integrase inhibitor. Clin Infect Dis. 2009; 48(7): 931-939.

[49] Eggink D, Berkhout B, Sanders RW. Inhibition of HIV-1 by fusion inhibitors. Curr Pharm Des. 2010; 16(33): 3716-3728.

[50] Tilton JC, Doms RW. Entry inhibitors in the treatment of HIV-1 infection. Antiviral Res. 2010; 85(1): 91-100.

[51] Sagar V, Pilakka-Kanthikeel S, Pottathil R, Saxena SK, Nair M. Towards nanomedicines for neuro AIDS. Rev Med Virol. 2014; 24(2): 103-124.

Section 2

Retrovirus Replication, Gene Expression, Latency and Reactivation

2

Role of Host Proteins in HIV-1 Early Replication

Lokeswara S. Balakrishna and Anand K. Kondapi

Abstract

After 33 years of the identification of HIV-1 infection, very little is known about the role of host cellular proteins. Till now considerable work has been done in the area of host-pathogen interactions facilitated by the viral proteins and host receptors. The role of the main receptor CD4 and co-receptors like CCR5, CXCR4 and their alternative receptors were well studied in disease progression. But the intracellular events during the host-pathogen interactions were poorly understood. Much data is available based on the global analysis of genome-wide RNA interference screens, yeast two-hybrid system and co-immunoprecipitation studies but their exact roles are not yet characterized. There are very few host proteins like APOBEC3G, LEDGF/p75, INI1, HMG I(Y), BAF which are well studied and characterized. Majority of the reported proteins are attributed to multiple functions. It will be useful to study such proteins to develop as future candidates in HIV-1 therapeutics.

Keywords: HIV, Reverse transcription, Host proteins, CD4, CCR5, CXCR4, Topoisomerase, PICs

1. Introduction

Host cell responses are key determinants of infection pertaining to infectious diseases. Different kind of host cell responses are exerted during the course of the infection, either a host defense response to restrict the invasion of pathogen or may promote the invasion. During the course of evolution, pathogens have acquired capability to protect themselves from host defense. This protection is mainly by modulating the key regulators of the host signal transduction mechanism. Unlike other pathogens, viruses are very small with a small genome which codes for the essential structural proteins and enzymes, a reason to consider them as primitive. These proteins are enough to takeover host cell and to control majority of the cellular processes. This takeover property may be due to its dependence on the host mechanisms to fulfill its needs to establish itself and for replication in the host. For this they have to suppress certain mechanisms and promote others. It is very important to know the host factors involved the host-pathogen interactions, establishment of the infection and pathogenicity. In retrovirus especially in HIV infection, host proteins requirement starts from the beginning of attachment of the virus to its target host cell, reverse transcription, integration, transcription and translation of viral proteins, carrying of viral proteins to plasma membrane and release of viral particles. Overall knowledge on host proteins involvement in any infection would increase the possibilities of invention of the inhibitors for the host proteins which may be candidates for the future drug inventions.

2. Binding of virus to cell surface

Viruses have remarkable specificity for the host species and the cell types that they can infect. Similar to other viruses, human immunodeficiency virus (HIV) also has host and exhibits target cell specificity. This feature is based on the properties of a cell which can fulfill the virus catch and grab mechanism to complete its infection cycle (from attachment to the production of progeny viruses is considered as an infection cycle).

HIV as a single particle is called as virion. This virion consists of an outer envelope and inner capsid. Outer envelope is host derived plasma membrane with host cell surface molecules and viral transmembrane glycoprotein called gp41, which connects outer surface gp120 glycoprotein expressed from *env* gene of virus. Capsid is a cone shaped structure made up of a viral protein p24 (named based on its molecular weight 24 kDa) which is a processed poly-protein product of virus gene called *gag*. Envelope glycoprotein is translated as a 160 kDa polyprotein, later processed into two subunits of 120 kDa and another is 41 kDa by a cellular endo-protease [1]. One gp120 and one gp41 collectively form a unit and a trimer of this unit enmeshed in the viral envelope [2, 3]. The gp41 is the transmembrane portion and gp120 is the extracellular region which works as an anchor to grab the specific host cells. The primary/preliminary target for this gp120 is CD4 (receptor) and a secondary target is a chemokine receptor (co-receptor) of host cells.

2.1. Role of CD4

Leucocyte differentiation antigen, CD4 is a cellular receptor for HIV-1, HIV-2 and Simian immune deficiency virus (SIV). These viruses share CD4 as the common primary receptor and the binding sites on these viruses are highly conserved [4]. The CD4 receptor is found on CD4 T-cells (high expression) macrophages and dendritic cells (DCs; low expression). A CD4 binding site present on each monomer of gp120. Recruitment of one CD4 molecule on a single gp120 in the trimeric anchor can induce conformational changes in all three glycoprotein monomers of the trimer [5]. This binding of gp120 to CD4 can be blocked by the host antibodies produced against the gp120 of the virus which are called neutralizing antibodies. But due to variation in gp120 among virus population, there is a lag time in producing enough antibodies to block virus attachment. Furthermore, the conformational change occurs in the gp120, which avoid recognition by the neutralizing antibodies, a process known as conformational masking. The conformational changes in gp120 allow it to bind to a second receptor on the CD4 cell surface [6].

The second docking area on the CD4+ cell surface is a chemokine receptor, a seven transmembrane (7TM) co-receptor namely C-C chemokine receptor type 5 (CCR5) or C-X-C chemokine receptor type 4 (CXCR4). The viral preference of using one co-receptor among others is called 'viral tropism'. The virus which can infect the macrophages predominantly uses CCR5 as their co-receptor. About 90% of all HIV infections involve the M-tropic HIV strain. CXCR4, also called fusin, which is a glycoprotein-linked chemokine receptor used by T-cell infecting (T-tropic) HIV to attach to the T cell. Indeed some HIV are reported to use co-receptors other than these two co-receptors like CCR1, CCR2b, CCR3, CCR8, CCR9, CXCR4, CX3CR1/V28, STRL-33/BONZO/CXCR6, GPR1, GPR15/BOB, APJ, ChemR23, RDC1, and Leukotriene B4 receptor though the mechanisms are unknown.

Once the HIV gp120 has attached to the CD4 molecule, it undergoes conformational changes which enables the binding of the gp120 to a co-receptor leads to the further structural rearrangements in the gp41 to fuse with the cell membrane and entry of the virions core into the cell's cytoplasm (Fig. 1). Once within a cell, virus is safe from neutralizing antibodies, but vulnerable to attack by CD8 cells (cytotoxic T-lymphocytes or CTLs).

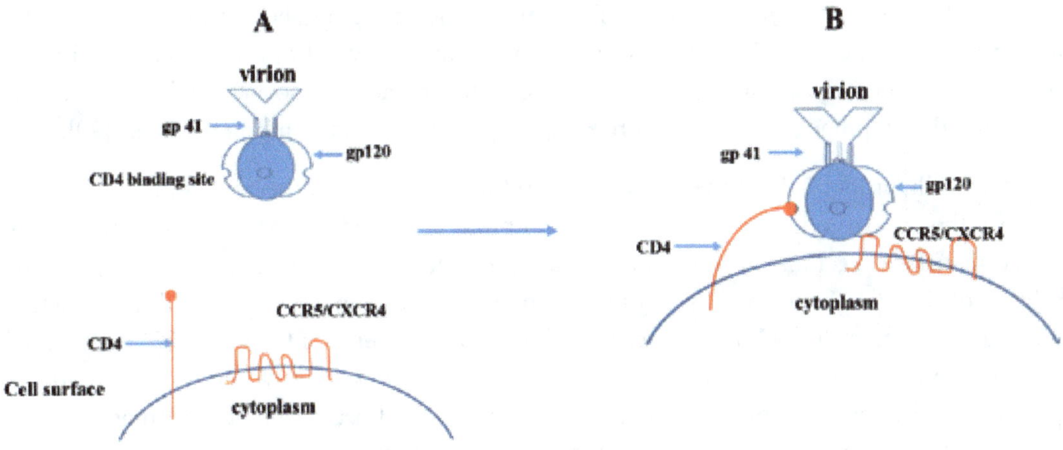

Figure 1. Virus and cell surface receptors interactions. In panel (A), schematic diagram represents virion particle and host cell surface receptors before attachment of the virion to the host. gp120, gp41 and CD4 binding region on gp120 of virus and CD4 and co-receptors (CCR5/CXCR4) were labeled. In panel (B), the attachment of CD4 in the CD4 binding site of gp120 and also binding of the co-receptor were shown.

In the absence of CD4, infection is inefficient and its significance *in vivo* is controversial. More than this binding and internalization, CD4 is involved in the signal transduction. Binding of gp120 to the CD4 induces rapid activation of the ERK/mitogen-activated protein (MAP) kinase pathway and stimulates expression of cytokine and chemokine genes by the binding of nuclear transcription factors (AP-1, NF-kB, and C/EBP) in both T-cell tropic and macrophage tropic strain. The activation of this signaling pathway requires functional CD4 receptors which is independent of binding to CXCR4 [7, 8]. Signal transduction by CD4 depends on its association with Lck, a src-family tyrosine kinase. Lck, interacts with CD4 with its unique NH2-terminal domain, and interacts with other intracellular signaling proteins with its SH2 and SH3 domains. Given the necessity of Lck kinase activity for T lymphocyte development and for mature T cell functions, perhaps Lck may function at different stages during T cell activation [9].

2.2. Role of chemokine receptors

Chemokine receptors are seven transmembrane (7TM) domain, G protein–coupled molecules that mediate the chemotaxis of T cells and phagocytic cells to the areas of inflammation [10]. Chemokine receptors have four domains exposed on the cell surface: the N terminus and three extracellular loops (E1, E2 and E3). Co-receptors take up different conformations on cell surface and on different cell types [11, 12], influencing their ability to support HIV infection. For X4 strains, E2 is critical. Deletion of the N terminus of CXCR4 affects some of the strains but not all [13], although, when present, participates in binding gp120 [14]. The gp120 of HIV-1 is structurally divided into five regions called Variable regions and represented with V1,V2 etc. The highly conserved region present between the variable region of gp120 functions as a co-receptor binding site [15]. *In vitro*, envelope glycoproteins in soluble form are even capable of co-receptor mediated signal transduction [16-18] which involves rapid phosphorylation of GPCRs at the carboxyl-terminal tail [19]. There are 21 potential phosphorylation sites in CXCR4 and only seven in CCR5. Chemokines, small low-molecular weight proteins, are the ligands that activate and signal through CCR5 and CXCR4 to mediate several cellular functions including development, leukocyte trafficking, angiogenesis, and immune response [20].

The extracellular loop2 (E2) which is present on the N terminus of the co-receptor, is responsible for the gp120 binding and HIV entry. After binding of the viral extracellular gp120 to the host cell surface, a cellular kinase called Focal adhesion kinase interacts with CCR5 [21]. The affinity of the gp120 and the co-receptors influences the strength of the signal transduction which can be correlated with the successes of post-fusion events of the virus [22]. This signal transduction from the co-receptor converts the host non permissive environment to the permissive environment for the virus establishment in the host. The binding capacity of the viral gp120 to the host co-receptor has been considered as the viral infectivity. Even though, the early and late events after co-receptor-gp120 interaction are not completely elucidated, yet these interactions have an important role in very early events of HIV-1 lifecycle.

2.3. CCR5

Virus using the CCR5 as co-receptor, infects the macrophages called M-tropic virus. This viral strain has ability to infect other cell types like dendritic cells and CD4 T-cells. Majority of the viral isolates utilize CCR5 co-receptors for their transmission. M-tropic HIV replicates in peripheral blood lymphocytes are less virulent and does not form syncytia. Syncytia are multinuclear cells which are result of cellular fusion.

The expression of CCR5 was observed on widely diverged cell types [23] and modulated by pro-inflammatory cytokines. A number of inflammatory CC-chemokines, like MIP-1, RANTES etc. [24] act as CCR5 agonists while, MCP-3 functions as antagonist. The cytosolic domain of the CCR5 bound with GPCRs (G-protein coupled receptors) transduces signals upon binding of ligands to CCR5. The signal transduction include different secondary messengers like cAMP, Ca2+, PI3-kinase, MAP kinases, as well as other tyrosine kinase cascades [7, 25-31]. The importance of the CCR5 mediated signaling [32] in HIV-1 infection was observed in CCR5 Δ 32 [33] mutant populations. Homozygous CCR5 Δ 32 hampers HIV's ability to infiltrate immune cells. Not only this, many other genetic mutations in CCR5 have effect on the progression and transmission of the HIV-1 infection. This mutation exerts resistance to many modes of HIV-1 infection [34].

Variation in the levels of chemokines was observed from person to person. In long-term non-progressors and seronegative individuals (people with repeated exposure to the virus but who do not become infected), unusually high levels of the CCR5 ligands were observed. These chemokines could function as natural competitive inhibitors to HIV-1 infection.[35].

2.4. CXCR4

T-tropic HIV uses CXCR4 as a co-receptor which belongs to the family of α-chemokine receptor. CXCR4 (known as fusin or X4) is also a GPCR with natural ligand CXCL12, known as Stromal Cell-Derived Factor 1 (SDF-1) [36, 37]. T-tropic virus can induce syncytium (SI) and are responsible for the rapid disease progression in HIV-positive individuals. During HIV-1 infections X4-tropic virus has the tendency of emergence and maintains higher viral loads and much lower CD4 cell counts in infected persons. Even though highly virulent, X4 infections are susceptible to antiretroviral therapy [7]. Mutational effect of CXCR4 in HIV-1 infection is not known due to its significant role in development and knockout mutant in mice for this gene is lethal at the embryonic stage [38].

CXCR4 also involved in cell death of $CD4^+$ T-cells which was induced by gp120 indicate an important *in vivo* role for CXCR4 mediated signaling. The interaction of gp120 with CXCR4 triggers a cell death pathway of Fas independent, mitochondrial dependent, cytochrome c mediated activation of caspase-9 and -3 [39]. Membrane fusion dependent $CD4^+$ T-cell death was observed in the virus strains of X4 and dual tropic (R5X4) [40].

However, majority of *in vivo* HIV-1 infection is mediated by M-tropic (R5) viruses could able to lyse their target cells and X4 viruses can kill $CXCR4^+$. It shows that, CD4 has no role in gp120 induced cell death. Moreover, cell lysis and syncytia formation were inhibited in the cells with high levels of CD4 expression. Interestingly, Glycol protein from non-infectious strains of X4 or R5X4 could not induce cell death [41].

Based on the capability to support infection of CD4+ cell lines, other than CCR5 and CXCR4 more than 14 potential co-receptors were identified *in vitro* [6] (Table 1). These receptors are members of (or closely related to) the chemokine receptor family. The significance of other co-receptors for HIV-1 replication *in vivo* and pathogenesis remains unclear.

Recently STRL-33 (CXCR6) functions as co-receptor for HIV-1 infection in primary T-cells [42], and in thymocytes, CCR8 were identified *in vitro* [43].

Co-receptors	Ligands	Role in viral replication	
		In vitro	*In vivo*
CCR1	MIP-1α, RANTES, MPIF-1, MCP-3	+	
CCR2b	MCP-1, MCP-2, MCP-3	+	
CCR3	Eotaxin, Eotaxin-2, MCP-3, MCP-4, RANTES	++	
CCR5	MIP-1α, MIP-1β, RANTES, MCP-2	++++	++++
CCR8	I-309	+	
CCR9	TECK	+	
CXCR4	SDF-1	+++	++
CX3CR1/V28	Fractalkine	+	
STRL-33/BONZO/CXCR6	CXCL16	+	
GPR1	?	+	
GPR15/BOB	?	+	
APJ	Apelin	+	
ChemR23	?	+	
RDC1	?	+	
Leukotriene B_4 receptor	Leukotriene B_4	+	

Table 1. HIV-1 receptors and cell tropism

3. CD4-independent infection

CD4 expression is not uniform in all hematopoietic cells. While, some cell types (T-cells) express high levels of CD4, others, including macrophages and dendritic cells (DCs), express barely detectable amounts. But T-cells as well as macrophages are susceptible to the HIV-1 infection *in vivo.* This susceptibility reveals the existence of alternative host cell surface receptors to which, HIV-1 may attach to cells by CD4-independent manner. Sugar groups present on the both virus and host cell surface (like mannose-specific macrophage endocytosis receptor) mediate the host pathogen interactions and helps in the HIV-1 infection [44]. Apart from the sugars, a cell surface protein (DC-SIGN) of dendritic cells [45, 46]. A closely related receptor to this (DC-SIGNR) on endothelial cells [47], Glycolipids namely galactocerebroside (GalC), galactosulfatide (sGalC) which express on neurons and glia in the brain, colon epithelial cell lines and, importantly, on macrophages [48-50] helps in HIV-1 infection. GalC supports suboptimal entry of particular HIV-1 strains without CD4, although infection requires a co-receptor [51]. Glycosaminoglycans like heparansulphate involved in the infection of HeLa cells [52]. Besides cell surface receptors, cell derived molecules incorporated onto virions such as integrin ICAM-1 (intercellular adhesion molecule-1) and LFA-1 (lymphocyte function-associated antigen-1) [53, 54] enhance the overall efficiency of virus entry.

Primary HIV-2 isolates generally infect CD4⁻ co-receptor⁺ cells more efficiently than HIV-1 [55, 56]. In these infections, CXCR4 plays a crucial role [56].

In glioma cell line (D-54 cells) binding of recombinant gp120 to the GalC or sGalC and a 180 kDa receptor activates signal transduction by a tyrosine-kinase which phosphorylates 130- and 115-kDa proteins [57]. It shows that not only main receptors but also the alternative receptors involved in signal transduction and host protein modifications in HIV-1 infections but the differences were observed in efficiency of infection and disease progression. It also conveys that HIV envelope glycoprotein can specifically bind to the different host cell surface receptors which can function as receptor or co-receptor and this feature provides flexibility to virus to infect wide variety of host cell types [56].

3.1. Fusion and internalization of viral particles

The HIV envelope glycoprotein is responsible not only for the virus attachment to the cell surface but also mediates viral entry. The two parts of the envelope; gp41 and gp120 trimer forms a functional unit, which under goes a series of structural changes (Fig. 2) upon binding to the CD4 and an appropriate chemokine receptors. These interactions promote conformational changes in gp120 and gp41, respectively. These changes exposes the fusion domain of the gp41 and allows to undergo fusion [58-60]. In the viral infection co-receptor has a crucial role in fusion. Fusion of viral particles in the absence of CD4 was observed but in the absence of co-receptor was not yet identified. Following attachment to the receptors, some virus particles enters into endosomes (Fig. 3) later the low pH of the endosomes promotes fusion [61]. However, HIV uses a co-receptor dependent and independent of pH [62].

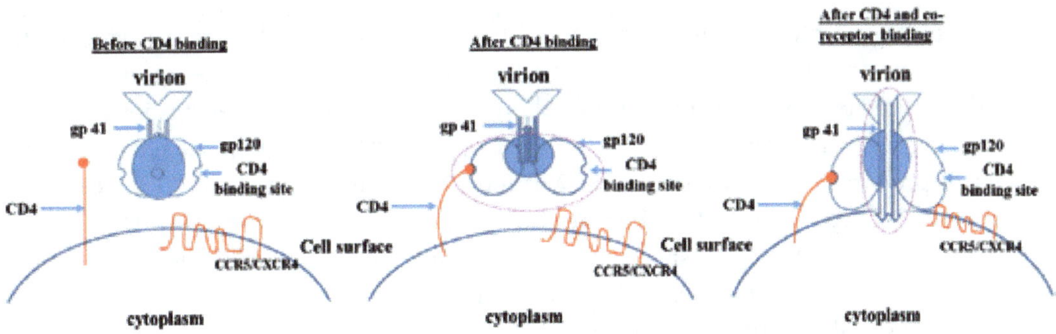

Figure 2. Schematic representation of structural changes in gp120 and gp41. Different conformational changes in gp120 and gp41 up on binding to the host cell surface receptors were represented schematically. From left, structures of the virion envelope gp120 before binding. Once gp120 bound to CD4 of host cell, structural change in the gp120 region in the pink dotted circle. The extreme right represents the changes in the gp120 and especially gp41 once receptor and co-receptor of the host cell interacts with gp120 of virus labeled in pink dotted circle.

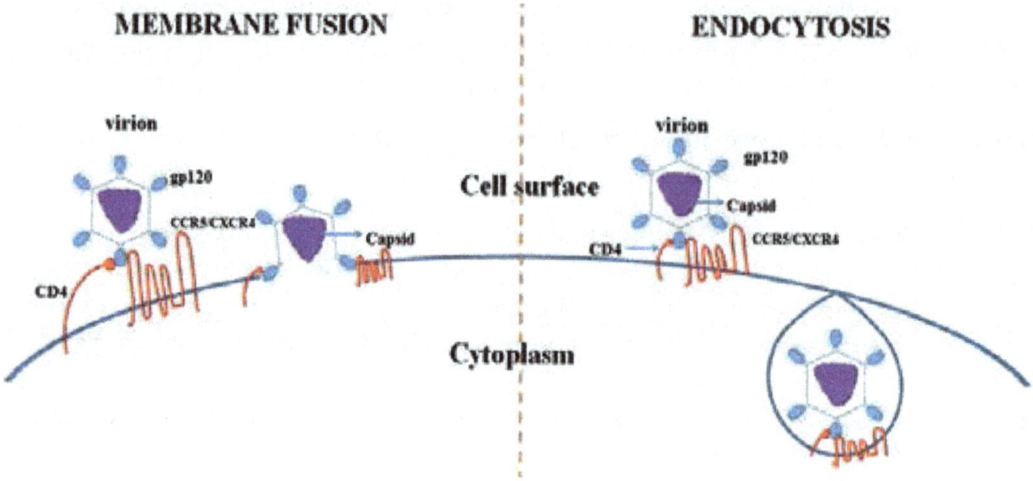

Figure 3. HIV-1 entry strategies. Two well reported strategies of HIV-1 entry into host cell. The direct membrane fusion upon binding to the CD4 and co-receptors like CCR5/CXCR4 and endocytosis mediated entry upon receptor and co-receptors binding were represented schematically.

CD4 is the primary receptor for HIVs, but virus penetration requires further interactions with chemokine receptor CCR5 or CXCR4. Earlier studies proposed that binding to co-receptor initiate fusion directly at the plasma membrane. The ability of formation of syncytia, and evidences of putative fusion events at the cell surface have supported the direct fusion at the plasma membrane [63]. However, recent evidence with endocytosis inhibitors and single-particle tracking revealed fusion and infection occur after endocytic uptake [64, 65]. Moreover, in macrophages HIV infection occurs through macropinocytosis [66, 67]. Based on the receptor density and mobility, the mode of viral entry to the host cell whether fusion on cell surface or after endocytosis will be determined [68].

3.2. Receptor-mediated endocytosis

Adhesion to the receptor initiates later events which enable viruses to enter the cytosol. The cortex is the potential barrier to prevent entry of large molecules [69]. Receptor-mediated signaling induced by envelope of the viruses allow viruses that undergo penetration at the cell surface and transit the cortex [70-72]. In HIV, Gαi (a heterotrimeric G protein subunit that inhibits the production of cAMP from ATP [73]) and CXCR4 on resting CD4+ve T cells activate cofilin which induces reorganization of the cortex that facilitates infection [71]. This evolvement of use of different mechanisms of viral invasion into host cell may have distinct advantages for virus, which provides broader range of cell types to infect or different ways to infect a same cell type [68] to bypass the host restriction mechanisms.

Immediately after its release into the cytoplasm, the viral core undergoes a partial and progressive disassembly, known as uncoating, that leads to the generation of sub viral particles called reverse-transcription complexes (RTCs) and pre-integration complexes (PICs). In HIV-1 the uncoating of capsid is coupled to the initiation of reverse transcription [74].

4. Reverse transcription of viral RNA

4.1. Uncoating of capsid

The events in retrovirus infection that occur between entry into the host cell and reverse transcription are less understood. Uncoating is the process of disintegration of the capsid and release of its components into the cytoplasm of the host cell. The uncoating of the HIV-1 capsid is thought to precede reverse transcription, whereas MLV capsid proteins remain associated with the reverse transcription and pre-integration complexes [75-77]. The uncoating of capsid is a temporarily regulated [78] and the host cell factors involved are poorly understood. There are few known host proteins function as restriction factors and block the HIV-1 infection. One of them is TRIM5α, which prevents retrovirus infection by disrupting an early, post entry events by associating with the retroviral capsid [79, 80]. In owl monkeys, RBCC domains of TRIM5 fused with cyclophilin A (CypA), a known capsid ligand, was identified. [81-87]. Mechanism of Cyp A in capsid uncoating or some other step in the post-entry phases of the HIV-1 life cycle not clearly understood [88]. TRIM5 has multiple roles in early infection. It can interfere with the uncoating process, intercept reverse transcription and viral genome transport to the nucleus by binding to the capsid proteins [89].

In vitro, uncoating of HIV-1 capsid required activated CD4+ lymphocytes. Two distinct cellular factors with molecular mass of approximately 60 and 160 kDa were found to be involved in capsid uncoating [90]. In contrary, cyclophilins mediated activation of T-cells by regulating the activity of calcineurin, a phosphatase necessary for T-cell activation was reported. [91]

4.2. Reverse transcription (Reverse transcriptase and cellular factors: regulators of HIV-1 reverse transcription)

Reverse transcription takes place in a complex organization called reverse transcription complex (RTC), is a nucleoprotein complex comprising viral RNA, a tRNA primer and newly synthesized DNA, along with these nucleic acids, viral factors and host factors present. Reverse transcription complex (RTC) is a composite organization of nucleoprotein complex comprising of viral RNA, tRNA primer and synthesized DNA where the reverse transcription takes place. RTC comprises of several other nucleic acids, host and viral factors.

The first host factors in association with reverse transcription which are packaged in virus particles is cellular $tRNA^{Lys3}$, which is bound to the aminoacyl-$tRNA^{Lys3}$, synthetase (LysRS) [92]. $tRNA^{Lys3}$ works as a primer by binding to the primer binding site (PBS) [93] of the HIV-1 genomic mRNA and is the first step in reverse transcription. Initiation of reverse transcription refers to the addition of the first five deoxynucleotides to the $tRNA^{Lys3}$ primer [94]. Several cellular factors which bound to Integrase (IN) and effects reverse transcription were identified [95] but their direct role in reverse transcription not yet illustrated. Some of the known cellular proteins are integrase interactor 1 (INI1, hSNF5) [96, 97], sin3A-associated protein (SAP18), histone deactylase 1 (HDAC1) [98] and survival motor neuron (SMN)-interacting protein 2 (Gemin2) [99].

INI1 is a component of the SWI/SNF chromatin remodeling complex of host cell [100] and is associated with virus. INI 1 Packaging into virus is specific for HIV-1 and is regulated by a direct interaction with the IN domain of the HIV-1 Gag-Pol protein [101]. INI1 associates with viral RTC/PIC [102] and stimulates IN [96] activity moreover involved in the regulation of reverse transcription.

Like INI1 of SWI/SNF complex, Sin3a-HDAC1 complex members Sin3a, Sap18, Sap30, and HDAC1 were found in virion [98]. HDAC1 is required for the initiation of reverse transcription and involved in a step between uncoating and reverse transcription which results in defective viral cDNA synthesis [98]. Another IN binding protein is Gemin2 is shown to be required for an early reverse transcription product (negative strand strong stop) or integration of viral DNA, suggesting that Gemin2 association with either the reverse transcription or pre-integration complexes [99]. It can recruit other cellular factors like DHX9 (RNA helicase A) [103], which intern associated with the SMN complex [104]. But the precise mechanism of how Gemin2 affects early replication remains to be determined.

Human antigen R (HuR) is a nuclear protein with nucleocytoplasmic shuttling capabilities [105]. It is a RNA binding protein with 3 RNA binding domains and exhibits high specificity and affinity for AU-rich elements (AREs) [106, 107]. HuR is required for optimal reverse transcription. While the mechanism behind this activity remains unclear, it appears to be due to an interaction with the RNase H domain of RT [108]. APOBEC3G (hA3G), is a host protein with negative effects on reverse transcription [109-112]. Furthermore, hA3G was a member of several ribonucleoproteins including DHX9 [103], hnRNP U [113], PABPC1 [114], YB-1 and SNRPA [115] which can affect HIV-1 replication including reverse transcription. A direct interaction between hA3G and HuR was reported [116]. By formation of protein complex, hA3G and HuR could regulate the functions of RTC in the cytoplasm.

A kinase anchor protein 1 (AKAP1) which bind the regulatory subunits of cAMP-dependent protein kinase A (PKA) and anchors them to various membranes throughout the cell [117]. Interaction between AKAP149 and HIV-1 RT was also reported [108]. Like HuR, it interacts with the RNase H region of RT but the mechanism of RT regulation was not yet clearly understood. DNA topoisomerases 1 (TOP1) is another host protein which interact with HIV-1 NC and participate in the initiation of the cDNA synthesis by enhancing the activity of HIV-1 RT [118, 119].

Several studies were conducted on the role of Topoisomerase II in HIV-1 infection. In eukaryotes two isoforms of topoisomerase II (Topo II) was identified. The smaller one is 170kDa topoisomerase II alpha (Topo IIα) and the bigger one is 180 kDa topoisomerase II beta (Topo IIβ) [120]. Recent evidence has proven their key role in viral infections [121, 122]. In response to HIV-1 infection, increased protein levels of Topo IIα and β were observed [123, 124]. Both of these isoforms were reported to undergo phosphorylation in HIV-1 infection [125-127]. These isoforms are phosphorylated by serine kinase present in the purified HIV-1 virion and are associated with pre-integration complexes (PICs) [126]. Topo II inhibitors abrogate HIV-1 replication cycle by interfering with the PICs formation [128]. Down regulation if these isoforms using siRNA resulting in impaired HIV-1 replication [124,129, 130] due to the incomplete reverse transcription. Co-localization studies reveled the association of these isoforms with HIV-1 reverse transcriptase [124]. Similar to this, another host protein HMGI(Y) reported to be with involvement in covalent strand transfer in HIV-1 reverse transcription. [131].

APOBEC3G (apolipoprotein B mRNA-editing enzyme catalytic polypeptide-like 3G, (A3G)) belong to a group of interferon-stimulated gene [132] and is an editing enzyme for nucleic acids. It blocks virus replication by deamination of viral minus-strand DNA, resulting in G-to-A hyper mutation. In addition to the deaminase activity, A3G has also been shown to directly inhibit HIV-1 reverse transcription by a non-editing mechanism [133, 134]. A3G may also reduce viral DNA synthesis and can inducing viral DNA degradation by interacting physically with HIV-1 reverse transcriptase [135]. Its action was blocked by the viral protein Vif and another host protein apoptosis signal-regulating kinase 1 (ASK1) by binding to Vif restores A3G function [136]. Cyclin-dependent kinase (CDK) 2 is a host protein which regulates the reverse transcriptase by phosphorylating on threonine which improves the increased efficiency and stability of reverse transcriptase and enhanced viral fitness. p21, a cell-intrinsic CDK inhibitor, counteracts the CDK2-dependent phosphorylation and significantly reduced the efficacy of viral reverse transcription [137].

4.3. Pre-integration complex formation

Immediately after disintegration of the capsid into the host cytosol, the viral +stand genomic RNA converted into double stranded DNA. This newly synthesized viral genomic DNA wrapped around the host and viral proteins in a protective manner and protected from nuclease degradation [138]. The viral single-stranded RNA genome is converted into a linear double-stranded DNA. The viral DNA intermediate then migrates to the cell nucleus and is covalently integrated into a host chromosome. The integration of reverse transcribed HIV-1 cDNA into a host cell chromosome is an essential step in the viral replication cycle [108, 109]. Retroviral integration *in vivo* is mediated by pre-integration complexes (PICs). PICs are be formed with viral and host cell proteins like high-mobility group protein A1 (HMGA1) and the barrier-to-auto-integration factor (BAF) which were well studied and identified their functions as cofactors for integration [139]. HMG I(Y) is another host protein required for the proper function of the PICs *in vitro* [131]. Other proteins like XRCC6, TFRC and HSP70 were identified as in association with viral DNA [140]. Topoisomerase IIα and β isoforms were also identified as nucleoprotein components of PICs, which suggesting their significant role in HIV-1 replication [128].

5. Integration of viral DNA into host genome

Integration of the viral DNA is mediated by PICs formation. These PICs are capable of performing integration *in vitro*. Even though specific (HMG I(Y)) and non-specific (bovine RNase A) could able to restore the activity, but only BAF can restore the native structure of the HIV-1 protein–DNA intasome from salt stripped PICs [141]. HIV-1 integration in host genome is not a random event. In majority of cases, it takes place in AT rich, euchromatin region. Even though PIC formation can protect the DNA and helps in the nuclear transport, it cannot guide the integration in a proper location in the host genome. A host protein called cellular lens epithelium-derived growth factor (LEDGF/p75) which binds both chromosomal DNA and HIV integrase [142], directs the integration to a location where active transcription takes place under its control [143]. The interaction between the integrase and INI1 stimulates the DNA-joining activity of the integrase and helps to target the viral DNA towards active genes [96, 102, 144]. Presence of a Topo II cleavage site in the HIV-1 promoter and also at 180 bp upstream of the HIV-1 integration site [145, 146] and association of these Topo II isoforms with HIV-1 PICs [128]. Based on the available information, it can be derived that, IN alone can carry out the integration reaction but for the selection of the proper location in the host genome for the integration and success full HIV-1 gene transcription, Integrase required the support of the host factors.

5.1. Transport of PICs into nucleus

Transportation of the PICs to the nucleus can be divided in two parts. One is transportation to the nuclear periphery and second is from periphery to inside the nucleus. After completion of the successful cell surface attachment, capsid internalization and degradation in cytoplasm which leads to the reverse transcription of the viral genome RNA to DNA and the formation of the pre-integration complex to protect the DNA from the host nucleases takes place. This nucleoprotein PICs should travel from the cytosol to the nucleus to form provirus which is an integrated viral DNA in host chromatin and can produce progeny virus [147, 148].

The PICs in the cytoplasm translocate to the perinuclear compartment by the cytoplasmic movements by the cytoskeleton. Actin and microtubule [149] selectively plays a role in this transport with the help of Myosin VI and Dynein [148]. The dynein complex proteins such as dynein light chain1 (DYNLL1), Tctex1 and Dynactin have been shown to be involved in this process. But very little is known about how HIV-1 targets cytoskeleton. Many of the viral elements found in association with the PIC have been proposed to be important for HIV-1 nuclear import.

Studies on the HIV-1 infection in dividing and non-dividing cells provided enough evidences to believe that PICs enter the nucleoplasm by crossing the nuclear envelope through nuclear pore complexes (NPCs), which form stable channels through the nuclear envelope and gate-keep the trafficking of molecules between the nucleus and cytoplasm [150]. The RTC/PIC with the size of~100-250nm [149, 151, 152] cannot cross the nuclear pore. So, only few important components of the PIC may enter into the nucleus [149, 152].

For nuclear transport, KPNs and NUPs functions as carrier proteins by binding to the integrase. KPN α adaptor proteins importin α1 (Rch1) [153] and importin α3 (KPNA4) [154], to which KPN β1 proteins bound additionally. Importin 7 [155, 156] and transportin 3 [157, 158] are recruited by the nuclear localization signal (NLS) present on IN. In both dividing and non-dividing cells Impα3 which interacts with IN is found to be essential nuclear import and replication [154]. In addition to these, IN can directly interact with the KPN β1, NUP153 [159], Pom121 [160] or hCG1 [161] who has possible interactions with IN and Vpr and can facilitate nuclear import [162].

6. Summary

After 33 years of the identification of HIV-1 infection, very little is known about the role of host cellular proteins. Till now considerable work has been done in the area of host–pathogen interactions facilitated by the viral proteins and host receptors. The role of the main receptor like CD4 and co-receptors like CCR5, CXCR4 and their alternative receptors were well studied with the role of their signaling in disease progression. But the intracellular events of the host–pathogen interactions were poorly understood. Much data is available based on the global wide analysis of genome-wide RNA interference screens, yeast two-hybrid system and co-immunoprecipitation studies but their exact roles were not yet characterized. There are very few host proteins like APOBEC3G, LEDGF/p75, INI1, HMG I(Y) and BAF, which were well studied and characterized. Majority of the reported proteins were attributed with multiple functions. It is very useful to study such proteins to develop as future candidates to HIV-1 therapy.

Author details

Lokeswara S. Balakrishna and Anand K. Kondapi*

*Address all correspondence to: akondapi@gmail.com

Department of Biotechnology and Bioinformatics, School of Life Sciences, University of Hyderabad, P O Central University, Hyderabad, Telangana, India

References

[1] Moulard, M. and E. Decroly, *Maturation of HIV envelope glycoprotein precursors by cellular endoproteases.* Biochim Biophys Acta, 2000. 1469(3): p. 121-32.

[2] Weiss, C.D., J.A. Levy, and J.M. White, *Oligomeric organization of gp120 on infectious human immunodeficiency virus type 1 particles.* J Virol, 1990. 64(11): p. 5674-7.

[3] Earl, P.L., R.W. Doms, and B. Moss, *Oligomeric structure of the human immunodeficiency virus type 1 envelope glycoprotein.* Proc Natl Acad Sci U S A, 1990. 87(2): p. 648-52.

[4] Sattentau, Q.J., et al., *The human and simian immunodeficiency viruses HIV-1, HIV-2 and SIV interact with similar epitopes on their cellular receptor, the CD4 molecule.* AIDS, 1988. 2(2): p. 101-5.

[5] Salzwedel, K. and E.A. Berger, *Cooperative subunit interactions within the oligomeric envelope glycoprotein of HIV-1: functional complementation of specific defects in gp120 and gp41.* Proc Natl Acad Sci U S A, 2000. 97(23): p. 12794-9.

[6] Clapham, P.R. and A. McKnight, *HIV-1 receptors and cell tropism.* Br Med Bull, 2001. 58: p. 43-59.

[7] Popik, W., J.E. Hesselgesser, and P.M. Pitha, *Binding of human immunodeficiency virus type 1 to CD4 and CXCR4 receptors differentially regulates expression of inflammatory genes and activates the MEK/ERK signaling pathway.* J Virol, 1998. 72(8): p. 6406-13.

[8] Popik, W. and P.M. Pitha, *Binding of human immunodeficiency virus type 1 to CD4 induces association of Lck and Raf-1 and activates Raf-1 by a Ras-independent pathway.* Mol Cell Biol, 1996. 16(11): p. 6532-41.

[9] Ravichandran, K.S., T.L. Collins, and S.J. Burakoff, *CD4 and signal transduction.* Curr Top Microbiol Immunol, 1996. 205: p. 47-62.

[10] Power, C.A. and T.N. Wells, *Cloning and characterization of human chemokine receptors.* Trends Pharmacol Sci, 1996. 17(6): p. 209-13.

[11] Baribaud, F., et al., *Antigenically distinct conformations of CXCR4.* J Virol, 2001. 75(19): p. 8957-67.

[12] Lee, B., et al., *Epitope mapping of CCR5 reveals multiple conformational states and distinct but overlapping structures involved in chemokine and coreceptor function.* J Biol Chem, 1999. 274(14): p. 9617-26.

[13] Picard, L., et al., *Role of the amino-terminal extracellular domain of CXCR-4 in human immunodeficiency virus type 1 entry.* Virology, 1997. 231(1): p. 105-11.

[14] Doranz, B.J., et al., *Identification of CXCR4 domains that support coreceptor and chemokine receptor functions.* J Virol, 1999. 73(4): p. 2752-61.

[15] Kwong, P.D., et al., *Structure of an HIV gp120 envelope glycoprotein in complex with the CD4 receptor and a neutralizing human antibody.* Nature, 1998. 393(6686): p. 648-59.

[16] Davis, C.B., et al., *Signal transduction due to HIV-1 envelope interactions with chemokine receptors CXCR4 or CCR5.* J Exp Med, 1997. 186(10): p. 1793-8.

[17] Hesselgesser, J., et al., *CD4-independent association between HIV-1 gp120 and CXCR4: functional chemokine receptors are expressed in human neurons.* Curr Biol, 1997. 7(2): p. 112-21.

[18] Weissman, D., et al., *Macrophage-tropic HIV and SIV envelope proteins induce a signal through the CCR5 chemokine receptor.* Nature, 1997. 389(6654): p. 981-5.

[19] Marchese, A., et al., *G protein-coupled receptor sorting to endosomes and lysosomes.* Annu Rev Pharmacol Toxicol, 2008. 48: p. 601-29.

[20] Viola, A. and A.D. Luster, *Chemokines and their receptors: drug targets in immunity and inflammation.* Annu Rev Pharmacol Toxicol, 2008. 48: p. 171-97.

[21] Cicala, C., et al., *Induction of phosphorylation and intracellular association of CC chemokine receptor 5 and focal adhesion kinase in primary human CD4+ T cells by macrophage-tropic HIV envelope.* J Immunol, 1999. 163(1): p. 420-6.

[22] Arthos, J., et al., *CCR5 signal transduction in macrophages by human immunodeficiency virus and simian immunodeficiency virus envelopes.* J Virol, 2000. 74(14): p. 6418-24.

[23] Rottman, J.B., et al., *Cellular localization of the chemokine receptor CCR5. Correlation to cellular targets of HIV-1 infection.* Am J Pathol, 1997. 151(5): p. 1341-51.

[24] Blanpain, C., et al., *CCR5 and HIV infection.* Receptors Channels, 2002. 8(1): p. 19-31.

[25] Furler, R.L. and C.H. Uittenbogaart, *Signaling through the P38 and ERK pathways: a common link between HIV replication and the immune response.* Immunol Res, 2010. 48(1-3): p. 99-109.

[26] Cohen, P.S., et al., *The critical role of p38 MAP kinase in T cell HIV-1 replication.* Mol Med, 1997. 3(5): p. 339-46.

[27] Francois, F. and M.E. Klotman, *Phosphatidylinositol 3-kinase regulates human immunodeficiency virus type 1 replication following viral entry in primary CD4+ T lymphocytes and macrophages.* J Virol, 2003. 77(4): p. 2539-49.

[28] Del Corno, M., et al., *HIV-1 gp120 and chemokine activation of Pyk2 and mitogen-activated protein kinases in primary macrophages mediated by calcium-dependent, pertussis toxin-insensitive chemokine receptor signaling.* Blood, 2001. 98(10): p. 2909-16.

[29] Lee, C., et al., *Macrophage activation through CCR5- and CXCR4-mediated gp120-elicited signaling pathways.* J Leukoc Biol, 2003. 74(5): p. 676-82.

[30] Popik, W. and P.M. Pitha, *Early activation of mitogen-activated protein kinase kinase, extracellular signal-regulated kinase, p38 mitogen-activated protein kinase, and c-Jun N-terminal kinase in response to binding of simian immunodeficiency virus to Jurkat T cells expressing CCR5 receptor.* Virology, 1998. 252(1): p. 210-7.

[31] Lopez-Ilasaca, M., *Signaling from G-protein-coupled receptors to mitogen-activated protein (MAP)-kinase cascades.* Biochem Pharmacol, 1998. 56(3): p. 269-77.

[32] Shen, R. and P.D. Smith, *Mucosal correlates of protection in HIV-1-exposed sero-negative persons.* Am J Reprod Immunol, 2014. 72(2): p. 219-27.

[33] Samson, M., et al., *Resistance to HIV-1 infection in caucasian individuals bearing mutant alleles of the CCR-5 chemokine receptor gene.* Nature, 1996. 382(6593): p. 722-5.

[34] Carrington, M., et al., *Genetics of HIV-1 infection: chemokine receptor CCR5 polymorphism and its consequences.* Hum Mol Genet, 1999. 8(10): p. 1939-45.

[35] De Clercq, E. and D. Schols, *Inhibition of HIV infection by CXCR4 and CCR5 chemokine receptor antagonists.* Antivir Chem Chemother, 2001. 12 Suppl 1: p. 19-31.

[36] Murphy, P.M., et al., *International union of pharmacology. XXII. Nomenclature for chemokine receptors.* Pharmacol Rev, 2000. 52(1): p. 145-76.

[37] Fredriksson, R., et al., *The G-protein-coupled receptors in the human genome form five main families. Phylogenetic analysis, paralogon groups, and fingerprints.* Mol Pharmacol, 2003. 63(6): p. 1256-72.

[38] Marechal, V., et al., *Opposite effects of SDF-1 on human immunodeficiency virus type 1 replication.* J Virol, 1999. 73(5): p. 3608-15.

[39] Roggero, R., et al., *Binding of human immunodeficiency virus type 1 gp120 to CXCR4 induces mitochondrial transmembrane depolarization and cytochrome c-mediated apoptosis independently of Fas signaling.* J Virol, 2001. 75(16): p. 7637-50.

[40] Lin, G., et al., *Identification of gp120 binding sites on CXCR4 by using CD4-independent human immunodeficiency virus type 2 Env proteins.* J Virol, 2003. 77(2): p. 931-42.

[41] LaBonte, J.A., N. Madani, and J. Sodroski, *Cytolysis by CCR5-using human immunodeficiency virus type 1 envelope glycoproteins is dependent on membrane fusion and can be inhibited by high levels of CD4 expression.* J Virol, 2003. 77(12): p. 6645-59.

[42] Sharron, M., et al., *Expression and coreceptor activity of STRL33/Bonzo on primary peripheral blood lymphocytes.* Blood, 2000. 96(1): p. 41-9.

[43] Lee, S., et al., *CCR8 on human thymocytes functions as a human immunodeficiency virus type 1 coreceptor.* J Virol, 2000. 74(15): p. 6946-52.

[44] Larkin, M., et al., *Oligosaccharide-mediated interactions of the envelope glycoprotein gp120 of HIV-1 that are independent of CD4 recognition.* AIDS, 1989. 3(12): p. 793-8.

[45] Curtis, B.M., S. Scharnowske, and A.J. Watson, *Sequence and expression of a membrane-associated C-type lectin that exhibits CD4-independent binding of human immunodeficiency virus envelope glycoprotein gp120.* Proc Natl Acad Sci U S A, 1992. 89(17): p. 8356-60.

[46] Geijtenbeek, T.B., et al., *DC-SIGN, a dendritic cell-specific HIV-1-binding protein that enhances trans-infection of T cells.* Cell, 2000. 100(5): p. 587-97.

[47] Pohlmann, S., et al., *DC-SIGNR, a DC-SIGN homologue expressed in endothelial cells, binds to human and simian immunodeficiency viruses and activates infection in trans.* Proc Natl Acad Sci U S A, 2001. 98(5): p. 2670-5.

[48] Fantini, J., et al., *Infection of colonic epithelial cell lines by type 1 human immunodeficiency virus is associated with cell surface expression of galactosylceramide, a potential alternative gp120 receptor.* Proc Natl Acad Sci U S A, 1993. 90(7): p. 2700-4.

[49] Harouse, J.M., et al., *Inhibition of entry of HIV-1 in neural cell lines by antibodies against galactosyl ceramide.* Science, 1991. 253(5017): p. 320-3.

[50] Seddiki, N., et al., *A monoclonal antibody directed to sulfatide inhibits the binding of human immunodeficiency virus type 1 (HIV-1) envelope glycoprotein to macrophages but not their infection by the virus.* Biochim Biophys Acta, 1994. 1225(3): p. 289-96.

[51] Delezay, O., et al., *Co-expression of CXCR4/fusin and galactosylceramide in the human intestinal epithelial cell line HT-29.* AIDS, 1997. 11(11): p. 1311-8.

[52] Mondor, I., S. Ugolini, and Q.J. Sattentau, *Human immunodeficiency virus type 1 attachment to HeLa CD4 cells is CD4 independent and gp120 dependent and requires cell surface heparans.* J Virol, 1998. 72(5): p. 3623-34.

[53] Paquette, J.S., et al., *Level of ICAM-1 surface expression on virus producer cells influences both the amount of virion-bound host ICAM-1 and human immunodeficiency virus type 1 infectivity.* J Virol, 1998. 72(11): p. 9329-36.

[54] Fortin, J.F., et al., *Role of the leukocyte function antigen-1 conformational state in the process of human immunodeficiency virus type 1-mediated syncytium formation and virus infection.* Virology, 1999. 257(1): p. 228-38.

[55] Reeves, J.D., et al., *Primary human immunodeficiency virus type 2 (HIV-2) isolates infect CD4-negative cells via CCR5 and CXCR4: comparison with HIV-1 and simian immunodeficiency virus and relevance to cell tropism in vivo.* J Virol, 1999. 73(9): p. 7795-804.

[56] Endres, M.J., et al., *CD4-independent infection by HIV-2 is mediated by fusin/CXCR4.* Cell, 1996. 87(4): p. 745-56.

[57] Schneider-Schaulies, J., et al., *HIV-1 gp120 receptor on CD4-negative brain cells activates a tyrosine kinase.* Virology, 1992. 191(2): p. 765-72.

[58] Chan, D.C., et al., *Core structure of gp41 from the HIV envelope glycoprotein.* Cell, 1997. 89(2): p. 263-73.

[59] Weissenhorn, W., et al., *Atomic structure of the ectodomain from HIV-1 gp41.* Nature, 1997. 387(6631): p. 426-30.

[60] Binley, J.M., et al., *Redox-triggered infection by disulfide-shackled human immunodeficiency virus type 1 pseudovirions.* J Virol, 2003. 77(10): p. 5678-84.

[61] Skehel, J.J., et al., *Membrane fusion by influenza hemagglutinin.* Cold Spring Harb Symp Quant Biol, 1995. 60: p. 573-80.

[62] Clapham, P.R. and A. McKnight, *Cell surface receptors, virus entry and tropism of primate lentiviruses.* J Gen Virol, 2002. 83(Pt 8): p. 1809-29.

[63] Stein, B.S., et al., *pH-independent HIV entry into CD4-positive T cells via virus envelope fusion to the plasma membrane.* Cell, 1987. 49(5): p. 659-68.

[64] Daecke, J., et al., *Involvement of clathrin-mediated endocytosis in human immunodeficiency virus type 1 entry.* J Virol, 2005. 79(3): p. 1581-94.

[65] Miyauchi, K., et al., *HIV enters cells via endocytosis and dynamin-dependent fusion with endosomes.* Cell, 2009. 137(3): p. 433-44.

[66] Marechal, V., et al., *Human immunodeficiency virus type 1 entry into macrophages mediated by macropinocytosis.* J Virol, 2001. 75(22): p. 11166-77.

[67] Carter, G.C., et al., *HIV-1 infects macrophages by exploiting an endocytic route dependent on dynamin, Rac1 and Pak1.* Virology, 2011. 409(2): p. 234-50.

[68] Grove, J. and M. Marsh, *The cell biology of receptor-mediated virus entry.* J Cell Biol, 2011. 195(7): p. 1071-82.

[69] Marsh, M. and R. Bron, *SFV infection in CHO cells: cell-type specific restrictions to productive virus entry at the cell surface.* J Cell Sci, 1997. 110 (Pt 1): p. 95-103.

[70] Wang, X., et al., *Integrin alphavbeta3 is a coreceptor for human cytomegalovirus.* Nat Med, 2005. 11(5): p. 515-21.

[71] Yoder, A., et al., *HIV envelope-CXCR4 signaling activates cofilin to overcome cortical actin restriction in resting CD4 T cells.* Cell, 2008. 134(5): p. 782-92.

[72] Taylor, M.P., O.O. Koyuncu, and L.W. Enquist, *Subversion of the actin cytoskeleton during viral infection.* Nat Rev Microbiol, 2011. 9(6): p. 427-39.

[73] Birnbaumer, L., *Expansion of signal transduction by G proteins. The second 15 years or so: from 3 to 16 alpha subunits plus betagamma dimers.* Biochim Biophys Acta, 2007. 1768(4): p. 772-93.

[74] Zhang, H., et al., *Morphologic changes in human immunodeficiency virus type 1 virions secondary to intravirion reverse transcription: evidence indicating that reverse transcription may not take place within the intact viral core.* J Hum Virol, 2000. 3(3): p. 165-72.

[75] Fassati, A. and S.P. Goff, *Characterization of intracellular reverse transcription complexes of human immunodeficiency virus type 1.* J Virol, 2001. 75(8): p. 3626-35.

[76] Fassati, A. and S.P. Goff, *Characterization of intracellular reverse transcription complexes of Moloney murine leukemia virus.* J Virol, 1999. 73(11): p. 8919-25.

[77] Miller, M.D., C.M. Farnet, and F.D. Bushman, *Human immunodeficiency virus type 1 preintegration complexes: studies of organization and composition.* J Virol, 1997. 71(7): p. 5382-90.

[78] Forshey, B.M., et al., *Formation of a human immunodeficiency virus type 1 core of optimal stability is crucial for viral replication.* J Virol, 2002. 76(11): p. 5667-77.

[79] Stremlau, M., et al., *The cytoplasmic body component TRIM5alpha restricts HIV-1 infection in Old World monkeys.* Nature, 2004. 427(6977): p. 848-53.

[80] Perron, M.J., et al., *TRIM5alpha mediates the postentry block to N-tropic murine leukemia viruses in human cells.* Proc Natl Acad Sci U S A, 2004. 101(32): p. 11827-32.

[81] Cowan, S., et al., *Cellular inhibitors with Fv1-like activity restrict human and simian immunodeficiency virus tropism.* Proc Natl Acad Sci U S A, 2002. 99(18): p. 11914-9.

[82] Besnier, C., Y. Takeuchi, and G. Towers, *Restriction of lentivirus in monkeys.* Proc Natl Acad Sci U S A, 2002. 99(18): p. 11920-5.

[83] Dodding, M.P., et al., *Capsid processing requirements for abrogation of Fv1 and Ref1 restriction.* J Virol, 2005. 79(16): p. 10571-7.

[84] Luban, J., et al., *Human immunodeficiency virus type 1 Gag protein binds to cyclophilins A and B.* Cell, 1993. 73(6): p. 1067-78.

[85] Nisole, S., et al., *A Trim5-cyclophilin A fusion protein found in owl monkey kidney cells can restrict HIV-1.* Proc Natl Acad Sci U S A, 2004. 101(36): p. 13324-8.

[86] Yap, M.W., et al., *Trim5alpha protein restricts both HIV-1 and murine leukemia virus.* Proc Natl Acad Sci U S A, 2004. 101(29): p. 10786-91.

[87] Thali, M., et al., *Functional association of cyclophilin A with HIV-1 virions.* Nature, 1994. 372(6504): p. 363-5.

[88] Grattinger, M., et al., *In vitro assembly properties of wild-type and cyclophilin-binding defective human immunodeficiency virus capsid proteins in the presence and absence of cyclophilin A.* Virology, 1999. 257(1): p. 247-60.

[89] Stremlau, M., et al., *Specific recognition and accelerated uncoating of retroviral capsids by the TRIM5alpha restriction factor.* Proc Natl Acad Sci U S A, 2006. 103(14): p. 5514-9.

[90] Auewarakul, P., et al., *Uncoating of HIV-1 requires cellular activation.* Virology, 2005. 337(1): p. 93-101.

[91] Schreiber, S.L. and G.R. Crabtree, *The mechanism of action of cyclosporin A and FK506.* Immunol Today, 1992. 13(4): p. 136-42.

[92] Cen, S., et al., *Incorporation of lysyl-tRNA synthetase into human immunodeficiency virus type 1.* J Virol, 2001. 75(11): p. 5043-8.

[93] Saadatmand, J., et al., *Interactions of reverse transcriptase sequences in Pol with Gag and LysRS in the HIV-1 tRNALys3 packaging/annealing complex.* Virology, 2008. 380(1): p. 109-17.

[94] Dobard, C.W., M.S. Briones, and S.A. Chow, *Molecular mechanisms by which human immunodeficiency virus type 1 integrase stimulates the early steps of reverse transcription.* J Virol, 2007. 81(18): p. 10037-46.

[95] Luban, J. and S.P. Goff, *The yeast two-hybrid system for studying protein-protein interactions.* Curr Opin Biotechnol, 1995. 6(1): p. 59-64.

[96] Kalpana, G.V., et al., *Binding and stimulation of HIV-1 integrase by a human homolog of yeast transcription factor SNF5.* Science, 1994. 266(5193): p. 2002-6.

[97] Rain, J.C., et al., *Yeast two-hybrid detection of integrase-host factor interactions.* Methods, 2009. 47(4): p. 291-7.

[98] Sorin, M., et al., *Recruitment of a SAP18-HDAC1 complex into HIV-1 virions and its requirement for viral replication.* PLoS Pathog, 2009. 5(6): p. e1000463.

[99] Hamamoto, S., et al., *Identification of a novel human immunodeficiency virus type 1 integrase interactor, Gemin2, that facilitates efficient viral cDNA synthesis in vivo.* J Virol, 2006. 80(12): p. 5670-7.

[100] Wang, W., et al., *Diversity and specialization of mammalian SWI/SNF complexes.* Genes Dev, 1996. 10(17): p. 2117-30.

[101] Yung, E., et al., *Specificity of interaction of INI1/hSNF5 with retroviral integrases and its functional significance.* J Virol, 2004. 78(5): p. 2222-31.

[102] Turelli, P., et al., *Cytoplasmic recruitment of INI1 and PML on incoming HIV preintegration complexes: interference with early steps of viral replication.* Mol Cell, 2001. 7(6): p. 1245-54.

[103] Roy, B.B., et al., *Association of RNA helicase a with human immunodeficiency virus type 1 particles.* J Biol Chem, 2006. 281(18): p. 12625-35.

[104] Pellizzoni, L., et al., *A functional interaction between the survival motor neuron complex and RNA polymerase II.* J Cell Biol, 2001. 152(1): p. 75-85.

[105] Fan, X.C. and J.A. Steitz, *Overexpression of HuR, a nuclear-cytoplasmic shuttling protein, increases the in vivo stability of ARE-containing mRNAs.* EMBO J, 1998. 17(12): p. 3448-60.

[106] Myer, V.E., X.C. Fan, and J.A. Steitz, *Identification of HuR as a protein implicated in AUUUA-mediated mRNA decay.* EMBO J, 1997. 16(8): p. 2130-9.

[107] Vakalopoulou, E., J. Schaack, and T. Shenk, *A 32-kilodalton protein binds to AU-rich domains in the 3′ untranslated regions of rapidly degraded mRNAs.* Mol Cell Biol, 1991. 11(6): p. 3355-64.

[108] Lemay, J., et al., *HuR interacts with human immunodeficiency virus type 1 reverse transcriptase, and modulates reverse transcription in infected cells.* Retrovirology, 2008. 5: p. 47.

[109] Bishop, K.N., et al., *APOBEC3G inhibits elongation of HIV-1 reverse transcripts.* PLoS Pathog, 2008. 4(12): p. e1000231.

[110] Guo, F., et al., *Inhibition of tRNA(3)(Lys)-primed reverse transcription by human APOBEC3G during human immunodeficiency virus type 1 replication.* J Virol, 2006. 80(23): p. 11710-22.

[111] Guo, F., et al., *The interaction of APOBEC3G with human immunodeficiency virus type 1 nucleocapsid inhibits tRNA3Lys annealing to viral RNA.* J Virol, 2007. 81(20): p. 11322-31.

[112] Li, X.Y., et al., *APOBEC3G inhibits DNA strand transfer during HIV-1 reverse transcription.* J Biol Chem, 2007. 282(44): p. 32065-74.

[113] Valente, S.T. and S.P. Goff, *Inhibition of HIV-1 gene expression by a fragment of hnRNP U.* Mol Cell, 2006. 23(4): p. 597-605.

[114] Afonina, E., M. Neumann, and G.N. Pavlakis, *Preferential binding of poly(A)-binding protein 1 to an inhibitory RNA element in the human immunodeficiency virus type 1 gag mRNA.* J Biol Chem, 1997. 272(4): p. 2307-11.

[115] Konig, R., et al., *Global analysis of host-pathogen interactions that regulate early-stage HIV-1 replication.* Cell, 2008. 135(1): p. 49-60.

[116] Gallois-Montbrun, S., et al., *Antiviral protein APOBEC3G localizes to ribonucleoprotein complexes found in P bodies and stress granules.* J Virol, 2007. 81(5): p. 2165-78.

[117] Feliciello, A., M.E. Gottesman, and E.V. Avvedimento, *The biological functions of A-kinase anchor proteins.* J Mol Biol, 2001. 308(2): p. 99-114.

[118] Takahashi, H., et al., *Human immunodeficiency virus type 1 reverse transcriptase: enhancement of activity by interaction with cellular topoisomerase I.* Proc Natl Acad Sci U S A, 1995. 92(12): p. 5694-8.

[119] Shoya, Y., et al., *Human topoisomerase I promotes HIV-1 proviral DNA synthesis: implications for the species specificity and cellular tropism of HIV-1 infection.* Proc Natl Acad Sci U S A, 2003. 100(14): p. 8442-7.

[120] Drake, F.H., et al., *Purification of topoisomerase II from amsacrine-resistant P388 leukemia cells. Evidence for two forms of the enzyme.* J Biol Chem, 1987. 262(34): p. 16739-47.

[121] Rainwater, R. and K. Mann, *Differential increase in topoisomerase II in simian virus 40-infected cells.* J Virol, 1990. 64(2): p. 918-21.

[122] Benson, J.D. and E.S. Huang, *Two specific topoisomerase II inhibitors prevent replication of human cytomegalovirus DNA: an implied role in replication of the viral genome.* J Virol, 1988. 62(12): p. 4797-800.

[123] Mandraju, R.K. and A.K. Kondapi, *Regulation of topoisomerase II alpha and beta in HIV-1 infected and uninfected neuroblastoma and astrocytoma cells: involvement of distinct nordihydroguaretic acid sensitive inflammatory pathways.* Arch Biochem Biophys, 2007. 461(1): p. 40-9.

[124] Lokeswara Balakrishna, S., N. Satyanarayana, and A.K. Kondapi, *Involvement of human topoisomerase II isoforms in HIV-1 reverse transcription.* Arch Biochem Biophys, 2013. 532(2): p. 91-102.

[125] Muller, W.E., et al., *Effect of nonviable preparations from human immunodeficiency virus type 1 on nuclear matrix-associated DNA polymerase alpha and DNA topoisomerase II activities.* J Acquir Immune Defic Syndr, 1990. 3(1): p. 1-10.

[126] Kondapi, A.K., et al., *A biochemical analysis of topoisomerase II alpha and beta kinase activity found in HIV-1 infected cells and virus.* Arch Biochem Biophys, 2005. 441(1): p. 41-55.

[127] Ponraj, K., et al., *HIV-1 associated Topoisomerase IIbeta kinase: a potential pharmacological target for viral replication.* Curr Pharm Des, 2013. 19(26): p. 4776-86.

[128] Kondapi, A.K., N. Satyanarayana, and A.D. Saikrishna, *A study of the topoisomerase II activity in HIV-1 replication using the ferrocene derivatives as probes.* Arch Biochem Biophys, 2006. 450(2): p. 123-32.

[129] Filion, L.G., et al., *Inhibition of HIV-1 replication by daunorubicin.* Clin Invest Med, 1993. 16(5): p. 339-47.

[130] Bouille, P., et al., *Antisense-mediated repression of DNA topoisomerase II expression leads to an impairment of HIV-1 replicative cycle.* J Mol Biol, 1999. 285(3): p. 945-54.

[131] Farnet, C.M. and F.D. Bushman, *HIV-1 cDNA integration: requirement of HMG I(Y) protein for function of preintegration complexes in vitro.* Cell, 1997. 88(4): p. 483-92.

[132] Strebel, K., J. Luban, and K.T. Jeang, *Human cellular restriction factors that target HIV-1 replication.* BMC Med, 2009. 7: p. 48.

[133] Malim, M.H., *APOBEC proteins and intrinsic resistance to HIV-1 infection.* Philos Trans R Soc Lond B Biol Sci, 2009. 364(1517): p. 675-87.

[134] Luo, K., et al., *Cytidine deaminases APOBEC3G and APOBEC3F interact with human immunodeficiency virus type 1 integrase and inhibit proviral DNA formation.* J Virol, 2007. 81(13): p. 7238-48.

[135] Wang, X., et al., *The cellular antiviral protein APOBEC3G interacts with HIV-1 reverse transcriptase and inhibits its function during viral replication.* J Virol, 2012. 86(7): p. 3777-86.

[136] Miyakawa, K., et al., *ASK1 restores the antiviral activity of APOBEC3G by disrupting HIV-1 Vif-mediated counteraction.* Nat Commun, 2015. 6: p. 6945.

[137] Leng, J., et al., *A cell-intrinsic inhibitor of HIV-1 reverse transcription in CD4(+) T cells from elite controllers.* Cell Host Microbe, 2014. 15(6): p. 717-28.

[138] Khiytani, D.K. and N.J. Dimmock, *Characterization of a human immunodeficiency virus type 1 pre-integration complex in which the majority of the cDNA is resistant to DNase I digestion.* J Gen Virol, 2002. 83(Pt 10): p. 2523-32.

[139] Lin, C.W. and A. Engelman, *The barrier-to-autointegration factor is a component of functional human immunodeficiency virus type 1 preintegration complexes.* J Virol, 2003. 77(8): p. 5030-6.

[140] Schweitzer, C.J., et al., *Proteomic analysis of early HIV-1 nucleoprotein complexes.* J Proteome Res, 2013. 12(2): p. 559-72.

[141] Chen, H. and A. Engelman, *The barrier-to-autointegration protein is a host factor for HIV type 1 integration.* Proc Natl Acad Sci U S A, 1998. 95(26): p. 15270-4.

[142] Llano, M., et al., *An essential role for LEDGF/p75 in HIV integration.* Science, 2006. 314(5798): p. 461-4.

[143] Ciuffi, A., et al., *A role for LEDGF/p75 in targeting HIV DNA integration.* Nat Med, 2005. 11(12): p. 1287-9.

[144] Wang, W., et al., *Purification and biochemical heterogeneity of the mammalian SWI-SNF complex.* EMBO J, 1996. 15(19): p. 5370-82.

[145] Howard, M.T. and J.D. Griffith, *A cluster of strong topoisomerase II cleavage sites is located near an integrated human immunodeficiency virus.* J Mol Biol, 1993. 232(4): p. 1060-8.

[146] Pommier, Y., et al., *DNA topoisomerases I & II cleavage sites in the type 1 human immunodeficiency virus (HIV-1) DNA promoter region.* Biochem Biophys Res Commun, 1994. 205(3): p. 1601-9.

[147] De Rijck, J., et al., *Lentiviral nuclear import: a complex interplay between virus and host.* Bioessays, 2007. 29(5): p. 441-51.

[148] Jayappa, K.D., Z. Ao, and X. Yao, *The HIV-1 passage from cytoplasm to nucleus: the process involving a complex exchange between the components of HIV-1 and cellular machinery to access nucleus and successful integration.* Int J Biochem Mol Biol, 2012. 3(1): p. 70-85.

[149] McDonald, D., et al., *Visualization of the intracellular behavior of HIV in living cells.* J Cell Biol, 2002. 159(3): p. 441-52.

[150] Hoelz, A., E.W. Debler, and G. Blobel, *The structure of the nuclear pore complex.* Annu Rev Biochem, 2011. 80: p. 613-43.

[151] Pereira, C.F., et al., *HIV taken by STORM: super-resolution fluorescence microscopy of a viral infection.* Virol J, 2012. 9: p. 84.

[152] Lelek, M., et al., *Superresolution imaging of HIV in infected cells with FlAsH-PALM.* Proc Natl Acad Sci U S A, 2012. 109(22): p. 8564-9.

[153] Gallay, P., et al., *HIV-1 infection of nondividing cells through the recognition of integrase by the importin/karyopherin pathway.* Proc Natl Acad Sci U S A, 1997. 94(18): p. 9825-30.

[154] Ao, Z., et al., *Importin alpha3 interacts with HIV-1 integrase and contributes to HIV-1 nuclear import and replication.* J Virol, 2010. 84(17): p. 8650-63.

[155] Fassati, A., et al., *Nuclear import of HIV-1 intracellular reverse transcription complexes is mediated by importin 7.* EMBO J, 2003. 22(14): p. 3675-85.

[156] Ao, Z., et al., *Interaction of human immunodeficiency virus type 1 integrase with cellular nuclear import receptor importin 7 and its impact on viral replication.* J Biol Chem, 2007. 282(18): p. 13456-67.

[157] Christ, F., et al., *Transportin-SR2 imports HIV into the nucleus.* Curr Biol, 2008. 18(16): p. 1192-202.

[158] Larue, R., et al., *Interaction of the HIV-1 intasome with transportin 3 protein (TNPO3 or TRN-SR2).* J Biol Chem, 2012. 287(41): p. 34044-58.

[159] Woodward, C.L., et al., *Integrase interacts with nucleoporin NUP153 to mediate the nuclear import of human immunodeficiency virus type 1.* J Virol, 2009. 83(13): p. 6522-33.

[160] Fouchier, R.A., et al., *Interaction of the human immunodeficiency virus type 1 Vpr protein with the nuclear pore complex.* J Virol, 1998. 72(7): p. 6004-13.

[161] Le Rouzic, E., et al., *Docking of HIV-1 Vpr to the nuclear envelope is mediated by the interaction with the nucleoporin hCG1.* J Biol Chem, 2002. 277(47): p. 45091-8.

[162] Matreyek, K.A. and A. Engelman, *Viral and cellular requirements for the nuclear entry of retroviral preintegration nucleoprotein complexes.* Viruses, 2013. 5(10): p. 2483-511.

3

Molecular Mechanisms Controlling HIV Transcription and Latency – Implications for Therapeutic Viral Reactivation

Michael D. Röling, Mateusz Stoszko and Tokameh Mahmoudi

Abstract

Persistence of transcriptionally silent replication competent HIV-1 is a major barrier to clearance of the virus from patients; current combinatorial antiretroviral therapies are successful in abrogating active viral replication, but are unable to eradicate latent HIV-1. A "shock and kill" strategy has been proposed as a curative approach in which latent virus is activated and infected cells are removed by immune clearance, while new rounds of infection are prevented by antiretroviral therapy. Much effort has been put toward understanding the molecular mechanisms maintaining HIV latency and the nature of reservoirs, to provide novel therapeutic targets. This has led to the development of latency reversal agents (LRAs), some of which are undergoing clinical trials. Targeting multiple mechanisms underlying HIV latency via a combination of LRAs is likely to result in more potent activation of the latent reservoir. Therefore, novel as well as synergistic combinations of therapeutic molecules are required to accomplish more potent latency reversal.

Keywords: HIV-1 latency, Latency reversal agents (LRAs), Combinatorial antiretroviral therapy

1. Introduction

Human immunodeficiency virus-1 (HIV-1) is a lentivirus, a subgroup of Retroviridae. Like all retroviruses, HIV-1 virions consist of an RNA genome with viral proteins encapsulated in a viral envelope. The viral proteins execute key steps to establish a productive infection by stably integrating into the host genome. Unlike most retroviruses, HIV-1 can also directly infect nondividing cells. HIV-1 preferably infects a subset of T-lymphocytes (CD4+ T-cells) that play a crucial role in the immune response. HIV-1 infection causes exhaustion and ultimately depletion of the host immune system, a syndrome termed acquired immuno-deficiency syndrome (AIDS). HIV-1 came into prominence with the outbreak of the AIDS epidemic in the 1980s. Major steps have been taken toward treating this viral infection. In particular, combinatorial antiretroviral therapy (cART) successfully abrogated HIV-1 replication. Thus, for compliant patients with access to c-ART, HIV infection has become a chronic rather than a lethal disease. However, cessation of antiretroviral therapy results in viral rebound in infected patients, even after years of cART. This is because in a small fraction of infected cells, HIV persists in a latent but replication-competent state. Latent HIV is unaffected by cART, but infection can rebound upon cART interruption. Therefore, HIV latency is the main challenge in developing a curative therapy for HIV.

The quest for an HIV-1 cure involves the development of either a sterilizing or a functional cure. A sterilizing cure would require complete removal of replication competent viral genetic material from the infected patient and thus the stable depletion of latently HIV-infected cells. A functional cure, on the other hand, requires the patient's immune system to suppress HIV-1 replication life-long in the absence of cART without disease progression, loss of CD4+ T cells and HIV transmission. The functional cure does not aim to eradicate the virus entirely from the patient. Both the sterilizing and functional cure strategies are currently the subject of major research efforts.

2. Clinical picture of HIV

The AIDS epidemic in the 1980s led to the identification of HIV as the causative agent. AIDS is a condition in which depletion of CD4+ T-cells overtime leads to the loss of the host immune system's ability to fight infections and cancers, eventually leading to death. As HIV was identified as the causative agent, cure efforts focused on disrupting the viral lifecycle. In the early 1990s, the first antiretroviral therapies – monotherapies – had limited success as they resulted in rebound of viremia due to the appearance of resistant viral strains. Resistant HIV required novel therapeutic strategies. Therefore, a combination of anti-retrovirals, targeting distinct steps of the viral life cycle was developed, so-called combinatorial antiretroviral therapy (cART). cART has proven to be extremely successful in lowering the amount of viral RNA in plasma below the limits of detection by standard laboratory techniques. Unfortunately, the therapy does not eradicate the virus as cessation of medication causes re-emergence of viral replication [1–3]. Thus, a fraction of the virus escapes the effects of cART. The source for this recurring viral replication is a small pool of latently infected cells that harbor integrated proviruses which, while silent, are not recognized by either the immune system nor are they subject to cART. Moreover, HIV can persist in the presence of cART in certain anatomical sites if drug penetrance is incomplete.

According to the World Health Organization (WHO), the number of HIV-infected individuals worldwide in late 2014 was estimated to be approximately 37 million [4]. The vast majority of infected people live in sub-Saharan Africa, where access to appropriate diagnostic centers and cART is limited. Estimates put new infections at 5,600 a day in 2014.

2.1. HIV-1 replication cycle and state-of-the-art antiretroviral therapy

HIV-1, as all viruses, is a parasite of the host cell and hijacks key cellular processes to establish a productive infection. To produce new virions, the virus goes through a viral replication cycle. HIV's replication cycle consists of entering the cell by docking at the cell surface receptor CD4 and co-receptors CCR5/CXCR5 and fusing to the cell, un-packaging of the genome, reverse transcription of the viral RNA genome into double-stranded DNA, which is the main component of the pre-integration complex, followed by integration of the double-stranded DNA genome into the host genome, transcription of the provirus, translation of viral proteins, and ultimately virion biogenesis followed by budding from host cell and maturation. Modern cART targets most steps in the HIV viral replication cycle (Figure 1). There are currently 28 approved agents for the treatment of HIV infection [5]. They fall into six mechanistic major classes, which act at different stages in the HIV replication cycle:

1. Fusion inhibitors: enfuvirtide (ENF, T-20), the only currently available fusion inhibitor, binds to the gp41 receptor site, preventing the fusion of the virus with the target cell.

2. C-C chemokine receptor type 5 (CCR5) antagonists: maraviroc (MVC) is currently the only available CCR5 antagonist. This drug is an entry inhibitor, specifically blocking the human chemokine receptor CCR5.

3. Nucleoside (nucleotide) reverse transcriptase inhibitors (NRTIs) block the addition of nucleosides to the DNA chain during reverse transcription of RNA.

4. Non-nucleoside reverse transcriptase inhibitors (NNRTIs) bind to and inhibit the enzyme reverse transcriptase (RT), preventing conversion of viral RNA to DNA during infection.

5. Integrase inhibitors (INIs): raltegravir (RAL), elvitegravir (EVG) and dolutegravir (DTG) are the only currently available drugs in this class. They target the HIV enzyme integrase (IN) that is required for insertion of viral genetic material into human DNA.

6. Protease inhibitors (PIs) bind to the catalytic site of HIV aspartic protease, blocking the processing of viral proteins (eg. Saquinavir).

These antivirals comprise the various current cART regimens that are used in the clinic. cART has proven to be extremely successful in suppressing viral replication in compliant patients. In fact, it has been argued that the theoretical potential of cART has already been reached [6]. Therefore, in the developed world with access to medication, HIV has become a chronic and not a lethal disease.

2.2. The burden of lifelong cART

Implementation of cART has provided long-term suppression of viral replication, improving the life expectancy and life quality of infected patients. Unfortunately, the economic burden of cART is debilitating. According to the Centers for Disease Control and Prevention (CDC), lifetime costs of treating HIV infection is estimated to be $379,668 per infected individual in the United States [7].

Moreover, patients on cART overtime can experience several side effects of cART such as: cardiovascular diseases (e.g., myocardial infarction); non-AIDS cancers (e.g., anal cancer, liver cancer, Hodgkin's disease); liver, kidney, and bone disease as well as neurologic complications, such as dementia [8]. Interestingly, most of these conditions are associated with the ageing process. Hence, it is thought, that HIV infection controlled by cART accelerates ageing. And importantly, HIV persists in a latent state that is not targeted by cART, rendering cART a therapeutic management of the disease as opposed to a curative treatment. Thus, there is much need to develop a curative therapy for HIV.

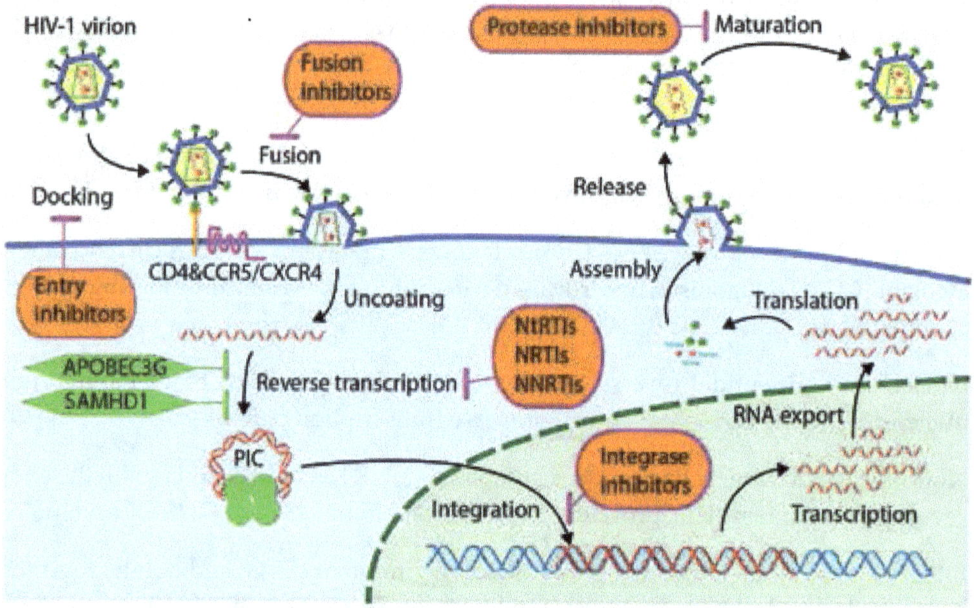

Figure 1. The viral replication cycle can be targeted pharmacologically at different stages

2.3. Clinical latency

The first step in finding a cure for HIV-1 infection is to identify the main source of cells that carry silenced, replication-competent HIV-1. Therefore, it is critical to define which cells or anatomical compartments constitute a reservoir of latent but replication-competent HIV-infected cells.

HIV-1 infects cells expressing the cell surface CD4 receptor and either of the co-receptors CCR5 or CXCR4. These cells include T helper cells, monocytes, macrophages, and dendritic cells. *In vivo,* HIV infects mostly activated CD4+ T-cells as quiescent and resting CD4+ T-cells are less permissive to infection due to low expression of CD4 and CCR5, and minimal metabolism [9–12]. The low metabolism is characterized by low levels of available dNTPs for reverse transcription and lack of energy sources [13–17]. Additionally, the cortical actin barrier in resting cells is thought to inhibit virus entry, reverse transcription and nuclear import [18,19]. However, the biggest pool of latently infected cells comprises resting memory CD4+ T-cells.

It is thought that these latent infections are predominantly generated while activated infected cells revert back to a resting memory state [20–22]. During this process, as the genome of the (partially) activated cell condenses and is silenced in transition to a memory state, so does the HIV genome [14,15]. There is also evidence for direct infection of resting cells by HIV, resulting in the generation of a latent infection [23]. Studying these cells in patients is challenging as the frequency of latently infected cells in suppressed patients is very low, estimated to be 1 latent cell per 1 million of uninfected cells [24,25]. Due to the long half-life of a latently infected resting memory CD4+ T-cells (estimated at 44 months), cART would take over 70 years in order to eradicate HIV from the infected patient [6,26,27].

Naive T-cells are also found to be latently infected; however, the frequency of such cells is even smaller than resting memory cells [28]. Interestingly, the naive T-cell reservoir may increase over time in suppressed individuals due to high proliferation of these cells compared to resting memory cells [29].

HIV is found also in cells of monocyte/macrophage lineage such as macrophages in brain and lung sections of infected individuals on anti-retroviral therapy [30,31]. However, proviral transcription occurs in these cells at low levels; therefore, it is debatable whether these cells are part of the latent reservoir [32,33].

Among the anatomical compartments affected by HIV-1, the central nervous system (CNS) and gut-associated lymphoid tissues (GALT) are two major sites [34–36]. The source of infection in the CNS is most likely infected monocytes, which are able to cross the blood–brain barrier as the virus itself cannot [37–39]. Approximately 5-10 times more HIV-1 RNA can be obtained from GALT than from blood cells in patients receiving cART [40,41], potentially indicative of lower penetrance of cART in cells within this anatomical site. However, the contribution of these compartments to rebound of viremia after cART cessation remains controversial [42,43].

2.4. Clinical proof-of-concepts for HIV-1 eradication

Thus far, only one patient, the so-called Berlin patient, was cured from HIV-1 after receiving treatment for acute myeloid leukemia [44,45]. HIV eradication in this patient was accomplished after several rounds of radio- and chemotherapy, total body irradiation, and two hematopoietic stem cell (HSC) transplantations from a donor bearing homozygous thirty-two base pair deletion in the CCR5 co-receptor gene (CCR5Δ32) were performed. The mutant CCR5 impedes viral entry of R5 tropic viruses in the first phase of the infection [46–49]. It is estimated that between 1% and 15% of the European Caucasian population harbor this mutation, while it occurs less frequently in African and Asian populations [47, 48]. In this patient, cART was ceased a day before the first transplant and after 7 years, no viremia or other indications of viral replication have been detectable [52].

Following the success of the case of the "Berlin patient", two HIV-1-positive patients, the "Boston patients", received HSCs transplants after developing Hodgkin's lymphoma [53]. Both patients carried heterozygous CCR5Δ32 mutation. While still under cART regimen, no viral production was observed which led to cessation of therapy. Unfortunately, after several months, strong viral rebound occurred in these patients. Follow-up analysis pointed to the likely presence of a small refractory source of cells, which is thought to have seeded the viral rebound; phylogenic studies revealed that only a few latent proviruses contributed to the viral rebound [53]. Several other similar studies have been conducted with infected patients suffering from either leukemias or lymphomas who received autologous or allogenic HSC transplantation alongside cART as a strategy to deplete the latent pool of cells. However, in most of these studies, viral rebound was detected following therapy interruption [54].

In another case, the Mississippi baby, an infant presumably infected *in utero,* received cART 30 h after birth. As newborns do not have resting memory CD4+ T-cells, it was reasoned that cART will prevent establishment of the latent reservoir – the main impediment in eradication strategies. One month after therapy, viremia reached undetectable levels and cART was stopped after 18 months. Unfortunately, 2 years post therapy interruption, rebound of viremia was detected (52, http://www.niaid.nih.gov/news/newsreleases/2014/pages/mississippibaby-hiv.aspx).

The immune system of rare "elite controllers" maintains low HIV-1 plasma levels, without the need of medication for many years. Although the capability of these patient to control viral replication is not completely understood, their circulating myeloid dendritic cells and CD8+ T-cells are more effective in depletion of infected CD4 T-cells [56–61]. Interestingly, the ARNS VISCONTI cohort showed that cessation of long-term cART, started during the acute phase of HIV-1 infection, resulted in post-treatment control (PST) of infection. Fourteen of the studied individuals were able to keep or even further reduce the viral reservoir. Furthermore, these individuals were able to maintain long-lasting, low level of viremia [62]. Recently, a perinatally infected baby displayed more than 11 years of HIV-1 remission. At 3 months of age, plasma HIV-RNA reached 2.1 x 10^6 copies/ml, and cART was administered for about 5–6 years. At 6.8 years of age, no HIV-1 RNA was detectable and cART was discontinued. After more than 12 years, plasma viremia still remains undetectable [63]. Therefore, this case provides the first evidence that early initiated, long-term cART can result in stable and durable HIV-1 remission.

Data from the Berlin and Boston patients provided a rationale for the creation of HIV-resistant cells. Since the CCR5Δ32 homozygous mutation is not lethal and not associated with abnormal immune functions [52], many approaches to silence the CCR5 gene have been or are under investigation [64–67]. These studies all employ genome editing technologies such as transcription activator-like effector nuclease (TALEN), clustered regularly interspaced short palindromic repeats (CRISPRs) or zinc-finger nucleases (ZNFs), which target the genome with high specificity and introduce deletions in the sequence of interest, in this case in the DNA sequence of CCR5 or/and CXCR4 co-receptors [64,65,68]. The rationale for this approach is based on the notion that cells bearing mutated CCR5 protein are not permissive to infection with R5 HIV-1 viruses, while cells with a mutated CXCR4 are resistant to C4 viruses. The double knock-out of both CCR5 and CXCR4 would allow resistance to infection regardless of viral tropism. However, the safety of such an approach remains to be elucidated. Uninfected HSCs isolated from infected individuals are engineered with either technology and then transfused back into patients. The ZNF approach targeting CCR5 has shown some promising results, although the sizes of cohorts used have been small. Gene-modified cells persisted in patients over 9 months, and cells seemed to expand and undergo trafficking to other tissues [66]. An increase in CD4+ T-cell counts was observed in all individuals. Importantly HIV-1 DNA in the blood decreased. The encouraging outcome of this study has resulted in phase II clinical trials.

Another gene therapy-based approach is the introduction of HIV-1 expression-dependent suicide genes encoding either toxic or pro-apoptotic proteins such as members of the Bcl-2 protein family. Constructs that are responsive to Tat and Rev viral proteins were tested [69]. While obtaining encouraging results, activity of such suicide genes only affects cells that are actively producing viruses, thus the latent pool of cells would still be unaffected.

Despite many attempts at HIV-1 cure, thus far only two cases, the "Berlin patient" and the early treated infant have resulted in eradication [44,45,63]. Due to safety and economic issues associated with transplantation and gene therapy approaches, broad use of such a therapeutic approach is not feasible for HIV cure. Moreover, the gene therapy approach provides a functional rather than sterilizing cure. Nevertheless, all these studies provided valuable insights into the biology of the latent reservoirs. They constitute a proof-of-concept for HIV-1 cure. Moreover, it seems that immediate initiation of cART contributes to restricting the establishment of the latent pool.

These studies highlight the need for more robust, cheaper, and feasible treatments in order to achieve HIV-1 eradication among all infected individuals. In 2004, the concept of so-called "shock and kill" or "kick and kill" therapy was proposed [70–72]. The aim is to specifically reactivate proviruses in latently infected cells ("shock") and eliminate the infected cells via viral cytophatic effects or/and render the cells susceptible to immune clearance ("kill"). New rounds of infection would be prevented by cART. "Shock and kill" therapy relies on the identification of potent and specific latency reversal agents (LRAs) alongside induction of an effective immune response against the reactivated latent pool of cells. The LRAs currently under investigation do not result in sufficient reactivation of latent HIV *in vivo*. Therefore, novel molecules that specifically reactivate latent HIV-1 are urgently needed.

3. Model systems and assays to detect and study HIV-1

To study the complex nature of HIV-1 latency, reliable model systems are required that recapitulate the nature and dynamics of the latent reservoir in vivo. Several cell lines of lymphocytic or monocytic lineage, primary-cell models, as well as animal models, are used to study HIV latency [73].

3.1. Cell lines

Immortalized cell lines of T-cell and monocytic origin are cost-effective and easy to use in the study of latent HIV. They allow fast read-outs in large scale for mechanistic molecular characterization of HIV gene expression. Therefore, cell lines are an attractive platform for screening and mechanistic characterization of LRAs. To generate a latent cell line, cells must first be latently infected with a HIV derived virus. Several different HIV derived viruses are used ranging from full length to minimal virus and can make use of a wide range of reporter constructs (e.g. GFP or luciferase). The viral Tat/TAR axis is of vital importance for the transcriptional regulation of HIV and can be included or excluded from the viral construct used. Latent infection of relevant cell lines derived from T-cells or monocytic lineage, depending on reservoir of interest generate cell lines that can be used to study the molecular mechanisms of HIV latency [74–78].

Ach-2 and U1 cells are characterized by low expression of HIV-1, which can be strongly upregulated upon TNFα or mitogens stimulation [74,79]. However, in these cell lines, latency results from mutations in Tat protein (U1 cell-line) or in RNA stem loop TAR (Ach-2) [76,77]. Therefore, these cell lines do not represent complexity of latency found *in vivo*, however, they do allow Tat/TAR-independent HIV-1 reactivation investigation.

A more appropriate system to study latency are J-Lat cell lines derived from Jurkat cells of T-lymphocytic origin [78, 80,81]. These cells have integrated replication-competent full-length or minimal proviral constructs with an intact promoter and Tat-TAR axis, a *GFP* reporter gene replaces the *Nef* sequence in full-length proviruses or is located downstream of *Tat* in minimal proviruses [78].

These cell lines have been extremely useful to delineate the molecular requirements of HIV transcription activation and silencing. Although useful for molecular analysis and screening platforms, the cell line model systems of HIV latency also present some limitations; first, clonal cell lines are derived from a single integration event, and therefore do not reflect the diverse distribution of integration sites in the host chromatin [82,83]. Consistently, results vary depending on the cell lines used, indicating possible clonal cell line effects [84]. Due to the above mentioned limitations and the considerable difference between cell line models and primary cells in terms of proliferative capacity, genomic stability and mechanisms involved in establishing and maintaining latency, generally latency models based on primary cells are preferable.

3.2. Primary cells

To more closely resemble infection *in vivo* and validate putative LRAs more accurately, several primary cell models have been developed. Depending on the cell status at infection, these models can be divided into two groups.

The first group relies on purification of CD4+ T-cells from healthy donors, that are then activated and subsequently infected. Depending on the method, CD4+T-cells are purified and stimulated with a-CD3/IL-2 [85], a-CD3/aCD-28 [86], a-CD3/aCD-28/IL-2 [87], or Ag-MDDC (antigen-loaded monocyte-derived dendritic cells; [88]), and infected with virus. Productively infected cells die due to virus-induced apoptosis or become latent by reverting back to a resting state. To limit infection to only one replication cycle, replication-defective viruses or antiretroviral drugs are also used. The rationale for these systems rely on the notion that a portion of activated, infected CD4+ T-cells transition to a quiescent state, shutting down general transcription and slowing down metabolism, resulting in latency [6,25,28,89–91]. Depending on the method used, different populations of latently infected cells are generated for use in reactivation studies. In the methods suggested by Sahu and Marini central memory T (T_{CM}) cells remain in culture, in Yang's protocol mainly effector memory T (T_{EM}) cells are produced, in Bosque and Planelles's method cells phenotype resembles central memory-like (T_{CM}). The main disadvantage of these methods is the time needed to obtain results, which varies from 1 to 4 months. Furthermore, they are labor-intensive and technically challenging.

The second group uses direct infection of resting memory CD4+ T-cells, which immediately after integration become latent. Cells are infected after purification and can be used after several days for reactivation studies [90, 91]. Stimulation of CCR7, CXCR3, or CCR6 receptors increases the susceptibility of resting memory CD4+ T-cells to infection without T-cell activation. In the methods of Swiggard and Lassen, central memory T (T_{CM}) and effector memory T (T_{EM}) cells are the source of latent HIV-1; in Saleh's method naïve resting memory T-cells, in addition to T_{CM} and T_{EM} cells, constitute the latent pool. The main advantage of these methods is the time needed to evaluate the potency of putative LRA, as results can be obtained within one week.

Depending on the protocol used, the amounts of cells that become latent differ from as little as 1% to up to 40%. In models where cells are activated, on average more latently infected cells are generated. Using these models, we can quantify the level of reactivation of HIV-1 in a reliable manner by measuring the production of the viral protein p24 by enzyme-linked immunosorbent assay (ELISA) or quantification of viral transcription by quantitative RT-PCR, or by detection of GFP/luciferase in case of reporter-based constructs.

A novel detection method distinguishes uninfected, productively infected, and latently infected cells using a dual reporter system. A modified HIV-1 derived genome containing GFP as a reporter of viral transcriptional activity and mCherry under an EF1a promoter as a reporter of infection (latent or productive) allows easy isolation of the different cell populations [23].

Ultimately, the golden standard for testing activity of LRAs are primary cells from infected individuals under cART obtained by leukophoresis, a process in which white blood cells are specifically isolated while other blood components are reverted back to the patients' circulatory system. The isolated cells are uninfected, latently infected, and infected with defective viruses. Large amounts of CD4+ T-cells are required and isolated from patients for testing LRAs.

The development of primary cell models greatly improved the quest for LRAs, yet results differ between each model system [84]. No *in vitro* models completely recapitulate the full range of latent cells *in vivo*; instead, only a small sub-fraction of latently infected cells is represented. Hence, the validation process of putative LRAs requires testing on cells derived from infected individuals [93].

3.3. Animal models of HIV-1 infection

The number of animal models available to study latency is limited. The toxicity of putative LRAs can be assessed with use of mouse and non-human primate (NHP) models [94]. Two mouse models have been developed and used in HIV latency studies: the humanized SCID

(SCID-hu) mouse, transplanted with human thymus and liver fragments, and the humanized blood, liver, and thymus (BLT) mouse which has a human immune system with full mucosal immunity [95–97]. Unfortunately, SCID-hu mice do not express human proteins involved in the viral replication cycle; therefore, the study of HIV-1 in these mice is restricted to events taking place within organs of human origin in this model. In addition, HIV-1 is not responsive to cART in these animals. BLT mice are a better model of HIV-1 infection, as they produce resting memory CD4+ T-cells of human origin. However, some components of cART do not repress replication in BLT mice [34].

NHP models employ the Simian immunodeficiency virus (SIV) infection in rhesus and pig tailed macaques to recapitulate HIV-1 infection in humans [98,99]. NHP models allow the monitoring of the spread of infection. Moreover, infection in this model can be controlled by antiretroviral therapy. NHP models are helpful in studying the first stages of latency establishment, as investigating this part of HIV-1 infection is extremely challenging in patients, as the pool of latently infected cells is established early during infection [100]. One caveat to the use of SIV-based NHP models of HIV latency is that the viral 5′LTR or promoter of SIV is considerably different in sequence from HIV-1 [101] and therefore latent SIV response to LRAs, which is a direct consequence of promoter-mediated transcription activation may vary substantially from latent HIV-1. In addition, animal models are far more expensive than cell-based systems. Nor do they fully reflect human infection or metabolism. Finally, ethical concerns are inherent to the use of NHP models of HIV latency.

3.4. Detection of the latent reservoir

The study of latent HIV infection requires accurate measurement of the size of the latent reservoir and the extent of reactivation following LRA treatment. Depending on the experimental aim, different detection methods can be employed. These methods generally rely on PCR, protein quantification, or reporter detection.

The quantitative viral outgrowth assay (QVOA) is a well-established method to estimate the latent pool. The assay relies on the use of serial dilutions of cells obtained from an infected individual in co-culture with uninfected cells that are permissive to infection. Viral proteins are detected by ELISA. Unfortunately, QVOA is time-consuming, costly, and might generate false-negative results as not all replication-competent proviruses are reactivated, and thus not detected [83].

The HIV reservoir can be approximated by detecting the number of viral DNA copies present in the cells. The recently introduced digital droplet PCR (ddPCR) improves on classic and nested qRT-PCR by simultaneously amplifying thousands of nanoliter reactions in combination with very sensitive detection system based on flow cytometry [94,102,103]. ddPCR is therefore superior to nested qRT-PCR in its ability to resolve rare events such as latent HIV-1. Although PCR based methods provide increased sensitivity for the detection of viral genetic material, these approaches also detect defective proviruses, which results in false-positive results.

Another recent PCR-based method for reservoir detection evades false positive results from defective proviruses. The *Tat/rev* induced limiting diluting assay (TILDA) relies on PCR amplification of multiply spliced RNA (msRNA) of *tat/rev* transcripts that are present in productively infected cells and absent in latent infection [104]. Small amounts of cells isolated from patients are divided into two equal parts and distributed in limiting dilution. One half is left unstimulated while the other is activated with PMA/Ionomycin. After 12 hours, cells are lysed and subjected to ultrasensitive nested RT-PCR. By employing statistical modeling, the frequency of cells that are expressing msRNA in both groups is estimated and based on the unstimulated group a threshold of activation can be set. Using the TILDA assay, the size of the reservoir is estimated at 24 cells per million, which is more than measured by QVOA but less than measured by PCR methods [24,83,104]. The assay more accurately estimates the true size of the latent reservoir, is highly sensitive, reproducible, fast, relatively inexpensive, and requires only 10 mL of patients' blood. However, a limitation on the TILDA assay is that it detects the presence of viral transcripts but not the production or release of infectious viral particles; therefore, it may still overestimate the true size of the reservoir, yet to a smaller extent than other PCR-based methods. Additionally, signal detection relies on amplification of highly variable region of the HIV-1 DNA; therefore, detection of all subspecies of HIV-1 might be challenging and require extra optimization steps.

Unfortunately, all current methods to detect latent HIV-1 have limitations. First, the pool of latently infected cells in patients is extremely low, resulting in a high noise-to-signal ratio. Furthermore, defective or hyper-mutated proviruses are detectable by PCR-based techniques, yet irrelevant for eradication strategies. Moreover, not all replication-competent proviruses are inducible in the first round of treatment, yet get reactivated upon subsequent rounds of stimulation [83]. Thus, assays to measure latency reversal are overestimating – in the case of PCR-based methods – or underestimating – in the case of QVOA – the latent pool. This poses a main problem in measuring efficiency of the reactivation of HIV-1. A captivating approach employing the use of a biomarker (e.g., gene), which responds to treatment in the same way as HIV-1, would allow more easily quantifiable assessments as to whether latent HIV in patient cells would be responsive to a particular treatment.

4. Molecular mechanisms of latency

Although replication-competent, latent HIV is transcriptionally silenced but susceptible to reactivation upon certain stimuli. Following integration into the host genome, transcription from the HIV genome is controlled by key cellular host factors, and subject to host cell gene regulation similar to endogenous genes. Since viral transcription initiation, elongation, and termination are tightly regulated by host proteins, HIV is also widely used as a model system to study gene regulation.

4.1. Host antiretroviral mechanisms thwart infection

Host defense mechanisms impede HIV-1 infection. Upon entering the cell, HIV's RNA genome is reverse transcribed into double-stranded DNA (dsDNA). This process requires freely available deoxynucleotide triphosphates (dNTPs). By limiting the pool of freely available dNTPs, the nucleotide scavenger SAMHD1 restricts viral replication in non-cycling myeloid cells and quiescent CD4+ T-cells [105–108]. Additionally, SAMHD1 has 3′–5′ exoribonucleases (RNAse) activity that specifically cleaves single-stranded RNA [109,110]. Interestingly, Vpx, encoded by HIV-2 and Simian immunodeficiency virus, is an accessory protein packaged into the virion, which induces SAMHD1 degradation [111].

Additionally, APOBEC3G limits viral replication by catalyzing the deamination of cytidine to uridine in the viral single-stranded DNA (ssDNA) genome during reverse transcription [112]. Interestingly, APOBEC3G is inactive in memory CD4+T-cells, which helps to explain why this cell type is more permissive to HIV-1 infection. Therefore, activated CD4+ T-cells are the main target cell type of HIV infection and of the main source of the latent reservoir.

4.2. Integration of HIV into the host genome required by host factors

The reverse-transcribed viral DNA genome is incorporated in the pre-integration complex (PIC). The PIC is imported into the nucleus. Host factors identified so far that affect viral integration are lens epithelium-derived growth factor (LEDGF/p75/PSIP1) and hepatoma-derived growth factor related protein 2 (HRP- 2/ HDGFRP2), through an integrase binding domain. In the absence of LEDGF, provirus integration is decreased 10-fold and HIV's pattern of integration is altered [113–115]. Simultaneous LEDGF and HRP-2 knockdown further decreases viral replication [116]. Nevertheless, knockdown of both factors does not completely abolish HIV-1 integration, indicating that IN alone and/or in cooperation with other host factors can still integrate the viral genome [117]. PIC nuclear import stimulates export to the cytoplasm of INI-1 and PML, disrupting this effect greatly improves integration efficiency [118–120]. Upon knockdown of transportin-3/TNPO3 and nuclear pore protein RanBP2/Nup35 HIV-1 integrates randomly [121]. Therefore, nuclear import affects the site of integration with a preference for open chromatin.

4.3. Pre-integration vs post-integration latency

Two states of latency can be defined based on the integration state of HIV: pre-integration latency and post-integration latency. Defects in integration or in a prior phase of the viral replication cycle (e.g., incomplete reverse transcription) might result in unintegrated viral DNA. The half-life of the linear pre-integration complex is approximately 1 day [122]. The linear unintegrated viral DNA can also be circularized, resulting in slightly extended half-life of the virus [123]. In quiescent cells, the pre-integrated virus can reside near the centromere for weeks [124]. Unintegrated virus can replicate, albeit very inefficiently [125]. The half-life of both forms of unintegrated virus is too short and replication inefficient to serve as the source required for the long-term persistence of latent HIV making pre-integration latency less clinically relevant.

Post-integration latency occurs when the HIV virus is stably integrated into the host genome, but a productive infection is not achieved. The site of integration and the abundance of transcription factors are crucial for determining whether an infection will be latent or productive. The site of integration will determine the chromatin environment (such as histone modifications), relative position to other genes (intronic insertion vs gene desert) and position within the nucleus of the provirus.

4.4. Integration biases

The site of integration greatly determines the transcriptional activity of the provirus. HIV preferentially integrates into active genes both in patient material and transformed cell lines [82,126–128]. Moreover, HIV-1 integrates in regions of genome that are in close proximity to nuclear envelope [129]. Latent integrations are in or close to alphoid repeat elements in heterochromatin, whereas productive integrations avoid insertion in or near heterochromatin [78]. Integration is associated with transcription-inducing histone modifications (i.e., H3 & H4 acetylation and H3K4 methylation) but not transcription-inhibiting modifications (i.e., H3K27 trimethylation and DNA CpG methylation) [130]. A comparison of integration sites in resting and activated CD4+ T-cells showed that in both cell types HIV integrates in active genes. However, in activated cells, insertions were enriched for gene dense, CpG island-rich and high G/C-content regions [131]. Latency in infected Jurkat cell lines correlated with integrations in gene deserts, centromeric heterochromatin, and highly expressed cellular genes [128]. Within the nucleus, HIV-1 is located mostly in decondensed chromatin at the nuclear periphery, while it disfavors heterochromatic regions [132]. Interestingly, latent proviruses were found to interact with a pericentromeric region of chromosome 12 in quiescent cells [133]. In a study of viremic progressors and viremic controllers, integration was enriched into, or in close proximity to, Alu repeats, local hotspots, and silent regions of the genome [134]. In addition, close proximity of the provirus to PML bodies is associated with latency, an association that is lost upon reactivation [135].

4.5. Integration relative to host genes affects transcriptional state of the provirus

Sense and antisense integration relative to host genes can greatly affect the transcriptional state of HIV. Integration in sense orientation can lead to promoter occlusion, whereas integration in antisense orientation can lead to collision of the transcriptional machinery. Promoter occlusion occurs when the transcriptional machinery is depleted from the viral promoter by a dominant host promoter that is transcribed and negatively affects proviral expression.

Indeed, chimeric transcripts of the host gene and in sense viral integrations were observed [136,137]. Additionally, Han et al. compared the effect of sense and antisense insertions of HIV relative to the active HPRT gene [138]. In this setting, sense integration enhanced viral expression whereas antisense integration (transcriptional collision) led to suppression. Sense integrations were shown to be modestly preferred in latent cells, a preference that was not present in productively infected cells [139]. Transcriptional interference and transcriptional collision are examples of host genes interference with viral expression. On the other hand, reactivation of HIV may lead to suppression of host gene expression [136]. Indeed, in a cell model with a latent integration into the HMBOX1 gene, the host gene was repressed upon viral reactivation [140].

4.6. Viral transcription starts at the 5'LTR

The provirus is flanked by a 5′ and 3′ long terminal repeats (LTRs). While transcription can be initiated from both LTRs, the 5′ LTR is dominant and serves as the HIV promoter, although 3′ transcription is activated when the 5′ LTR is defective [141]. Transcriptional interference has been proposed as the mechanism by which the 5′ LTR exerts its dominance over the 3′ [142]. Interestingly, low-level antisense transcription takes place at the 3′ LTR, a mechanism by which latency can be maintained [143–147]. Sense transcription results in at least 40 coding transcripts due to alternative splicing of the HIV-1 genome [148]. Finally, both LTRs also act as a source of negative sense transcription, which could potentially affect the expression of neighboring genes [149,150].

4.7. The 5′ LTR contains numerous putative transcription factor binding sites

HIV-1 encodes a potent trans-activating protein – Tat – that drives viral expression during productive infection. However, initially, before sufficient levels of Tat are expressed, the provirus relies on host factors to initiate transcription. The 5′ LTR contains three regions – U3, R, and U5 (Figure 2) [151]. The R region, immediately next to the transcription start site (TSS), contains the trans-activation response (TAR) element, an important regulator of HIV expression. The U3 region contains the core promoter (nucleotides –78 to –1 upstream of TSS), a core enhancer (nucleotides –105 to –79), and a modulator region (nucleotides –454 to –104) [152,153]. The core promoter contains three Sp1 binding sites in tandem, a TATA box, and an initiator element at the transcription start site. The core enhancer contains two NF-kB-binding sites. The modulator region – so-called because early experiments with deletion upstream of the LTR caused increased activity of the LTR – was shown by later experiments to contain binding sites for both repressive and activating factors including nuclear factor of activated T-cells NFAT, STAT5, NF-kB p65/p50 heterodimers, lymphocyte enhancer factor (LEF-1), CCAAT/enhancer binding protein (C/EBP) factors, AP-1, and activating transcription factor/cyclic AMP response element binding (ATF/CREB) factors (Figure 2) [152,154–162]. It is well established that these transcription factors have binding sites within HIV-1 sequence. Moreover, they are strong activators of HIV-1 transcription of which NF-κB is considered the most critical [163–166]. In addition to the presence of these sites, bioinformatic tools indicate that this region of the HIV LTR contains a tightly clustered distribution of multiple transcription factor consensus binding elements [167].

4.8. Positive host factors bind to the 5′ LTR

Initial transcription of HIV-1 is entirely dependent on host factors. Nuclear factor kappa-light-chain-enhancer of activated B cells (NF-kB) is a hetero dimer comprised of p50 and p65 subunits involved in T-cell activation. NF-kB acts as a transcription factor and is a potent activator of HIV-1 transcription initiation and elongation. It interacts and functions cooperatively with numerous proteins. Independent of Tat, NF-kB can reactivate HIV to high expression levels [168]. Mutated NF-kB-binding sites on the LTR inhibit basal transcription and Tat transactivation [169]. NF-kB, Sp1, and other factors (LEF-1, Ets1, and TFE-3) bind to sites near NF-kB sites and synergistically activate HIV transcription, even in the presence of repressive

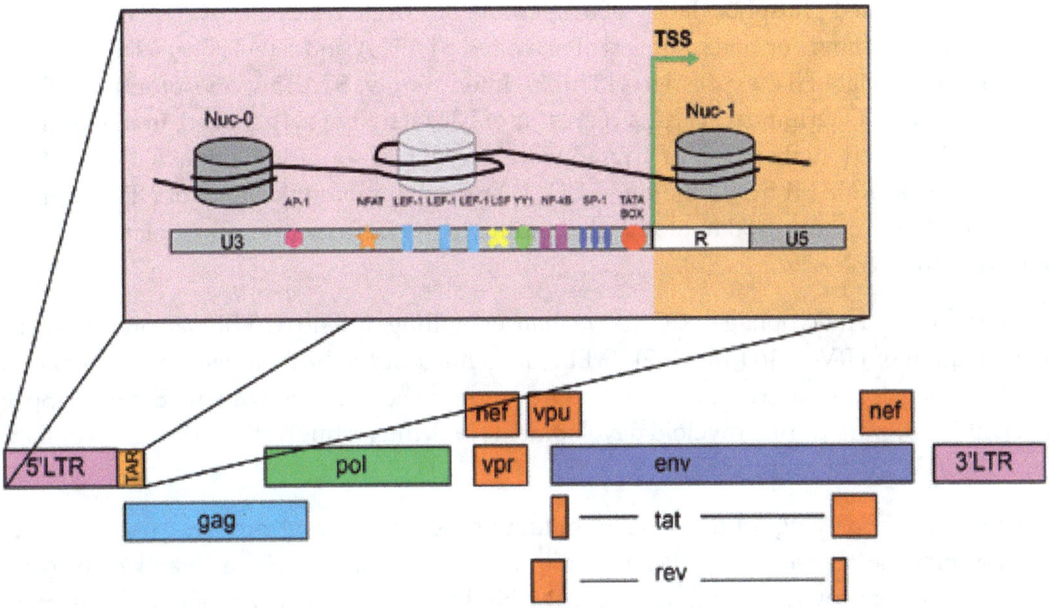

Figure 2. The genome of HIV-1

chromatin structures [170,171]. NF-kB and AP-1, a heterodimer of proteins from the c-Fos, c-Jun, ATF, and JDP families, cooperatively trans-activated LTR activity through the ERK1/ERK2 mitogen-activated protein kinase (MAPK) pathway [161]. Acetylation of Lys310 in NF-kB p65 subunit is an activating mark that is removed by NAD+-dependent protein deacetylases SIRT1 and SIRT2 [172]. Tat positively affects NF-kB by inhibiting SIRT1 and stimulating degradation of IkB, a protein that sequesters NF-kB in the cytoplasm [169,173]. The viral nucleocapsid (NC) protein enhances NF-kB-mediated activity by interacting with the LTR [174]. p65 recruits THIIH which is part of the preinitiation complex and its subunit CDK7 with kinase activity activates CDK9, resulting in increased HIV transcription [175,176]. The cell surface receptor OX40, bound by its ligand gp34, activates transcription from 5′ LTR, in a manner dependent on the presence of NF-kB-binding sites on the LTR [177]. The transcription factor E2F-1, a regulator of S-phase gene expression, inhibits LTR transcription through the recruitment of p50 at the NF-kB-binding sites on the LTR [178].

Members of the SV40-promoter (Sp) specific transcription factor family regulate LTR activity. Sp1 and Sp4 are activators of HIV-1 [179]. Expression of Sp transcription factors changes during monocytic maturation, suggesting differences in susceptibility to LTR activation during differentiation [180].

Nuclear factor of activated T-cells (NFAT) can induce LTR activity in T-cells [155]. NFAT recruits HATs through CBP/p300, which results in reactivation of HIV-1 transcription [181]. The Janus kinase (JAK)/signal transducers and activators of transcription (STAT5) can stimulate or inhibit HIV transcription. STAT5 binds to its binding sites in the U3 enhancer region on the LTR where it promotes transcription [156]. In response to a broad range of cytokines (e.g., IL-2, IL-7, IL-15) and granulocyte-macrophage colony-stimulating factor (GM-

CSF) JAK-mediated phosphorylation of a C-terminal tyrosine residue activates STAT5A and STAT5B. Homodimers or heterodimers of activated STAT5A and STAT5B translocate to the nucleus to stimulate HIV expression [182,183]. Interestingly, STAT5Δ, an isoform of STAT5 truncated on the C-terminus, acts as a repressor of LTR activity [184]. Indeed, in the promonocytic cell line U1 high levels of STAT5Δ are present. Upon stimulation with GM-CSF, STAT5Δ blocks RNAPII from binding to LTR U3 region, inhibiting activity of HIV promoter [185]. STAT5Δ promotes p50 homodimers binding to the LTR, contributing to latency maintenance [186].

In monocytes and macrophages, CCAAT/enhancer binding protein (C/EBP) factors are crucial for activation of HIV-1 [160,187–189]. C/EBP, a member of the bZIP superfamily, contains a DNA-binding domain and a leucine zipper for homo- and heterodimerizations. Similar to Sp-1, levels of C/EBP vary during myeloid development [190]. Interestingly, the HIV-1 LTR contains several C/EBP binding sites [159].

Some studies employing mutagenesis of binding sites for activator protein-1 (AP-1) within proviral genome showed that AP-1 transcription factor is the crucial activator of proviral transcription, as proviruses with altered AP-1-binding sites were less prone to reactivation even if treated with strong activator such as phorbol 12-myristate 13-acetate – PMA [191]. Furthermore, the latent pool was bigger in cells infected with a virus carrying a deletion in AP-1 sites, implicating that the AP-1 protein is necessary for successful provirus transcription [192]. Heterodimeric protein AP-1 is formed upon phosphorylation od c-Jun N-terminal kinase (JNK) in JNK/MAPK pathway [193]. It is well established that activation of TLR signaling induces nuclear localization of NF-kB and AP-1 mediated via JNK pathway [194–196].

In addition to the already mentioned host factors, the potent viral trans-activating protein Tat and to a lesser extent the multifunctional viral protein, viral protein R (Vpr), positively affect viral transcription. Productive infection requires the presence of Tat. Exogenous expression of Tat rescues HIV from latency [197]. A defective Tat mutant (C22G) is incapable of full-length viral expression [198]. Additionally, the Tat mutant (H13L) is more prone to establish latency [197]. Tat recruits the positive transcription elongation factor b (P-TEFb), which shifts RNAPII promoter proximal pausing to transcriptional elongation leading to a productive infection [199,200]. P-TEFb consists of CDK9, a serine/threonine kinase, and CyclinT1. The N-terminal cystein-rich region of Tat (Cy22-Cy37) binds to CycT1 through Zn2+-mediated interactions [201–203].

Vpr is a multifunctional HIV-1 protein that plays a role in nuclear import of the PIC and cell cycle arrest in proliferating cells. Vpr also activates LTR activity through multiple mechanisms. Vpr recruits p300 to the 5′ LTR increasing acetylation, resulting in HIV-1 transcription [204]. Moreover, Vpr interacts with Sp1 and TFIIB, part of the transcription initiation complex, stimulating proviral transcription [204–206].

4.9. Repressive host factors at the 5′ LTR

Not all host transcription factors have an activating effect on LTR activity (Figure 3). YY1 and LSF recognize binding sequences in the LTR and repress transcription through epigenetic modification [207]. C-promoter binding factor-1 (CBF-1) also represses HIV through epigenetic silencing [208,209]. c-Myc recruits an epigenetic silencing factor to repress HIV-1 [210].

Transcription factors initiate LTR activity, but full-length transcripts are not produced because transcription elongation is inhibited. DRB sensitivity-inducing factor (DSIF), a heterodimer composed of hSpt4 and hSpt5 proteins, induces capping of RNA from newly initiated transcription complexes [211]. The subunit hSpt5 interacts directly with nascent RNA as it appears from the RNAPII exit site and recruits negative elongation factor (NELF) (Figure 3) [212–214]. Escape of transcripts from the promoter proximal pause site is prevented by NELF, which induces termination of transcription over several hundred bases [215]. Moreover, the binding sequence of NELF subunit E recognizes a homologous sequence on TAR, increasing association of NELF with the LTR, which results in transcription silencing. Indeed, experiments where NELF is knocked down show higher basal HIV transcription and reactivation from latency [216–218].

A novel, RNA interference independent, mechanism mediated by microprocessor and termination factors causes transcriptional silencing and chromatin remodeling at the HIV-1 promoter [219]. Microprocessor binds to TAR, which is then cleaved by Drosha into two RNAs, a 5′-end and 3′-end product. The 5′ is further processed in an Rrp6-dependent manner into a transcription repressing RNA species. The 3′ RNA recruits termination factor Xrn2 and Setx, which induces RNAPII pausing and premature termination of transcription [219].

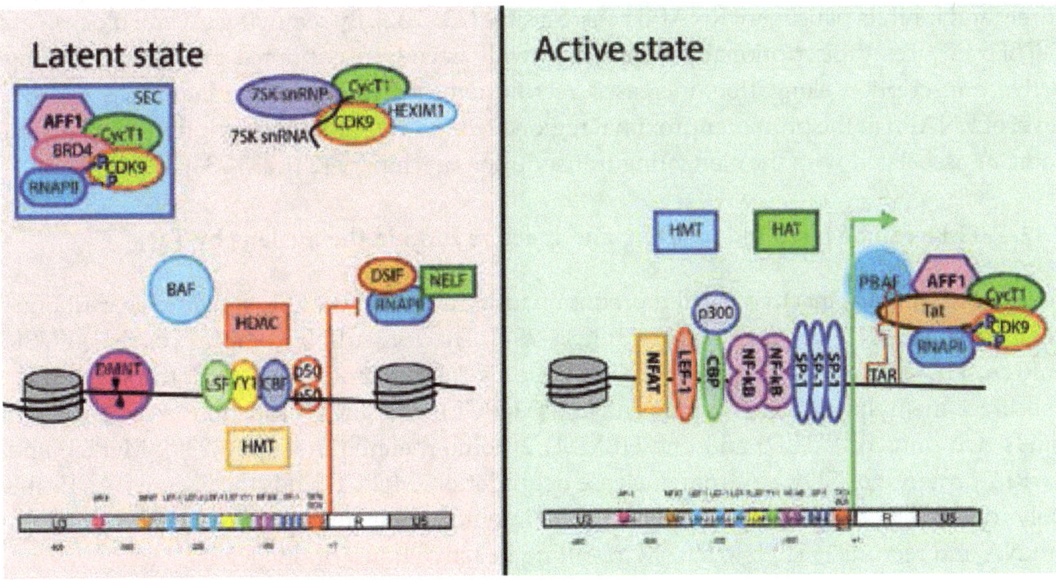

Figure 3. Molecular mechanisms in latent and productive HIV-1 infection

4.10. Host factors induce transcriptional initiation, but not elongation

While some host transcription factors recruit RNAPII, in the absence of Tat, transcription elongation does not occur resulting in the generation of short abortive transcripts by promoter proximal pausing [220,221]. These ~60nt transcripts include TAR, which has a stem-loop structure and binds near the HIV 5′LTR, inhibiting RNA-polymerase. TAR directly binds Tat, which recruits transcriptional elongation complex to the LTR [222].

4.11. Tat-dependent transcription leads to productive infection

If cells become activated or due to leaky transcription, Tat can be produced. Tat binding to P-TEFb induces significant conformational changes in P-TEFb, allowing Tat and CycT1 to cooperatively recognize and stably bind TAR [200,223].

Tat-P-TEFb phosphorylates NELF-E resulting in the dissociation of NELF from TAR and the paused RNAPII complex [214,216,218,224]. CDK9 phosphorylates RNAPII at the carboxyl terminal domain (CTD) at Ser2 and Ser5 residues of the 52 heptad repeats, which regulates progression to the elongation phase of transcription [225–227]. The phosphorylation status determines regular and alternative RNA splicing and the 3′ end recruitment of polyadenylation factors [228,229]. Ser2 phosphorylation of the RNAPII CTD recruits splicing-associated c-Ski-interacting protein, SKIP, and stimulates elongation transcription and alternative splicing of the Tat-specific splice site through interactions with U5snRNP proteins and tri-snRNP110K [230].

Phosphorylation of hSpt5, a subunit of DSIF, by CDK9 converts it into a positive elongation factor that prevents nascent RNA from breaking of from the transcription complex prematurely and inhibits pausing of RNAPII at arrest sites [231,232]. By removing several blocks Tat-P-TEFb induces transcriptional elongation as well as co-transcriptional processing. During active transcription elongation, increased recruitment of RNAPII to TSS maintains a stable level of RNAPII at the promoter proximal region [218]. Throughout transcription, Tat-P-TEFb remains associated with the elongating transcription machinery [231,233,234].

4.12. P-TEFb can be recruited in active and inactive form in the nucleus by Tat

In activated T-cells, inactive P-TEFb predominantly resides in the 7SK small nuclear ribonucleoprotein (snRNP) complex (Figure 3) [235–237]. The 7SK snRNP complex consists of 7SK snRNA, HEXIM1 (or its homolog HEXIM2), the La-related protein 7 (LARP7), and the 7SK-specific 5′ methylphosphate capping enzyme (MePCE). The snRNA functions as a scaffold: it binds two units of P-TEFb and one HEXIM1/2 homo-/heterodimers [238,239]. MePCE and LARP7 protect the 7SK RNA from nuclease degradation, MePCE binds the 5′ end, LARP7 the polyuridine 3′ end [240–242]. Tat disrupts the interaction between pTEFb and HEXIM1/7SK snRNA and recruits P-TEFb to 5′ LTR, resulting in active transcription [226].

BRD4 can also recruit P-TEFb from 7SK snRNP [241,243], to promote transcription. Due to similarities in their C-terminal P-TEFb interacting domains [244], Tat and BRD4 compete for P-TEFb [245,246]. In a latent model, knockdown of BRD4 results in Tat-dependent reactivation of HIV-1 [247].

Bromodomain and extra-terminal domain family of proteins (BET) play an important role in repression of the HIV-1 transcription. BET proteins are responsible for the recruitment of P-

TEFb to transcribed genes [246,248]. BRD4 competes with viral protein Tat for binding site on pTEFb, and it represses HIV-1 transcription [245,246]. Knockdown of BRD2 indicates this protein contributes to the maintenance of latency. These results are consistent with the notion that BRD2 is binding to remodeling factors such as HDACs [249,250].

P-TEFb can be recruited to transcription complexes by other factors. CTIP2 recruits P-TEFb by binding HEXIM1 and negatively regulates the complex by repressing the CDK9 kinase activity of P-TEFb [251]. Phosphorylation of HEXIM1 at Tyr271 and Tyr 274 decreases retention of P-TEFb in the 7SK RNP [252]. Additionally, through the binding of nascent RNA, SRSF2 and P-TEFb are released from the 7SK complex and induce transcription elongation in a manner similar to TAR/Tat-mediated recruitment of P-TEFb [253].

4.13. P-TEFb is a subunit of the super elongation complex

P-TEFb is required for activation of HIV transcription but does not explain the maximum observed viral expression; therefore, additional factors are necessary [254,255]. P-TEFb is an integral part of the super elongation complex (SEC) (Figure 3), which is a potent activator of transcriptional elongation of host genes [234,256]. It is composed of one of two scaffold proteins, AF4/FMR2 proteins AFF1 or AFF4. Translocations of AFF1 and AFF4 resulting in fusion proteins are commonly found in mixed lineage leukemia (MLL) [257–259]. The resultant fusion proteins cause aberrant recruitment of SEC to MML-specific genes [260]. AFF1 and AFF4 recruit many other proteins to the SEC [261], such as ELL family of elongation stimulatory factors ELL1 and ELL2, which inhibit RNAPII pausing and synergistically improve Tat-transactivation with P-TEFb [256]. Moreover, knockdown of ELL2 strongly suppresses viral expression. [203,210,230,252]. Tat and AFF4 inhibit the polyubiquitination-mediated degradation of ELL2, increasing available levels of SEC. [256,262].

4.14. Tat can be extensively post-translationally modified – "Tat code"

Modifications on numerous amino residues of Tat regulate the interaction with a wide variety of host proteins. In comparison to the histone code which is used to explain the multiple modification on histone tails and their function, a "Tat-code" has been proposed [34]. Tat is phosphorylated on Ser16 and Ser 46 by CDK2, modifications which result in transcription inhibition [263]. Acetylation of Lys28 increases affinity for P-TEFb binding and is removed by HDAC6 [264–266]. Tat dissociates from TAR and binds acetyltransferase PCAF which acetylates Tat at Lys50 and Lys51 [264,265,267–270]. Acetylated Lys50 allows recruitment of the PBAF (SWI/SNF B) chromatin remodeling complex to the LTR [267,271–273]. SIRT1 deacetylates Tat at Lys50 as part of a late phase of transcriptional regulation, striping Tat of acetyl groups allowing its reuse in subsequent rounds of transcriptional cycles [274]. Mono-methyl-transferase Set7/9 and LSD1, respectively, methylate and demethylate Lys51. Demethylated Lys51 of Tat enhances HIV-1 transcription [275,276]. Hdm2 polyubiquitinates Lys71, activating Tat [277].

4.15. Nucleosome positioning at the 5′ LTR controls viral expression

Regardless of integration position, the latent 5′ LTR typically contains two nucleosomes, Nuc-0 and Nuc-1, at fixed positions [278]. Nuc-1 blocks transcription elongation as it is positioned just downstream of the TSS. Nuc-1 is displaced upon virus reactivation [278–280]. Nucleosomes can be altered by chromatin remodeling complexes. A third unstable or loosely positioned nucleosome is located in between nuc-0 and nuc-1 [281] (Figures 2 and 3A).

BCL11B, together with the chromatin remodeling complex NuRD, strongly represses HIV-1 transcription [282]. BCL11B is specifically expressed in T-cells and neurons. Interestingly, the NuRD complex consists of several proteins with histone deacetylase activities – i.e., HDAC1 and HDAC2 [283,284].

The ATP-dependent chromatin remodeler BAF (SWI/SNF-A) was discovered by our group to be essential to both the establishment and maintenance of HIV latency (Figure 3). The BAF complex utilizes energy from ATP to push Nuc-1 from an energetically favorable position upstream of the TSS to a suboptimal region, downstream of TSS, resulting in a transcriptional block [281]. siRNA depletion of the BAF complex de-repressed proviral transcription. Furthermore, in siRNA-mediated BAF knockdown, latency establishment occurred less frequently than in the presence of the functional complex. The PIC through LEDGF interacts with INI-1 a subunit of BAF, allowing nucleosomes to be deposited at the provirus, contributing to latency establishment [118].

4.16. Epigenetic modifications regulate latency

Epigenetic modifications of nucleosomes such as histone-acetylation and -methylation and of DNA such as DNA-methylation play an important role in regulating the proviral transcription. Nucleosomes are the basic units of organization of chromatin and consist of a combination of histone subunits. Histones have an amino acids tail that can be extensively modified. Two broadly studied modifications that regulate expression effects are histone-acetylation and histone-methylation

Histone-acetylation by histone acetyl transferases (HATs) induces chromatin loosening, while histone deacetylases (HDACs) reverse the effect by removing the acetyl group (Figure 3). HATs such as p300/CREB-binding protein (p300/CBP) and p300/CBP-associated factor (P/CAF) can be recruited to activate the HIV LTR [158,285]. HDAC1, HDAC2, HDAC3, and HDAC6 repress HIV [286–289]. Numerous host factors recruit HDACs to the LTR. A negative regulator of P-TEFb, CTIP2 in cooperation with COUP-TF and Sp1 also recruits HDAC1 and HDAC2 to the HIV LTR in microglial cells [290,291]. Host factors LSF and YY1 co-operatively bind to the LTR, where YY1 recruits HDAC1 to deacetylate Nuc-1[207]. CBF-1 and c-Myc also repress HIV through the recruitment of HDAC1 [208–210].

Methylation of histones by histone methyltransferases (HMT) can act as an activating or repressing mark depending on the histone tail residue modified (e.g., methylation of lysine 4 on histone 3 (H3K4) is activating whereas H3K9, H3K27, and H4K20 methylation is repressive). HMTs modify specific histone residues, e.g., EZH2 (H3K27me3), SUV39H1 (H3K9me3), G9a (H3K9me2), and G9a like protein, GLP (H3K9me2). The repressive methyl groups deposited by these HMTs contribute to the maintenance of latency [292–295]. Moreover, EZH2 is suspected to recruit additional repressive proteins such as HDACs and other HMTs [294].

DNA methylation at CpG dinucleotides represses transcription by disrupting the binding of transcription activators to their binding sites or indirectly through the binding methyl-CpG binding proteins (MeCPs). In cell line models of latency, the HIV-1 LTR contains two CpG islands that are hypermethylated (Figure 3) [296]. Methyl-CpG binding domain protein 2 (MDB2) and HDAC-2 bind to the second CpG island on the HIV LTR and are displaced from there when cells are treated with cytosine-methylation inihibitor 5-aza-2′deoxycytidine [296]. In memory CD4+ T-cells from long-term aviremic and viremic patients, an increase in HIV LTR DNA methylation was observed in the aviremic patients [297]. The methylation of the HIV LTR in long-term non-progressors and elite controllers is increased compared to the LTR of aviremic patients on cART [298]. In contrast, this difference was not found in the first CpG island of resting memory CD4+ T-cells from aviremic patients, indicating that the mechanism by which DNA-methylation regulates latency deserves further exploration.

4.17. Viral and host non-coding RNAs regulate viral expression

Non-coding RNAs exert post transcriptional control on gene expression. Small non-coding RNAs (<200 nt) and in particular microRNAs (miRNAs) are well established to have regulatory function. The study of long non-coding RNAs (lncRNA, >200 nt) is an emerging field because of their epigenetic regulatory potential. Both viral and host miRNAs and lncRNAs affect replication of HIV-1 [146,299–301].

RNA interference (RNAi) is a post-transcriptional gene silencing mechanism. miRNAs post-transcriptionally suppress or silence gene expression as part of the RNA-induced silencing complex (RISC) forming a protein–RNA complex. Pri-miRNAs are generated by RNAPII and are subsequently processed by microprocessor into pre-miRNAs in the nucleus. Following export to the cytoplasm, they are cleaved by Dicer and incorporated into RISC. RISC generally binds in the 3′-untranslated region (3′UTR) of a target mRNA. The bound transcript is degraded or transcription is impeded depending on the level of homology, resulting in translational repression. The RNAi affects the infectivity of monocytes and macrophages [302]. Comparisons of productively infected, suppressed, and uninfected patients found difference in miRNA profiles, but it is very unlikely that the observed effects are due to viral activity because the number of infected cells is low in elite controllers or under cART [303–305]. Knockdown of Dicer or Drosha, a component of microprocessor, stimulates HIV-1 replication, indicating that miRNA generally are responsible for suppression of proviral transcription [299,300]. However, phenotypic effects are hard to interpret due to the pleiotropic side effects of microprocessor depletion. RNAi affects infectivity by targeting transcripts of key host factors and viral proteins involved in HIV-1 repression. In resting T-cells, the polycistronic miRNA cluster miR-17/92 is suppressed by HIV, resulting in PCAF upregulation [299]. Additionally, CycT1 is negatively regulated by miR27b [306]. Moreover, during differentiation from monocytes to macrophages, expression of miRNA198 and miR27b decreases relieving suppression of CycT1 [307,308]. In infected cells Tat, and possibly Vpr, inhibit RNAi [309–311]. In resting, but not activated, CD4+ T-cells a cluster of five miRNAs (miR-28, miR-125b, miR-150, miR-223, and miR-382) were found to be upregulated. They all target viral mRNAs for degradation; therefore, these miRNAs are contributing to latency maintenance [312]. However, further studies are required as results thus far are inconsistent [313–319].

The viral protein Nef is targeted by miR29a which interferes with HIV replication [300,320]. TRIM32 activates HIV-1 expression through the NF-kB pathway and is downregulated by miRNA-155 [321]. Tat-induced upregulation of miR34a and miR217 inhibits SIRT1 expression, which in turn results in high abundance of NF-kB, enhancing proviral transcription [322,323]. miRNA-182 has a positive effect on LTR activation by Tat [324]. miR-1236 restricts viral replication by repressing Vpr (HIV-1)-binding protein expression, VprBP [325].

HIV-1-derived miRNAs (vmiRNAs) were predicted *in silico* [326]. Applying deep sequencing technologies vmiRNAs were observed in cell line model systems of latency [327,328]. The TAR-derived miRNA-TAR5p and miR-TAR3p are asymmetrically processed and both repress LTR activity [329]. The *Nef*-derived miR-N367 inhibits viral promotor activity [330]. Nevertheless, relevance of vmiRNAs is debatable as no vmiRNAs were detected in PBMCs or macrophages of infected patients [331].

lncRNAs can modulate gene expression through different proposed mechanisms: (1) affecting mRNAs through sequence recognition, (2) recruiting proteins to DNA, (3) blocking host factors by assuming a secondary structure, (4) functioning as a scaffold for protein complexes. An anti-sense lncRNA of HIV-1 inhibits viral replication[146]. The non-coding repressor of NFAT (NRON) inhibits LTR activity in a NFAT-dependent manner [301].

4.18. Stochastic gene expression

The current model of HIV latency proposes that resting memory CD4+ T-cells are deprived of host factors that are necessary for viral expression. An alternative model proposes that expression is highly stochastic. Due to fluctuations in chromatin state and availability of the transcription factors, the latent and productive state co-exist [332]. In support, clonal lines (containing the same integration) showed binominal distributions of viral expression [333]. Transcriptional bursts of 2–10 mRNA transcripts were estimated to be the source of HIV-1 gene expression [334]. Tat-controlled positive feedback extends the expression reactivation [335]. The sensitivity to reactivation is also stochastic, as cells derived from patients remained latent during a first round of activation and were reactivatable in the next round of activation [83]. Moreover, molecules that increase gene expression fluctuations synergistically enhance HIV-1 reactivation [336].

5. HIV cure

Mechanistic insight into the complex nature of latent HIV-1 infection provides a rationale for eradication strategies. Therefore, identification of molecules that inhibit activity of repressors or potentiate HIV-1 activators alongside with immune system boosting are important objectives in eradication strategies.

5.1. Shocking the virus: screening for Latency Reversal Agents (LRAs)

The initial step of LRA discovery is screening drug libraries with cell-line-based models. Positive hits are evaluated further using primary-cell-based models as they better recapitulate the nature of latent reservoirs. If effective and not toxic, putative LRAs should undergo reactivation studies using primary cells derived from HIV-1-positive individuals that are on cART as well as toxicology studies in animal models, in case of novel molecules. It is advantageous to include molecules that are already approved drugs in such putative LRAs libraries, employing them into clinical practice would be time and resources effective. Moreover, in order to easily diffuse through cell membranes, ideal LRAs are small molecules, with molecular weight below 900 daltons, although clinical practice shows that most effective compounds do not exceed 500 daltons [337,338].

The first attempts to reactivate proviral DNA failed, due to the use of agents (e.g., IL-2 or a monoclonal antibody against CD3 receptor) which resulted in global T-cell activation. Indeed, viral p24 and plasma HIV-1 RNA levels increased, but the toxicity of such treatment left this approach useless [339–341]. Therefore, there is a need for more specific agents, which are able to reactivate proviral transcription without T-cell activation.

5.2. HDAC inhibitors (HDACis)

Histone deacetylase inhibitors (HDACis) are a very promising class of LRAs which include valporic acid (VPA), Vorinostat (SAHA), Romidepsin, Panobinostat, Givinostat, Droxinostat, or Entinostat. Some (Vorinostat (SAHA), Romidepsin, Panobinostat) are undergoing clinical trials [94,342–344].

The focus on HDACis is due to their ability to loosen up the compact chromatin structure at the latent proviral promoter. Inhibition of HDACs results in an increase of histone acetylation level by HATs. HDACs 1, 2, and 3 are of particular interest as they considerably contribute to HIV-1 repression [287]. Fortunately, HDACis are already used in clinical therapies, e.g., VPA is used in epilepsy and bipolar disorders, Vorinostat and Romidepsin are used to treat cutaneous T-cell lymphoma (CTCL) while Panabinostat is used in patients with multiple myeloma. In a very promising study by Archin et al., a single treatment with Vorinostat resulted in an increase in proviral RNA [345]. Unfortunately, the follow-up study with additional, multiple-dose rounds of treatment showed that increase on HIV-1 transcription is neither sustained nor elevated [346]. It is possible that other mechanisms maintaining latency compensate histone acetylation, in order to restrain proviral transcription. Alternatively such low concentrations of Vorinostat result in activation of pTEF-b instead of HDAC inhibition [347]. Since HDACs are involved in general regulation of gene expression; they have pleiotropic effects causing toxicities. Therefore, their use must be strictly controlled and monitored in order to provide maximal safety [348]. Nevertheless, HDACis are still under much interest. Especially, finding more specific HDAC inhibitors is very appealing, as current drugs are inhibiting a wide range of different HDACs, contributing to high toxicity [349].

5.3. BET inhibitors (BETi's)

Since BET proteins repress the HIV-1 promoter, it is worth to use their inhibitors in latency reversal strategies. Treatment with BET protein inhibitor JQ1 reactivates HIV-1 transcription in Tat-independent fashion [247]. Furthermore, BET inhibitor activity was positively tested in more relevant primary model system of latency [249]. Unfortunately, JQ-1 is not clinically available, due to its short half-life.

5.4. HMT inhibitors (HMTis)

Several histone methyltransferases (HMTs) such as EZH2, SUV39H1, and G9a interact with 5′ LTR contributing to maintenance of latency by deposition of repressive methyl groups on nucleosomal proteins [292–295]. Moreover, EZH2 recruits additional repressive proteins such as HDACs and other HMTs [294]. Several inhibitors of these proteins were tested in cell lines or primary cells from HIV-1 positive patients. Among which, Chaetocin (SUV39H1 inhibitor) and BIX-01294 (G9a inhibitor) were most potent [292,350]. However, high toxicity, due to pleiotropic effects, makes them unsuitable for clinical practice. Therefore, identification of novel compounds that are able to inhibit the activity of HMTs is needed.

5.5. DNMT inhibitors (DNMTis)

Inhibition of DNA methyltransferases (DNMTs) with 5-aza-2′ deoxycytidine (aza-CdR or Decitabine) leads to modest reactivation of latent HIV-1. This activity can be further enhanced with PKC agonists [351]. However, 5′ LTR methylation in patients material remains controversial [352]. Thus, further investigation of provirus methylation *in vivo* is needed.

5.6. Toll-like receptors (TLRs) stimulation

TLRs recently gained more attention, as theirs agonists are strong reactivators of HIV-1 [353–357]. The main role of these receptors is to activate an immune response against bacterial or viral infections [358]. Stimulating TLRs (as adjuvants in immunization) as well as opportunistic bacterial infections elevate plasma HIV-RNA and improve immune function [359–363].

Vaccine adjuvant – CPG 7909 (TLR 9 agonist) is able to decrease plasma HIV-1 RNA via activation of HIV-specific CD8+ T-cells in peripheral blood [359]. More recently, in SIV-positive rhesus monkeys undergoing cART were treated with GS-9620, a TLR7 agonist, reversible CD8 cytotoxic T-cells activation alongside with modest CD4 T-cell activation were observed. Moreover, elevated plasma viremia was observed as well as decrease in HIV-1 DNA in blood, colon, and lymph nodes. Interestingly, viral load returned back to undetectable levels when GS-9620 was no longer administrated. More strikingly, when cART was stopped, GS-9620-treated monkeys had 0.5 log lower viral set-point than untreated, infected animals. Additionally, in cells isolated from HIV-positive individuals transcription of HIV-1 was observed. However, some variability between samples was noticed. Clinical trials with the use of this compound are planned [364,365].

5.7. Super elongation complex stimulation

Treatment of cell lines and cells isolated from aviremic patients on cART with hexamethylene bisacetamide (HMBA), an anticancer drug that transiently activates PI3K/Akt pathway, results in phosphorylation of HEXIM1. P-TEFb is subsequently released and interacts with RNAP II, resulting in latency reversal [366–368]. Moreover, HMBA provides CDK9 recruitment to the viral promoter by interaction with SP1, which enhances transcription from proviral DNA. Furthermore, Klichko et al. showed that treatment with HMBA resulted in a decrease of CD4 receptor expression without affecting transcription of CCR5 and CXCR4 co-receptors [369]. Moreover, HMBA does not trigger activation of T-cells. Studies on P-TEFb's role in HIV-1 latency indicate that this heterocomplex might be an interesting target for inclusion in "shock and kill" therapies.

5.8. PKC pathway activation

Another interesting approach is the use of molecules that are able to selectively activate the protein kinase C (PKC) pathway. PKC pathway agonists trigger nuclear localization of NF-kB, NFAT, and AP-1 transcription factors. Therefore, PKC agonists are one of the most potent activators of HIV-1 transcription. Currently, two PKC agonists are being scrutinized clinically: prostratin and bryostatin, due to their safety and specificity toward HIV-1 reactivation. The latter is a clinically available drug [370]. Moreover, these two compounds prevent *de novo* infections, as they downregulate viral receptor and co-receptors CD4, CCR5 and CXCR4 in PBMCs [371]. A rather controversial molecule that reactivates HIV-1 transcription via NF-κB pathway is arsenic trioxide (As_2O_3). In the Jurkat model system of latency, As_2O_3 activates NF-κB leading to HIV-1 replication. Moreover, it synergizes with prostratin, tumor necrosis factor alpha (TNFα), and VPA [372]. Interestingly, arsenic is already used in clinical practice to treat acute promyelocytic leukemia (APL). Therefore, it would be interesting to test this compound in more relevant models of HIV-1 latency such as primary cells infected *ex vivo* and in cells derived from aviremic patients.

The use of PKC agonists raises concerns about their safety in a clinical setting. The protein kinase enzyme family consists of several isoenzymes that play important roles in signal transduction cascades [373]. As activation of latent HIV-1 is mediated via PKCα and PKCθ, the identification of more specific agonists of PKCα and PKCθ is needed. Alternatively, lowering the concentration of a specific agonists might decrease toxicity and contribute to latency reversal [374].

5.9. JNK/MAPK pathway activation

Studies employing mutagenesis of binding sites for activator protein-1 (AP-1) within the proviral genome showed that the AP-1 transcription factor is a crucial activator of proviral transcription, as proviruses with altered AP-1 binding sites were less prone to reactivation even if treated with a strong activator such as phorbol 12-myristate 13-acetate – PMA [191]. Furthermore, the latent pool of cells infected by virus with deletion in AP-1 sites was bigger, implicating that AP-1 is necessary for provirus transcription [192]. Heterodimeric protein AP-1 is formed upon phosphorylation of c-Jun N-terminal kinase (JNK) in JNK/MAPK pathway [193]. It is well established that activation of TLR signaling induces nuclear localization of NF-kB and AP-1 mediated via JNK pathway [194,196,376,377].

Virtual screening followed by validation of positive hits in cell line model systems for HIV-1 latency discovered 8-methoxy-6-methylquinolin-4-ol (MMQO) as a specific activator of the JNK-AP-1 pathway, which is able to reactivate HIV-1 from its latent state. Interestingly, MMQO inhibits IL-2 and TNFa expression, contributing to maintenance of resting state of CD4+ T-cells [378]. The recently synthetized panel of inhibitors of farnesyl transferase (FTase) are able to moderately reactivate HIV-1 transcription via JNK pathway. Interestingly, strong synergy with other LRAs, such as Vorinostat or TNF-a, was observed for these molecules in latency reversal [379,380].

5.10. Canonical Wnt signaling pathway activation

Recently, our group showed that treatment with Wnt3A/Rsp (natural stimulators of Wnt pathway) and lithium (inhibitor of Wnt repressor protein GSK3) leads to latency reversal in latent cell lines and enhances the latency reversal potential of HDAC inhibitors in CD4+ T primary cells obtained from patient volunteers when co-treated [381]. This observation shows a functional role for three LEF1 binding sites in the 5′ LTR contains, which are downstream targets of the classical Wnt pathway [381,382]. It would be very interesting to find more potent and selective inducers of Wnt pathway, as lithium exhibits many pleiotropic, toxic effects [383,384].

5.11. Chromatin loosening

It was discovered by our group that a main player in the establishment and maintenance of latency is the BAF complex (SWI/SNF-A), which belongs to ATP-dependent chromatin remodelers' family. Interestingly, Dykhuizen et al. [378] screened a library of compounds that would be able to mimic BRG-1 knock out. In their study, they found 20 compounds that were transcriptionally mimicking BAF complex disruption. We showed that several of those molecules were able to decrease the frequency of latency establishment and reactivate HIV-1 in cell line and primary cells models of latency [386, in press]. Moreover, they synergize with other LRAs – SAHA and prostratin. Two most potent inhibitors – caffeic acid phenethyl ester (CAPE) and pyrimethamine (PYR) did not activate T-cells derived from healthy donors and cells obtained from aviremic patients. Moreover, PYR is a registered drug used in malaria treatment. Therefore, these inhibitors are promising molecules to include in eradication strategies.

5.12. Multifunctional LRAs

In vitro treatment with cocaine leads to increase in HIV- replication in PBMCs as well as increased viral load in mouse models of HIV infection [387–389]. Interestingly, in *ex vivo* infected primary CD4+ cocaine treatment resulted in downregulation of miR125-b expression, which led to enhanced replication of HIV-1 [314]. In primary human macrophages and myeloid cell systems of latency, cocaine increased replication of HIV-1. Cocaine treatment activates NF-κB and leads to phosphorylation of mitogen- and stress-activated kinase 1 (MSK1). Furthermore, pMSK1 phosphorylates RELA (p65), a subunit of NF-κB promoting the interaction of NF-κB with p300 and recruitment of P-TEFb to the proviral 5′ LTR [390]. Moreover, treatment with cocaine results in histone H3 phosphorylation, thus increasing accessibility of HIV-1 promoter for transcription factors [390]. Therefore, cocaine not only reverses latency via NF-κB pathway but also causes epigenetic changes on 5′ LTR as well as blocks repressive miRNA.

Oral bacteria secrete short-chain fatty acids (SCFAs) including butyric acid, propionic acid, isovaleric acid, and isobutyric acid that are capable of HIV-1 and herpesviruses latency reversal [384,385]. Some of these molecules are known HDACis (e.g. Butyric acid) [393]. Moreover, SCFAs not only promotes histone acetylation, but also inhibit repressive histone formation and DNA methylation. Furthermore, they activate P-TEFb resulting in increased elongation of transcription from 5′ LTR. [345,385,386].

5.13. Immune clearance of reactivated cells – "Kill"

The majority of chronic patients are facing immune exhaustion, characterized by low cytokine secretion, smaller proliferative capacity, and low cytopathic potential of CD8+ T-cells [394,395]. Therefore, the first line of action would be reviving normal immune activity. Indeed, inhibition of programmed cell death protein 1 (PD-1) leads to restoration of immune functions in mouse models of HIV-1 infection [396]. However, these results were obtained in viremic animals. Nevertheless, an IgG4 antibody targeting PD-1 receptor is undergoing clinical trials to assess safety, immunotherapeutic activity, and the ability of treatment to reduce pool of latently infected cells [397].

In so-called "elite controllers", CD8+ T-cells effectively restrain infection without intervention of cART, by killing CD4+ T-cells that are actively producing HIV-1 particles [398,399]. The immune system can be boosted by specific amplification of HIV-1-specific CD8+ T-cells. These observations again aroused the idea of developing a vaccine. Indeed, rhesus monkeys vaccinated with CMV vectors resulted in broad cellular immune response to SIV [400–402]. However, safety issues related to the use of such vectors remain to be elucidated. Another platform being investigated to increase immune response against HIV-1 are Ad26 vectors, as it was shown that vaccinated rhesus monkeys were protected against infection with SIV as well as viral loads were lowered after vaccination [403,404].

A very interesting group of immunoglobulins to include in eradication strategies are broadly neutralizing monoclonal antibodies (mAbs or bNAbs) isolated from chronically infected patients. New generations of bNAbs exert higher potency and wider range of activity against many HIV-1 subtypes. It was shown that a combination of bNAbs is potent enough to transiently suppress viremia in rhesus monkeys as well as to reduce the amount of HIV-1 DNA in the blood, lymph nodes, and gastrointestinal mucosa [403,405,406].

6. Future perspectives and challenges

A reservoir of latent HIV is the main obstacle in finding a functional and sterilizing cure. Several challenges need to be addressed in order to overcome this obstacle. Defining the latent reservoir is impeded by the rare occurrence of a latent infection in a high background of defective proviral integration. Although HIV prefers integration in or near transcriptionally active genes which leaves ample room for variation in chromatin environment and available host transcription factors. This puts considerable demands on LRAs. LRAs should be effective, yet specific, without being toxic. As LRAs act via pathways involved in distinct cellular processes, pleiotropic effects are to be expected. Furthermore, recent studies on material obtained from HIV-1-positive suppressed patients revealed that currently available LRAs are not strong enough to reactivate the whole pool of latent proviruses, even after multiple rounds of stimulation. One of the concerns arising from "shock and kill" therapy is whether putative LRAs are strong enough to drive virus production to a level at which the immune system will be able to recognize and destroy HIV-1-producing cells. Indeed, trials aiming at testing HDAC inhibitors are inconsistent in showing depletion of latently infected cells while showing increased proviral transcription [407–412]. A complementary strategy would be to use multiple LRAs in combination to broadly and potentially synergistically reactivate the diversely integrated latent proviruses. Synergism between LRAs was already identified, e.g., Vorinostat and Prostratin [84]. Therefore, the quest for identification and characterization of novel compounds which are able to reactivate HIV-1 transcription as well as identifying combinations of drugs that can synergize to reverse latency is needed. Currently, no cell model is able to recapitulate the complexities of latency *in vivo*. A better system that more closely resembles the *in vivo* situation would greatly aid the understanding of molecular mechanisms underlying latency and the screening of new LRA. Moreover, as HIV-1 persists in a silent state, it contributes to a low level of inflammation, which over time leads to immune exhaustion. Furthermore, depletion of cells harboring latent provirus requires antigen-specific CTLs stimulation [399]. Most likely successful eradication therapies will be based on the combination of LRAs coupled with boosting HIV-1-specific immune response. A "shock and kill" approach in combination with immune therapies provides hope for reversing HIV-1 infection.

Author details

Michael D. Röling[#], Mateusz Stoszko[#], Tokameh Mahmoudi[*]

*Address all correspondence to: t.mahmoudi@erasmusmc.nl

Department of Biochemistry, Erasmus Medical Centre, Rottrerdam, The Netherlands

[#]These authors contributed equally to this work.

References

[1] Carpenter CCJ, Fischl MA, Hammer SM, Hirsch MS, Jacobsen DM, Katzenstein DA, et al. Antiretroviral Therapy for HIV Infection in 1998. JAMA. American Medical Association; 1998;280:78.

[2] Dinoso JB, Kim SY, Wiegand AM, Palmer SE, Gange SJ, Cranmer L, et al. Treatment intensification does not reduce residual HIV-1 viremia in patients on highly active antiretroviral therapy. Proc. Natl. Acad. Sci. U. S. A. 2009;106:9403–8.

[3] Davey RT, Bhat N, Yoder C, Chun TW, Metcalf JA, Dewar R, et al. HIV-1 and T cell dynamics after interruption of highly active antiretroviral therapy (HAART) in patients with a history of sustained viral suppression. Proc. Natl. Acad. Sci. U. S. A. 1999;96(26):15109-14.

[4] (WHO) WHO. WHO | Data and statistics [Internet]. World Health Organization; 2014 [cited 2015 Jul 29]. Available from: http://www.who.int/hiv/data/en/

[5] Pau AK, George JM. Antiretroviral Therapy. Infect. Dis. Clin. North Am. Elsevier Inc; 2014;28:371–402.

[6] Siliciano JD, Kajdas J, Finzi D, Quinn TC, Chadwick K, Margolick JB, et al. Long-term follow-up studies confirm the stability of the latent reservoir for HIV-1 in resting CD4+ T cells. Nat. Med. 2003;9:727–8.

[7] Centers for Disease Control and Prevention - CDC. CDC - HIV Cost-effectiveness - Ongoing Research - Prevention Research - HIV/AIDS [Internet]. 2015 [cited 2015 Jul 29]. Available from: http://www.cdc.gov/hiv/prevention/ongoing/costeffectiveness/

[8] Deeks SG. HIV infection, inflammation, immunosenescence, and aging. Annu. Rev. Med. 2011;62:141–55.

[9] Meyerhans A, Vartanian JP, Hultgren C, Plikat U, Karlsson A, Wang L, et al. Restriction and enhancement of human immunodeficiency virus type 1 replication by modulation of intracellular deoxynucleoside triphosphate pools. J. Virol. 1994;68:535–40.

[10] Zhou Y, Zhang H, Siliciano JD, Siliciano RF. Kinetics of human immunodeficiency virus type 1 decay following entry into resting CD4+ T cells. J. Virol. 2005;79:2199–210.

[11] Bukrinsky MI, Stanwick TL, Dempsey MP, Stevenson M. Quiescent T lymphocytes as an inducible virus reservoir in HIV-1 infection. Science. 1991;254:423–7.

[12] Zack JA, Arrigo SJ, Weitsman SR, Go AS, Haislip A, Chen IS. HIV-1 entry into quiescent primary lymphocytes: molecular analysis reveals a labile, latent viral structure. Cell. 1990;61:213–22.

[13] Ganesh L, Burstein E, Guha-Niyogi A, Louder MK, Mascola JR, Klomp LWJ, et al. The gene product Murr1 restricts HIV-1 replication in resting CD4+ lymphocytes. Nature. 2003;426:853–7.

[14] Pan X, Baldauf H-M, Keppler OT, Fackler OT. Restrictions to HIV-1 replication in resting CD4+ T lymphocytes. Cell Res. Shanghai Institutes for Biological Sciences, Chinese Academy of Sciences; 2013;23:876–85.

[15] Donahue D a, Wainberg M a. Cellular and molecular mechanisms involved in the establishment of HIV-1 latency. Retrovirology. Retrovirology; 2013;10:11.

[16] Bleul CC, Wu L, Hoxie JA, Springer TA, Mackay CR. The HIV coreceptors CXCR4 and CCR5 are differentially expressed and regulated on human T lymphocytes. Proc. Natl. Acad. Sci. U. S. A. 1997;94:1925–30.

[17] Stevenson M, Stanwick TL, Dempsey MP, Lamonica CA. HIV-1 replication is controlled at the level of T cell activation and proviral integration. EMBO J. 1990;9:1551–60.

[18] Wang W, Guo J, Yu D, Vorster PJ, Chen W, Wu Y. A dichotomy in cortical actin and chemotactic actin activity between human memory and naive T cells contributes to their differential susceptibility to HIV-1 infection. J. Biol. Chem. 2012;287:35455–69.

[19] Spear M, Guo J, Wu Y. The trinity of the cortical actin in the initiation of HIV-1 infection. Retrovirology. 2012;9:45.

[20] Siliciano RF, Greene WC. HIV latency. Cold Spring Harb. Perspect. Med. 2011;1:a007096.

[21] Ho DD, Neumann AU, Perelson AS, Chen W, Leonard JM, Markowitz M. Rapid turnover of plasma virions and CD4 lymphocytes in HIV-1 infection. Nature. 1995;373:123–6.

[22] Wei X, Ghosh SK, Taylor ME, Johnson VA, Emini EA, Deutsch P, et al. Viral dynamics in human immunodeficiency virus type 1 infection. Nature. 1995;373:117–22.

[23] Chavez L, Calvanese V, Verdin E. HIV Latency Is Established Directly and Early in Both Resting and Activated Primary CD4 T Cells. PLoS Pathog. 2015;11:e1004955.

[24] Eriksson S, Graf EH, Dahl V, Strain MC, Yukl SA, Lysenko ES, et al. Comparative analysis of measures of viral reservoirs in HIV-1 eradication studies. PLoS Pathog. Public Library of Science; 2013;9:e1003174.

[25] Chun T, Carruth L, Finzi D, Shen X, DiGiuseppe J, Taylor H, et al. Quantification of latent tissue reservoirs and total body viral load in HIV-1 infection. Lett. to Nat. 1997;246:170–170.

[26] Finzi D, Blankson J, Siliciano JD, Margolick JB, Chadwick K, Pierson T, et al. Latent infection of CD4+ T cells provides a mechanism for lifelong persistence of HIV-1, even in patients on effective combination therapy. Nat. Med. 1999;5:512–7.

[27] Strain MC, Günthard HF, Havlir D V, Ignacio CC, Smith DM, Leigh-Brown AJ, et al. Heterogeneous clearance rates of long-lived lymphocytes infected with HIV: intrinsic stability predicts lifelong persistence. Proc. Natl. Acad. Sci. U. S. A. 2003;100:4819–24.

[28] Chomont N, El-Far M, Ancuta P, Trautmann L, Procopio F a, Yassine-Diab B, et al. HIV reservoir size and persistence are driven by T cell survival and homeostatic proliferation. Nat. Med. 2009;15:893–900.

[29] Wightman F, Solomon A, Khoury G, Green JA, Gray L, Gorry PR, et al. Both CD31(+) and CD31⁻ naive CD4(+) T cells are persistent HIV type 1-infected reservoirs in individuals receiving antiretroviral therapy. J. Infect. Dis. 2010;202:1738–48.

[30] Costiniuk CT, Jenabian M-A. The lungs as anatomical reservoirs of HIV infection. Rev. Med. Virol. 2014;24:35–54.

[31] Salemi M, Lamers SL, Yu S, de Oliveira T, Fitch WM, McGrath MS. Phylodynamic analysis of human immunodeficiency virus type 1 in distinct brain compartments provides a model for the neuropathogenesis of AIDS. J. Virol. 2005;79:11343–52.

[32] garashi T, Brown CR, Endo Y, Buckler-White A, Plishka R, Bischofberger N, et al. Macrophage are the principal reservoir and sustain high virus loads in rhesus macaques after the depletion of CD4+ T cells by a highly pathogenic simian immunodeficiency virus/HIV type 1 chimera (SHIV): Implications for HIV-1 infections of humans. Proc. Natl. Acad. Sci. U. S. A. 2001;98:658–63.

[33] Gartner S, Markovits P, Markovitz DM, Kaplan MH, Gallo RC, Popovic M. The role of mononuclear phagocytes in HTLV-III/LAV infection. Science. 1986;233:215–9.

[34] Van Lint C, Bouchat S, Marcello A. HIV-1 transcription and latency: an update. Retrovirology. Retrovirology; 2013;10:67.

[35] Eisele E, Siliciano RF. Redefining the viral reservoirs that prevent HIV-1 eradication. Immunity. Elsevier Inc.; 2012;37:377–88.

[36] Tyagi M, Bukrinsky M. Human immunodeficiency virus (HIV) latency: the major hurdle in HIV eradication. Mol. Med. 2012;18:1096–108.

[37] Schnell G, Price RW, Swanstrom R, Spudich S. Compartmentalization and clonal amplification of HIV-1 variants in the cerebrospinal fluid during primary infection. J. Virol. 2010;84:2395–407.

[38] Churchill MJ, Gorry PR, Cowley D, Lal L, Sonza S, Purcell DFJ, et al. Use of laser capture microdissection to detect integrated HIV-1 DNA in macrophages and astrocytes from autopsy brain tissues. J. Neurovirol. 2006;12:146–52.

[39] Wiley CA, Schrier RD, Nelson JA, Lampert PW, Oldstone MB. Cellular localization of human immunodeficiency virus infection within the brains of acquired immune deficiency syndrome patients. Proc. Natl. Acad. Sci. U. S. A. 1986;83:7089–93.

[40] Yukl SA, Gianella S, Sinclair E, Epling L, Li Q, Duan L, et al. Differences in HIV burden and immune activation within the gut of HIV-positive patients receiving suppressive antiretroviral therapy. J. Infect. Dis. 2010;202:1553–61.

[41] Chun T-W, Nickle DC, Justement JS, Meyers JH, Roby G, Hallahan CW, et al. Persistence of HIV in gut-associated lymphoid tissue despite long-term antiretroviral therapy. J. Infect. Dis. 2008;197:714–20.

[42] Lerner P, Guadalupe M, Donovan R, Hung J, Flamm J, Prindiville T, et al. The gut mucosal viral reservoir in HIV-infected patients is not the major source of rebound plasma viremia following interruption of highly active antiretroviral therapy. J. Virol. 2011;85:4772–82.

[43] Yilmaz A, Verhofstede C, D'Avolio A, Watson V, Hagberg L, Fuchs D, et al. Treatment intensification has no effect on the HIV-1 central nervous system infection in patients on suppressive antiretroviral therapy. J. Acquir. Immune Defic. Syndr. 2010;55:590–6.

[44] Symons J, Vandekerckhove L, Hütter G, Wensing AMJ, Van Ham PM, Deeks SG, et al. Dependence on the CCR5 coreceptor for viral replication explains the lack of rebound of CXCR4-predicted HIV variants in the Berlin patient. Clin. Infect. Dis. 2014;59:596–600.

[45] Hütter G, Nowak D, Mossner M, Ganepola S, Müssig A, Allers K, et al. Long-term control of HIV by CCR5 Delta32/Delta32 stem-cell transplantation. N. Engl. J. Med. 2009;360:692–8.

[46] Deng H, Liu R, Ellmeier W, Choe S, Unutmaz D, Burkhart M, et al. Identification of a major co-receptor for primary isolates of HIV-1. Nature. 1996;381:661–6.

[47] Dragic T, Litwin V, Allaway GP, Martin SR, Huang Y, Nagashima KA, et al. HIV-1 entry into CD4+ cells is mediated by the chemokine receptor CC-CKR-5. Nature. 1996;381:667–73.

[48] Libert F, Cochaux P, Beckman G, Samson M, Aksenova M, Cao A, et al. The deltaccr5 mutation conferring protection against HIV-1 in Caucasian populations has a single and recent origin in Northeastern Europe. Hum. Mol. Genet. 1998;7:399–406.

[49] Yukl SA, Boritz E, Busch M, Bentsen C, Chun T-W, Douek D, et al. Challenges in detecting HIV persistence during potentially curative interventions: a study of the Berlin patient. PLoS Pathog. Public Library of Science; 2013;9:e1003347.

[50] Martinson JJ, Chapman NH, Rees DC, Liu YT, Clegg JB. Global distribution of the CCR5 gene 32-basepair deletion. Nat. Genet. 1997;16:100–3.

[51] Sabeti PC, Walsh E, Schaffner SF, Varilly P, Fry B, Hutcheson HB, et al. The case for selection at CCR5-Delta32. PLoS Biol. Public Library of Science; 2005;3:e378.

[52] Yukl SA, Boritz E, Busch M, Bentsen C, Chun T-W, Douek D, et al. Challenges in detecting HIV persistence during potentially curative interventions: a study of the Berlin patient. PLoS Pathog. Public Library of Science; 2013;9:e1003347.

[53] Henrich TJ, Hu Z, Li JZ, Sciaranghella G, Busch MP, Keating SM, et al. Long-term reduction in peripheral blood HIV type 1 reservoirs following reduced-intensity conditioning allogeneic stem cell transplantation. J. Infect. Dis. 2013;207:1694–702.

[54] Passaes CP, Sáez-Cirión A. HIV cure research: advances and prospects. Virology. 2014;454-455:340–52.

[55] Persaud D, Gay H, Ziemniak C, Chen YH, Piatak M, Chun T-W, et al. Absence of detectable HIV-1 viremia after treatment cessation in an infant. N. Engl. J. Med. 2013;369:1828–35.

[56] Huang J, Burke PS, Cung TDH, Pereyra F, Toth I, Walker BD, et al. Leukocyte immunoglobulin-like receptors maintain unique antigen-presenting properties of circulating myeloid dendritic cells in HIV-1-infected elite controllers. J. Virol. 2010;84:9463–71.

[57] Lewin SR, Rouzioux C. HIV cure and eradication: how will we get from the laboratory to effective clinical trials? AIDS. 2011;25:885–97.

[58] Autran B, Descours B, Avettand-Fenoel V, Rouzioux C. Elite controllers as a model of functional cure. Curr. Opin. HIV AIDS. 2011;6:181–7.

[59] Okulicz JF, Lambotte O. Epidemiology and clinical characteristics of elite controllers. Curr. Opin. HIV AIDS. 2011;6:163–8.

[60] Blankson JN. Control of HIV-1 replication in elite suppressors. Discov. Med. 2010;9:261–6.

[61] Shacklett BL. Understanding the 'lucky few': the conundrum of HIV-exposed, seronegative individuals. Curr. HIV/AIDS Rep. 2006;3:26–31.

[62] Sáez-Cirión A, Bacchus C, Hocqueloux L, Avettand-Fenoel V, Girault I, Lecuroux C, et al. Post-treatment HIV-1 controllers with a long-term virological remission after the interruption of early initiated antiretroviral therapy ANRS VISCONTI Study. PLoS Pathog. Public Library of Science; 2013;9:e1003211.

[63] Frange P, Faye A, Avettand-Fenoel V, Bellaton E, Deschamps D, Angin M, et al. HIV-1 virological remission for more than 11 years after interruption of early initiated antiretroviral therapy in a perinatally-infected child. 8th Int. AIDS Soc. Conf. HIV Pathog. Treat. Prev. (IAS 2015) Vancouver. 2015. p. MOAA0105LB.

[64] Ru R, Yao Y, Yu S, Yin B, Xu W, Zhao S, et al. Targeted genome engineering in human induced pluripotent stem cells by penetrating TALENs. Cell Regeneration; 2013;2(1):5.

[65] Mandal PK, Ferreira LMR, Collins R, Meissner TB, Boutwell CL, Friesen M, et al. Efficient Ablation of Genes in Human Hematopoietic Stem and Effector Cells using CRISPR/Cas9. Cell Stem Cell. Elsevier Inc.; 2014;15:643–52.

[66] Tebas P, Stein D, Tang WW, Frank I, Wang SQ, Lee G, et al. Gene Editing of CCR5 in Autologous CD4 T Cells of Persons Infected with HIV. N. Engl. J. Med. 2014;370:901–10.

[67] Allers K, Schneider T. CCR5Δ32 mutation and HIV infection: basis for curative HIV therapy. Curr. Opin. Virol. 2015;14:24–9.

[68] Vrins CLJ, Out R, van Santbrink P, van der Zee A, Mahmoudi T, Groenendijk M, et al. Znf202 affects high density lipoprotein cholesterol levels and promotes hepatosteatosis in hyperlipidemic mice. PLoS One. 2013;8:e57492.

[69] Huelsmann PM, Hofmann AD, Knoepfel SA, Popp J, Rauch P, Di Giallonardo F, et al. A suicide gene approach using the human pro-apoptotic protein tBid inhibits HIV-1 replication. BMC Biotechnol. 2011;11:4.

[70] Hamer DH. Can HIV be Cured? Mechanisms of HIV persistence and strategies to combat it. Curr. HIV Res. 2004;2:99–111.

[71] Deeks SG, Autran B, Berkhout B, Benkirane M, Cairns S, Chomont N, et al. Towards an HIV cure: a global scientific strategy. Nat. Rev. Immunol. Nature Publishing Group; 2012;12:607–14.

[72] Barouch DH, Deeks SG. immunologic strategies for HIV-1 remission and eradication. Science (80-.). 2014;345:169–74.

[73] Archin NM, Margolis DM. Emerging strategies to deplete the HIV reservoir. Curr. Opin. Infect. Dis. 2014;27:29–35.

[74] Folks TM, Clouse KA, Justement J, Rabson A, Duh E, Kehrl JH, et al. Tumor necrosis factor alpha induces expression of human immunodeficiency virus in a chronically infected T-cell clone. Proc. Natl. Acad. Sci. U. S. A. 1989;86:2365–8.

[75] Folks TM, Justement J, Kinter a, Dinarello C a, Fauci a S. Cytokine-induced expression of HIV-1 in a chronically infected promonocyte cell line. Science. 1987;238:800–2.

[76] Emiliani S, Van Lint C, Fischle W, Paras P, Ott M, Brady J, et al. A point mutation in the HIV-1 Tat responsive element is associated with postintegration latency. Proc. Natl. Acad. Sci. U. S. A. 1996;93:6377–81.

[77] Emiliani S, Fischle W, Ott M, Van Lint C, Amella CA, Verdin E. Mutations in the tat Gene Are Responsible for Human Immunodeficiency Virus Type 1 Postintegration Latency in the U1 Cell Line. J. Virol. 1998;72:1666–70.

[78] Jordan A, Bisgrove D, Verdin E. HIV reproducibly establishes a latent infection after acute infection of T cells in vitro. EMBO. 2003;22:1868–77.

[79] Folks TM, Justement J, Kinter A, Dinarello CA, Fauci AS. Cytokine-induced expression of HIV-1 in a chronically infected promonocyte cell line. Science. 1987;238:800–2.

[80] Weiss A, Wiskocil RL, Stobo JD. The role of T3 surface molecules in the activation of human T cells: a two-stimulus requirement for IL 2 production reflects events occurring at a pre-translational level. J. Immunol. 1984;133:123–8.

[81] Schneider U, Schwenk HU, Bornkamm G. Characterization of EBV-genome negative 'null' and 'T' cell lines derived from children with acute lymphoblastic leukemia and leukemic transformed non-Hodgkin lymphoma. Int. J. Cancer. 1977;19:621–6.

[82] Schröder ARW, Shinn P, Chen H, Berry C, Ecker JR, Bushman F. HIV-1 integration in the human genome favors active genes and local hotspots. Cell. 2002;110:521–9.

[83] Ho Y-C, Shan L, Hosmane NN, Wang J, Laskey SB, Rosenbloom DIS, et al. Replication-competent noninduced proviruses in the latent reservoir increase barrier to HIV-1 cure. Cell; 2013;155:540–51.

[84] Spina C a, Anderson J, Archin NM, Bosque A, Chan J, Famiglietti M, et al. An in-depth comparison of latent HIV-1 reactivation in multiple cell model systems and resting CD4+ T cells from aviremic patients. PLoS Pathog. 2013;9:e1003834.

[85] Sahu GK, Lee K, Ji J, Braciale V, Baron S, Cloyd MW. A novel in vitro system to generate and study latently HIV-infected long-lived normal CD4+ T-lymphocytes. Virology. 2006;355:127–37.

[86] Bosque A, Planelles V. Induction of HIV-1 latency and reactivation in primary memory CD4+ T cells. Blood. 2009;113:58–65.

[87] Yang H, Xing S, Shan L, Connell KO, Dinoso J, Shen A, et al. Small-molecule screening using a human primary cell model of HIV latency identifies compounds that reverse latency without cellular activation. 2009;119(11):3473-86..

[88] Marini A, Harper JM, Romerio F. An in vitro system to model the establishment and reactivation of HIV-1 latency. J. Immunol. 2008;181:7713–20.

[89] Chun TW, Finzi D, Margolick J, Chadwick K, Schwartz D, Siliciano RF. In vivo fate of HIV-1-infected T cells: quantitative analysis of the transition to stable latency. Nat. Med. 1995;1:1284–90.

[90] Pierson T, McArthur J, Siliciano RF. Reservoirs for HIV-1: mechanisms for viral persistence in the presence of antiviral immune responses and antiretroviral therapy. Annu. Rev. Immunol. 2000;18:665–708.

[91] Finzi D, Hermankova M, Pierson T, Carruth LM, Buck C, Chaisson RE, et al. Identification of a reservoir for HIV-1 in patients on highly active antiretroviral therapy. Science. 1997;278:1295–300.

[92] Swiggard WJ, Baytop C, Yu JJ, Li C, Schretzenmair R, Doherty UO, et al. Human Immunodeficiency Virus Type 1 Can Establish Latent Infection in Resting CD4 + T Cells in the Absence of Activating Stimuli. J..... 2005;79:14179–88.

[93] De Crignis E, Mahmoudi T. HIV eradication: combinatorial approaches to activate latent viruses. Viruses. 2014;6:4581–608.

[94] Archin NM, Sung JM, Garrido C, Soriano-Sarabia N, Margolis DM. Eradicating HIV-1 infection: seeking to clear a persistent pathogen. Nat. Rev. Microbiol. Nature Publishing Group; 2014;12:750–64.

[95] Denton PW, Olesen R, Choudhary SK, Archin NM, Wahl a., Swanson MD, et al. Generation of HIV Latency in Humanized BLT Mice. J. Virol. 2012;86:630–4.

[96] Melkus MW, Estes JD, Padgett-Thomas A, Gatlin J, Denton PW, Othieno F a, et al. Humanized mice mount specific adaptive and innate immune responses to EBV and TSST-1. Nat. Med. 2006;12:1316–22.

[97] Denton PW, Estes JD, Sun Z, Othieno F a., Wei BL, Wege AK, et al. Antiretroviral pre-exposure prophylaxis prevents vaginal transmission of HIV-1 in humanized BLT mice. PLoS Med. 2008;5:0079–89.

[98] North TW, Higgins J, Deere JD, Hayes TL, Villalobos A, Adamson L, et al. Viral sanctuaries during highly active antiretroviral therapy in a nonhuman primate model for AIDS. J. Virol. 2010;84:2913–22.

[99] Van Rompay KK a. Evaluation of antiretrovirals in animal models of HIV infection. Antiviral Res. 2010;85:159–75.

[100] Hatziioannou T, Ambrose Z, Chung NPY, Piatak M, Yuan F, Trubey CM, et al. A macaque model of HIV-1 infection. Proc. Natl. Acad. Sci. U. S. A. 2009;106:4425–9.

[101] Krebs FC, Hogan TH, Quiterio S, Gartner S, Wigdahl B. Lentiviral LTR-directed Expression, Sequence Variation, and Disease Pathogenesis Reviews. In: CL K, B F, B H, B K, F M, PA M, et al., editors. HIV Seq. Compend. 2001. Los Alamos: Theoretical Biology and Biophysics Group, Los Alamos National Laboratory: Los Alamos, NM; 2001. p. 1–42.

[102] Bruner KM, Hosmane NN, Siliciano RF. Towards an HIV-1 cure: measuring the latent reservoir. Trends Microbiol. Elsevier Ltd; 2015;23:192–203.

[103] Hindson BJ, Ness KD, Masquelier D a., Belgrader P, Heredia NJ, Makarewicz AJ, et al. High-throughput droplet digital PCR system for absolute quantitation of DNA copy number. Anal. Chem. 2011;83:8604–10.

[104] Procopio FA, Fromentin R, Kulpa DA, Brehm JH, Bebin A-G, Strain MC, et al. A Novel Assay to Measure the Magnitude of the Inducible Viral Reservoir in HIV-infected Individuals. EBioMedicine. Elsevier; 2015;

[105] Descours B, Cribier A, Chable-Bessia C, Ayinde D, Rice G, Crow Y, et al. SAMHD1 restricts HIV-1 reverse transcription in quiescent CD4+ T-cells. Retrovirology. Retrovirology; 2012;9:87.

[106] Goldstone DC, Ennis-Adeniran V, Hedden JJ, Groom HCT, Rice GI, Christodoulou E, et al. HIV-1 restriction factor SAMHD1 is a deoxynucleoside triphosphate triphosphohydrolase. Nature. Nature Publishing Group; 2011;480:379–82.

[107] Kim B, Nguyen L a., Daddacha W, Hollenbaugh J a. Tight interplay among SAMHD1 protein level, cellular dNTP levels, and HIV-1 proviral DNA synthesis kinetics in human primary monocyte-derived macrophages. J. Biol. Chem. 2012;287:21570–4.

[108] Lahouassa H, Daddacha W, Hofmann H, Ayinde D, Logue EC, Dragin L, et al. SAMHD1 restricts the replication of human immunodeficiency virus type 1 by depleting the intracellular pool of deoxynucleoside triphosphates. Nat. Immunol. 2012;13:621–621.

[109] Ryoo J, Choi J, Oh C, Kim S, Seo M, Diaz-griffero F, et al. The ribonuclease activity of SAMHD1 is required for HIV-1 restriction. Nat. Med. 2014;20:936–41.

[110] Choi J, Ryoo J, Oh C, Hwang S, Ahn K. SAMHD1 specifically restricts retroviruses through its RNase activity. Retrovirology. BioMed Central; 2015;12:1–12.

[111] Hofmann H, Logue EC, Bloch N, Daddacha W, Polsky SB, Schultz ML, et al. The Vpx lentiviral accessory protein targets SAMHD1 for degradation in the nucleus. J. Virol. 2012;86:12552–60.

[112] Sheehy AM, Gaddis NC, Choi JD, Malim MH. Isolation of a human gene that inhibits HIV-1 infection and is suppressed by the viral Vif protein. Nature. 2002;418:646–50.

[113] Schrijvers R, De Rijck J, Demeulemeester J, Adachi N, Vets S, Ronen K, et al. LEDGF/p75-independent HIV-1 replication demonstrates a role for HRP-2 and remains sensitive to inhibition by LEDGINs. PLoS Pathog. 2012;8.

[114] Ciuffi A, Llano M, Poeschla E, Hoffmann C, Leipzig J, Shinn P, et al. A role for LEDGF/p75 in targeting HIV DNA integration. Nat. Med. 2005;11:1287–9.

[115] Shun MC, Raghavendra NK, Vandegraaff N, Daigle JE, Hughes S, Kellam P, et al. LEDGF/p75 functions downstream from preintegration complex formation to effect gene-specific HIV-1 integration. Genes Dev. 2007;21:1767–78.

[116] Engelman A, Cherepanov P. The lentiviral integrase binding protein LEDGF/p75 and HIV-1 replication. PLoS Pathog. 2008;4.

[117] Wang H, Jurado KA, Wu X, Shun M-C, Li X, Ferris AL, et al. HRP2 determines the efficiency and specificity of HIV-1 integration in LEDGF/p75 knockout cells but does not contribute to the antiviral activity of a potent LEDGF/p75-binding site integrase inhibitor. Nucleic Acids Res. 2012;40:11518–30.

[118] Lesbats P, Botbol Y, Chevereau G, Vaillant C, Calmels C, Arneodo A, et al. Functional coupling between HIV-1 integrase and the SWI/SNF chromatin remodeling complex for efficient in vitro integration into stable nucleosomes. PLoS Pathog. 2011;7:e1001280.

[119] Maroun M, Delelis O, Coadou G, Bader T, Ségéral E, Mbemba G, et al. Inhibition of early steps of HIV-1 replication by SNF5/Ini1. J. Biol. Chem. 2006;281:22736–43.

[120] Turelli P, Doucas V, Craig E, Mangeat B, Klages N, Evans R, et al. Cytoplasmic recruitment of INI1 and PML on incoming HIV preintegration complexes: Interference with early steps of viral replication. Mol. Cell. 2001;7:1245–54.

[121] Ocwieja KE, Brady TL, Ronen K, Huegel A, Roth SL, Schaller T, et al. HIV integration targeting: A pathway involving transportin-3 and the nuclear pore protein RanBP2. PLoS Pathog. 2011;7:19–21.

[122] Pierson TC, Kieffer TL, Ruff CT, Buck C, Gange SJ, Siliciano RF. Intrinsic Stability of Episomal Circles Formed during Human Immunodeficiency Virus Type 1 Replication. J. Virol. 2002;76:4138–44.

[123] Pierson TC, Zhou Y, Kieffer TL, Christian T, Buck C, Siliciano RF, et al. Molecular Characterization of Preintegration Latency in Human Immunodeficiency Virus Type 1 Infection. J. Virol. 2002;76:8518–31.

[124] Zamborlini A, Lehmann-Che J, Clave E, Giron M-L, Tobaly-Tapiero J, Roingeard P, et al. Centrosomal pre-integration latency of HIV-1 in quiescent cells. Retrovirology. 2007;4:63.

[125] Trinité B, Ohlson EC, Voznesensky I, Rana SP, Chan CN, Mahajan S, et al. An HIV-1 replication pathway utilizing reverse transcription products that fail to integrate. J. Virol. 2013;87:12701–20.

[126] Han Y, Lassen K, Monie D, Ahmad R, Shimoji S, Liu X, et al. Resting CD4 + T Cells from Human Immunodeficiency Virus Type 1 (HIV-1) -Infected Individuals Carry Integrated HIV-1 Genomes within Actively Transcribed Host Genes. J. Virol. 2004;78:6122.

[127] Liu H, Dow EC, Arora R, Kimata JT, Bull LM, Arduino RC, et al. Integration of human immunodeficiency virus type 1 in untreated infection occurs preferentially within genes. J. Virol. 2006;80:7765–8.

[128] Lewinski MK, Bisgrove D, Shinn P, Chen H, Hoffmann C, Hannenhalli S, et al. Genome-Wide Analysis of Chromosomal Features Repressing Human Immunodeficiency Virus Transcription. J Virol. 2005;79:6610–9.

[129] Marini B, Kertesz-farkas A, Ali H, Lucic B, Lisek K, Manganaro L, et al. Nuclear architecture dictates HIV-1 integration site selection. Nature. 2015;521(7551):227-31

[130] Wang GP, Ciuffi A, Leipzig J, Berry CC, Bushman FD. HIV integration site selection: analysis by massively parallel pyrosequencing reveals association with epigenetic modifications. Genome Res. 2007;17:1186–94.

[131] Brady T, Agosto LM, Malani N, Berry CC, O'Doherty U, Bushman F. HIV integration site distributions in resting and activated CD4+ T cells infected in culture. AIDS. 2009;23:1461–71.

[132] Albanese A, Arosio D, Terreni M, Cereseto A. HIV-1 pre-integration complexes selectively target decondensed chromatin in the nuclear periphery. PLoS One. 2008; 3(6): e2413.

[133] Dieudonné M, Maiuri P, Biancotto C, Knezevich A, Kula A, Lusic M, et al. Transcriptional competence of the integrated HIV-1 provirus at the nuclear periphery. EMBO J. 2009;28:2231–43.

[134] Cohn LB, Silva IT, Oliveira TY, Rosales RA, Parrish EH, Learn GH, et al. HIV-1 Integration Landscape during Latent and Active Infection. Cell. Elsevier Inc.; 2015;160:420–32.

[135] Lusic M, Marini B, Ali H, Lucic B, Luzzati R, Giacca M. Proximity to PML nuclear bodies regulates HIV-1 latency in CD4+ T cells. Cell Host Microbe. Elsevier Inc.; 2013;13:665–77.

[136] Gallastegui E, Millán-Zambrano G, Terme J-M, Chávez S, Jordan A. Chromatin reassembly factors are involved in transcriptional interference promoting HIV latency. J. Virol. 2011;85:3187–202.

[137] Lenasi T, Contreras X, Peterlin BM. Transcriptional interference antagonizes proviral gene expression to promote HIV latency. Cell Host Microbe. 2008;4:123–33.

[138] Han Y, Lin YB, An W, Xu J, Yang H-C, O'Connell K, et al. Orientation-dependent regulation of integrated HIV-1 expression by host gene transcriptional readthrough. Cell Host Microbe. 2008;4:134–46.

[139] Shan L, Yang H-C, Rabi SA, Bravo HC, Shroff NS, Irizarry R a, et al. Influence of host gene transcription level and orientation on HIV-1 latency in a primary-cell model. J. Virol. 2011;85:5384–93.

[140] De Marco A, Biancotto C, Knezevich A, Maiuri P, Vardabasso C, Marcello A. Intragenic transcriptional cis-activation of the human immunodeficiency virus 1 does not result in allele-specific inhibition of the endogenous gene. Retrovirology. 2008;5:98.

[141] Klaver B, Berkhout B. Comparison of 5′ and 3' long terminal repeat promoter function in human immunodeficiency virus. J. Virol. 1994;68:3830–40.

[142] Cullen BR, Lomedico PT, Ju G. Transcriptional interference in avian retroviruses—implications for the promoter insertion model of leukaemogenesis. Nature. 1984;307:241–5.

[143] Michael NL, Vahey MT, d'Arcy L, Ehrenberg PK, Mosca JD, Rappaport J, et al. Negative-strand RNA transcripts are produced in human immunodeficiency virus type 1-

infected cells and patients by a novel promoter downregulated by Tat. J. Virol. 1994;68:979–87.

[144] Landry S, Halin M, Lefort S, Audet B, Vaquero C, Mesnard J-M, et al. Detection, characterization and regulation of antisense transcripts in HIV-1. Retrovirology. 2007;4:71.

[145] Ludwig LB, Ambrus JL, Krawczyk K a, Sharma S, Brooks S, Hsiao C-B, et al. Human Immunodeficiency Virus-Type 1 LTR DNA contains an intrinsic gene producing antisense RNA and protein products. Retrovirology. 2006;3:80.

[146] Kobayashi-Ishihara M, Yamagishi M, Hara T, Matsuda Y, Takahashi R, Miyake A, et al. HIV-1-encoded antisense RNA suppresses viral replication for a prolonged period. Retrovirology. 2012;9:38.

[147] Saayman S, Ackley A, Turner A-MW, Famiglietti M, Bosque A, Clemson M, et al. An HIV-encoded antisense long noncoding RNA epigenetically regulates viral transcription. Mol. Ther. 2014;22:1164–75.

[148] Karn J, Stoltzfus CM. Regulation of HIV-1 Gene Expression. Cold Spring Harb. Perspect. Med. 2012;1–17.

[149] Peeters a, Lambert PF, Deacon NJ. A fourth Sp1 site in the human immunodeficiency virus type 1 long terminal repeat is essential for negative-sense transcription. J. Virol. 1996;70:6665–72.

[150] Bentley K, Deacon N, Sonza S, Zeichner S, Churchill M. Mutational analysis of the HIV-1 LTR as a promoter of negative sense transcription. Arch. Virol. 2004;149:2277–94.

[151] Gaynor Richard. Cellular transcription factors involved in the regulation of HIV-1 gene expression. Aids. 1992. p. 347–63.

[152] Rosen C a, Sodroski JG, Haseltine W a. Location of cis-acting regulatory sequences in the human T-cell leukemia virus type I long terminal repeat. Proc. Natl. Acad. Sci. U. S. A. 1985;82:6502–6.

[153] Siekevitz M, Josephs SF, Dukovich M, Peffer N, Wong-Staal F, Greene WC. Activation of the HIV-1 LTR by T cell mitogens and the trans-activator protein of HTLV-I. Science. 1987;238:1575–8.

[154] Pereira LA, Bentley K, Peeters A, Churchill MJ, Deacon NJ. SURVEY AND SUMMARY A compilation of cellular transcription factor interactions with the HIV-1 LTR promoter. 2000;28:663–8.

[155] Kinoshita S, Su L, Amano M, Timmerman L a, Kaneshima H, Nolan GP. The T cell activation factor NF-ATc positively regulates HIV-1 replication and gene expression in T cells. Immunity. 1997;6:235–44.

[156] Selliah N, Zhang M, DeSimone D, Kim H, Brunner M, Ittenbach RF, et al. The??c-cytokine regulated transcription factor, STAT5, increases HIV-1 production in primary CD4 T cells. Virology. 2006;344:283–91.

[157] Nabel G, Baltimore D. An inducible transcription factor activates expression of human immunodeficiency virus in T cells. Nature. 1987;326:711–3.

[158] Gerritsen ME, Williams a J, Neish a S, Moore S, Shi Y, Collins T. CREB-binding protein/p300 are transcriptional coactivators of p65. Proc. Natl. Acad. Sci. U. S. A. 1997;94:2927–32.

[159] Tesmer VM, Rajadhyaksha a, Babin J, Bina M. NF-IL6-mediated transcriptional activation of the long terminal repeat of the human immunodeficiency virus type 1. Proc. Natl. Acad. Sci. 1993;90:7298–302.

[160] Henderson a J, Zou X, Calame KL. C/EBP proteins activate transcription from the human immunodeficiency virus type 1 long terminal repeat in macrophages/monocytes. J. Virol. 1995;69:5337–44.

[161] Yang X, Chen Y, Gabuzda D. ERK MAP kinase links cytokine signals to activation of latent HIV-1 infection by stimulating a cooperative interaction of AP-1 and NF-??B. J. Biol. Chem. 1999;274:27981–8.

[162] Krebs FC, Mehrens D, Pomeroy S, Goodenow MM, Wigdahl B. Human immunodeficiency virus type 1 long terminal repeat quasispecies differ in basal transcription and nuclear factor recruitment in human glial cells and lymphocytes. J. Biomed. Sci. 1998;5:31–44.

[163] Coiras M, López-Huertas MR, Rullas J, Mittelbrunn M, Alcamí J. Basal shuttle of NF-kappaB/I kappaB alpha in resting T lymphocytes regulates HIV-1 LTR dependent expression. Retrovirology. 2007;4:56.

[164] Colin L, Van Lint C. Molecular control of HIV-1 postintegration latency: implications for the development of new therapeutic strategies. Retrovirology. 2009;6:111.

[165] Cron RQ, Bartz SR, Clausell A, Bort SJ, Klebanoff SJ, Lewis DB. NFAT1 enhances HIV-1 gene expression in primary human CD4 T cells. Clin. Immunol. 2000;94:179–91.

[166] McKernan LN, Momjian D, Kulkosky J. Protein Kinase C: One Pathway towards the Eradication of Latent HIV-1 Reservoirs. Adv. Virol. 2012;2012:805347.

[167] Mahmoudi T. The BAF complex and HIV latency. Transcription. 2012;3:171–6.

[168] West MJ, Lowe AD, Karn J. Activation of Human Immunodeficiency Virus Transcription in T Cells Revisited: NF- κ B p65 Stimulates Transcriptional Elongation Activation of Human Immunodeficiency Virus Transcription in T Cells Revisited: NF-, B p65 Stimulates Transcriptional Elonga. J. Virol. 2001;75:8524–37.

[169] Li X, Josef J, Marasco W a. Hiv-1 Tat can substantially enhance the capacity of NIK to induce IkappaB degradation. Biochem. Biophys. Res. Commun. 2001;286:587–94.

[170] Pazin MJ, Sheridan PL, Cannon K, Cao Z, Keck JG, Kadonaga JT, et al. NF-κB-mediated chromatin reconfiguration and transcriptional activation of the HIV-1 enhancer in vitro. Genes Dev. 1996;10:37–49.

[171] Steger DJ, Workman JL. Stable co-occupancy of transcription factors and histones at the HIV-1 enhancer. EMBO J. 1997;16:2463–72.

[172] Rothgiesser KM, Erener S, Waibel S, Lüscher B, Hottiger MO. SIRT2 regulates NF-κB dependent gene expression through deacetylation of p65 Lys310. J. Cell Sci. 2010;123:4251–8.

[173] Kwon H-S, Ott M. The ups and downs of SIRT1. Trends Biochem. Sci. 2008;33:517–25.

[174] Zhang JL, Sharma PL, Crumpacker CS. Enhancement of the basal-level activity of HIV-1 long terminal repeat by HIV-1 nucleocapsid protein. Virology. 2000;268:251–63.

[175] Kim YK, Bourgeois CF, Pearson R, Tyagi M, West MJ, Wong J, et al. Recruitment of TFIIH to the HIV LTR is a rate-limiting step in the emergence of HIV from latency. EMBO J. 2006;25:3596–604.

[176] Larochelle S, Amat R, Glover-Cutter K, Sansó M, Zhang C, Allen JJ, et al. Cyclin-dependent kinase control of the initiation-to-elongation switch of RNA polymerase II. Nat. Struct. Mol. Biol. 2012;19:1108–15.

[177] Takahashi Y, Tanaka Y, Yamashita A. OX40 Stimulation by gp34/OX40 ligand enhances prodcuttive human immunodeficiency virus type1 infection. J. Virol. 2001;75:674806757.

[178] Kundu M, Srinivasan a, Pomerantz RJ, Khalili K. Evidence that a cell cycle regulator, E2F1, down-regulates transcriptional activity of the human immunodeficiency virus type 1 promoter. J. Virol. 1995;69:6940–6.

[179] Majello B, De Luca P, Hagen G, Suske G, Lania L. Different members of the Sp1 multigene family exert opposite transcriptional regulation of the long terminal repeat of HIV-1. Nucleic Acids Res. 1994;22:4914–21.

[180] Millhouse S, Krebs FC, Yao J, McAllister JJ, Conner J, Ross H, et al. Sp1 and related factors fail to interact with the NF-kappaB-proximal G/C box in the LTR of a replication competent, brain-derived strain of HIV-1 (YU-2). J. Neurovirol. 1998;4:312–23.

[181] Garcia-Rodriguez C, Rao A. Nuclear Factor of Activated T Cells (NFAT)-dependent Transactivation Regulated by the Coactivators p300/CREB-binding Protein (CBP). J. Exp. Med. 1998;187:2031–6.

[182] Wang FX, Xu Y, Sullivan J, Souder E, Argyris EG, Acheampong E a., et al. IL-7 is a potent and proviral strain-specific inducer of latent HIV-1 cellular reservoirs of infected individuals on virally suppressive HAART. J. Clin. Invest. 2005;115:128–37.

[183] Shuai K, Liu B. Regulation of JAK-STAT signalling in the immune system. Nat. Rev. Immunol. 2003;3:900–11.

[184] Bovolenta C, Camorali L, Lorini a L, Ghezzi S, Vicenzi E, Lazzarin a, et al. Constitutive activation of STATs upon in vivo human immunodeficiency virus infection. Blood. 1999;94:4202–9.

[185] Crotti A, Lusic M, Lupo R, Lievens PMJ, Liboi E, Della Chiara G, et al. Naturally occurring C-terminally truncated STAT5 is a negative regulator of HIV-1 expression. Blood. 2007;109:5380–9.

[186] Della Chiara G, Crotti A, Liboi E, Giacca M, Poli G, Lusic M. Negative regulation of HIV-1 transcription by a heterodimeric NF-??B1/p50 and C-terminally truncated STAT5 complex. J. Mol. Biol. Elsevier Ltd; 2011;410:933–43.

[187] Henderson AJ, Connor RI, Calame KL. C/EBP activators are required for HIV-1 replication and proviral induction in monocytic cell lines. Immunity. 1996;5:91–101.

[188] Henderson a J, Calame KL. CCAAT/enhancer binding protein (C/EBP) sites are required for HIV-1 replication in primary macrophages but not CD4(+) T cells. Proc. Natl. Acad. Sci. U. S. A. 1997;94:8714–9.

[189] Lee ES, Zhou H, Al LEEET, Irol J V. Endothelial Cells Enhance Human Immunodeficiency Virus Type 1 Replication in Macrophages through a C / EBP-Dependent Mechanism. Society. 2001;75:9703–12.

[190] Combates J, Kwon P, Rzepka W, Cohen D. Involvement of the Transcription Factor NF-1L6 in Phorbol Ester Induction of P-glycoprotein in U937 Cells. Cell. 1997;8:213–9.

[191] Colin L, Vandenhoudt N, de Walque S, Van Driessche B, Bergamaschi A, Martinelli V, et al. The AP-1 binding sites located in the pol gene intragenic regulatory region of HIV-1 are important for viral replication. PLoS One. Public Library of Science; 2011;6:e19084.

[192] Duverger A, Wolschendorf F, Zhang M, Wagner F, Hatcher B, Jones J, et al. An AP-1 binding site in the enhancer/core element of the HIV-1 promoter controls the ability of HIV-1 to establish latent infection. J. Virol. 2013;87:2264–77.

[193] Hess J, Angel P, Schorpp-Kistner M. AP-1 subunits: quarrel and harmony among siblings. J. Cell Sci. 2004;117:5965–73.

[194] Challacombe SJ, Naglik JR. The effects of HIV infection on oral mucosal immunity. Adv. Dent. Res. 2006;19:29–35.

[195] Long J, Wang Y, Wang W, Chang BHJ, Danesh FR. Identification of microRNA-93 as a novel regulator of vascular endothelial growth factor in hyperglycemic conditions. J. Biol. Chem. 2010;285:23457–65.

[196] Novis CL, Archin NM, Buzon MJ, Verdin E, Round JL, Lichterfeld M, et al. Reactivation of latent HIV-1 in central memory CD4+ T cells through TLR-1/2 stimulation. Retrovirology. 2013;10:119.

[197] Pearson R, Kim YK, Hokello J, Lassen K, Friedman J, Tyagi M, et al. Epigenetic silencing of human immunodeficiency virus (HIV) transcription by formation of restrictive chromatin structures at the viral long terminal repeat drives the progressive entry of HIV into latency. J. Virol. 2008;82:12291–303.

[198] Kim YK, Mbonye U, Hokello J, Karn J. T-cell receptor signaling enhances transcriptional elongation from latent HIV proviruses by activating P-TEFb through an ERK-dependent pathway. J. Mol. Biol. 2011;410:896–916.

[199] Herrmann CH, Rice a P. Lentivirus Tat proteins specifically associate with a cellular protein kinase, TAK, that hyperphosphorylates the carboxyl-terminal domain of the large subunit of RNA polymerase II: candidate for a Tat cofactor. J. Virol. 1995;69:1612–20.

[200] Wei P, Garber ME, Fang SM, Fischer WH, Jones K a. A novel CDK9-associated C-type cyclin interacts directly with HIV-1 Tat and mediates its high-affinity, loop-specific binding to TAR RNA. Cell. 1998;92:451–62.

[201] Bieniasz PD, Grdina T a, Bogerd HP, Cullen BR. Recruitment of a protein complex containing Tat and cyclin T1 to TAR governs the species specificity of HIV-1 Tat. EMBO J. 1998;17:7056–65.

[202] Fujinaga K, Cujec TP, Peng J, Garriga J, Price DH, Graña X, et al. The ability of positive transcription elongation factor B to transactivate human immunodeficiency virus transcription depends on a functional kinase domain, cyclin T1, and Tat. J. Virol. 1998;72:7154–9.

[203] Garber ME, Wei P, KewalRamani VN, Mayall TP, Herrmann CH, Rice AP, et al. The interaction between HIV-1 Tat and human cyclin T1 requires zinc and a critical cysteine residue that is not conserved in the murine CycT1 protein. Genes Dev. 1998;12:3512–27.

[204] Felzien LK, Woffendin C, Hottiger MO, Subbramanian R a, Cohen E a, Nabel GJ. HIV transcriptional activation by the accessory protein, VPR, is mediated by the p300 co-activator. Proc. Natl. Acad. Sci. U. S. A. 1998;95:5281–6.

[205] Wang L, Mukherjee S, Jia F, Narayan O, Zhao LJ. Interaction of virion protein Vpr of human immunodeficiency virus type 1 with cellular transcription factor Sp1 and trans-activation of viral long terminal repeat. J. Biol. Chem. 1995;270:25564–9.

[206] Agostini I, Navarro JM, Rey F, Bouhamdan M, Spire B, Vigne R, et al. The human immunodeficiency virus type 1 Vpr transactivator: cooperation with promoter-bound activator domains and binding to TFIIB. J. Mol. Biol. 1996;261:599–606.

[207] He G, Margolis DM. Counterregulation of chromatin deacetylation and histone deacetylase occupancy at the integrated promoter of human immunodeficiency virus type 1 (HIV-1) by the HIV-1 repressor YY1 and HIV-1 activator Tat. Mol. Cell. Biol. 2002;22:2965–73.

[208] Tyagi M, Karn J. CBF-1 promotes transcriptional silencing during the establishment of HIV-1 latency. EMBO J. 2007;26:4985–95.

[209] Tyagi M, Pearson RJ, Karn J. Establishment of HIV latency in primary CD4+ cells is due to epigenetic transcriptional silencing and P-TEFb restriction. J. Virol. 2010;84:6425–37.

[210] Jiang G, Espeseth A, Hazuda DJ, Margolis DM. c-Myc and Sp1 contribute to proviral latency by recruiting histone deacetylase 1 to the human immunodeficiency virus type 1 promoter. J. Virol. 2007;81:10914–23.

[211] Wen Y, Shatkin AJ. Transcription elongation factor hSPT5 stimulates mRNA capping Transcription elongation factor hSPT5 stimulates mRNA capping. Genes Dev. 1999; 13(14): 1774–1779.

[212] Cheng B, Price DH. Analysis of factor interactions with RNA polymerase II elongation complexes using a new electrophoretic mobility shift assay. Nucleic Acids Res. 2008;36:1–10.

[213] Missra A, Gilmour DS. Interactions between DSIF (DRB sensitivity inducing factor), NELF (negative elongation factor), and the Drosophila RNA polymerase II transcription elongation complex. Proc. Natl. Acad. Sci. U. S. A. 2010;107:11301–6.

[214] Yamaguchi Y, Inukai N, Narita T, Wada T, Handa H. Evidence that Negative Elongation Factor Represses Transcription Elongation through Binding to a DRB Sensitivity-Inducing Factor / RNA Polymerase II Complex and RNA. Mol. Cell. Biol. 2002;22:2918–27.

[215] Renner DB, Yamaguchi Y, Wada T, Handa H, Price DH. A Highly Purified RNA Polymerase II Elongation Control System. J. Biol. Chem. 2001;276:42601–9.

[216] Zhang Z, Klatt A, Gilmour DS, Henderson AJ. Negative elongation factor NELF represses human immunodeficiency virus transcription by pausing the RNA polymerase II complex. J. Biol. Chem. 2007;282:16981–8.

[217] Natarajan M, Schiralli Lester GM, Lee C, Missra A, Wasserman G a., Steffen M, et al. Negative elongation factor (NELF) coordinates RNA polymerase II pausing, premature termination, and chromatin remodeling to regulate HIV transcription. J. Biol. Chem. 2013;288:25995–6003.

[218] Jadlowsky JK, Wong JY, Graham AC, Dobrowolski C, Devor RL, Adams MD, et al. Negative Elongation Factor Is Required for the Maintenance of Proviral Latency but Does Not Induce Promoter-Proximal Pausing of RNA Polymerase II on the HIV Long Terminal Repeat. Mol. Cell. Biol. 2014;34:1911–28.

[219] Wagschal A, Rousset E, Basavarajaiah P, Contreras X, Harwig A, Laurent-Chabalier S, et al. Microprocessor, Setx, Xrn2, and Rrp6 co-operate to induce premature termination of transcription by RNAPII. Cell. 2012;150:1147–57.

[220] Kao SY, Calman a F, Luciw P a, Peterlin BM. Anti-termination of transcription within the long terminal repeat of HIV-1 by tat gene product. Nature. 1987;330:489–93.

[221] Laspia MF, Rice a P, Mathews MB. HIV-1 Tat protein increases transcriptional initiation and stabilizes elongation. Cell. 1989;59:283–92.

[222] Dingwall C, Ernberg I, Gait MJ, Green SM, Heaphy S, Karn J, et al. HIV-1 tat protein stimulates transcription by binding to a U-rich bulge in the stem of the TAR RNA structure. EMBO J. 1990;9:4145–53.

[223] Tahirov TH, Babayeva ND, Varzavand K, Cooper JJ, Sedore SC, Price DH. Crystal structure of HIV-1 Tat complexed with human P-TEFb. Nature. 2010;465:747–51.

[224] Fujinaga K, Irwin D, Huang Y, Taube R, Kurosu T, Peterlin BM. Dynamics of human immunodeficiency virus transcription: P-TEFb phosphorylates RD and dissociates negative effectors from the transactivation response element. Mol. Cell. Biol. 2004;24:787–95.

[225] Komarnitsky P, Cho EJ, Buratowski S. Different phosphorylated forms of RNA polymerase II and associated mRNA processing factors during transcription. Genes Dev. 2000;14:2452–60.

[226] Kim YK, Bourgeois CF, Isel C, Churcher MJ, Karn J. Phosphorylation of the RNA polymerase II carboxyl-terminal domain by CDK9 is directly responsible for human immunodeficiency virus type 1 Tat-activated transcriptional elongation. Mol. Cell. Biol. 2002;22:4622–37.

[227] Czudnochowski N, Bösken C a., Geyer M. Serine-7 but not serine-5 phosphorylation primes RNA polymerase II CTD for P-TEFb recognition. Nat. Commun. 2012;3:842.

[228] Ahn SH, Kim M, Buratowski S. Phosphorylation of serine 2 within the RNA polymerase II C-terminal domain couples transcription and 3J end. Process. Mol. Cell. 2004;13:67–76.

[229] Lenasi T, Peterlin BM, Barboric M. Cap-binding protein complex links pre-mRNA capping to transcription elongation and alternative splicing through positive transcription elongation factor b (P-TEFb). J. Biol. Chem. 2011;286:22758–68.

[230] Brиs V, Gomes N, Pickle L, Jones K a. A human splicing factor, SKIP, associates with P-TEFb and enhances transcription elongation by HIV-1 Tat. Genes Dev. 2005;19:1211–26.

[231] Bourgeois CF, Kim YK, Churcher MJ, West MJ, Karn J. Spt5 cooperates with human immunodeficiency virus type 1 Tat by preventing premature RNA release at terminator sequences. Mol. Cell. Biol. 2002;22:1079–93.

[232] Yamada T, Yamaguchi Y, Inukai N, Okamoto S, Mura T, Handa H. P-TEFb-mediated phosphorylation of hSpt5 C-terminal repeats is critical for processive transcription elongation. Mol. Cell. 2006;21:227–37.

[233] Keen NJ, Churcher MJ, Karn J. Transfer of Tat and release of TAR RNA during the activation of the human immunodeficiency virus type-1 transcription elongation complex. EMBO J. 1997;16:5260–72.

[234] Sobhian B, Laguette N, Yatim A, Nakamura M, Levy Y, Kiernan R, et al. HIV-1 Tat Assembles a Multifunctional Transcription Elongation Complex and Stably Associates with the 7SK snRNP. Mol. Cell. Elsevier Ltd; 2010;38:439–51.

[235] Nguyen VT, Kiss T, Michels a a, Bensaude O. 7SK small nuclear RNA binds to and inhibits the activity of CDK9/cyclin T complexes. Nature. 2001;414:322–5.

[236] Yang Z, Zhu Q, Luo K, Zhou Q. The 7SK small nuclear RNA inhibits the CDK9/cyclin T1 kinase to control transcription. Nature. 2001;414:317–22.

[237] Yik JHN, Chen R, Nishimura R, Jennings JL, Link AJ, Zhou Q. Inhibition of P-TEFb (CDK9/cyclin T) kinase and RNA polymerase II transcription by the coordinated actions of HEXIM1 and 7SK snRNA. Mol. Cell. 2003;12:971–82.

[238] Li Q, Price JP, Byers SA, Cheng D, Peng J, Price DH. Analysis of the large inactive P-TEFb complex indicates that it contains one 7SK molecule, a dimer of HEXIM1 or HEXIM2, and two P-TEFb molecules containing Cdk9 phosphorylated at threonine 186. J. Biol. Chem. 2005;280:28819–26.

[239] Peterlin BM, Brogie JE, Price DH. 7SK snRNA: A noncoding RNA that plays a major role in regulating eukaryotic transcription. Wiley Interdiscip. Rev. RNA. 2012;3:92–103.

[240] Jeronimo C, Forget D, Bouchard A, Li Q, Chua G, Poitras C, et al. Systematic Analysis of the Protein Interaction Network for the Human Transcription Machinery Reveals the Identity of the 7SK Capping Enzyme. Mol. Cell. 2007;27:262–74.

[241] Krueger BJ, Varzavand K, Cooper JJ, Price DH. The mechanism of release of P-TEFb and HEXIM1 from the 7SK snRNP by viral and cellular activators includes a conformational change in 7SK. PLoS One. 2010;5.

[242] Markert A, Grimm M, Martinez J, Wiesner J, Meyerhans A, Meyuhas O, et al. The La-related protein LARP7 is a component of the 7SK ribonucleoprotein and affects transcription of cellular and viral polymerase II genes. EMBO Rep. 2008;9:569–75.

[243] Schröder S, Cho S, Zeng L, Zhang Q, Kaehlcke K, Mak L, et al. Two-pronged binding with bromodomain-containing protein 4 liberates positive transcription elongation factor b from inactive ribonucleoprotein complexes. J. Biol. Chem. 2012;287:1090–9.

[244] Bisgrove D a, Mahmoudi T, Henklein P, Verdin E. Conserved P-TEFb-interacting domain of BRD4 inhibits HIV transcription. Proc. Natl. Acad. Sci. U. S. A. 2007;104:13690–5.

[245] Jang MK, Mochizuki K, Zhou M, Jeong H-S, Brady JN, Ozato K. The bromodomain protein Brd4 is a positive regulatory component of P-TEFb and stimulates RNA polymerase II-dependent transcription. Mol. Cell. 2005;19:523–34.

[246] Yang Z, Yik JHN, Chen R, He N, Jang MK, Ozato K, et al. Recruitment of P-TEFb for stimulation of transcriptional elongation by the bromodomain protein Brd4. Mol. Cell. 2005;19:535–45.

[247] Zhu J, Gaiha GD, John SP, Pertel T, Chin CR, Gao G, et al. Reactivation of latent HIV-1 by inhibition of BRD4. Cell Rep. 2012;2:807–16.

[248] Wu S-Y, Chiang C-M. The double bromodomain-containing chromatin adaptor Brd4 and transcriptional regulation. J. Biol. Chem. 2007;282:13141–5.

[249] Boehm D, Calvanese V, Dar RD, Xing S, Schroeder S, Martins L, et al. BET bromodomain-targeting compounds reactivate HIV from latency via a Tat-independent mechanism. Cell Cycle. 2013;12:452–62.

[250] Denis G V, McComb ME, Faller D V, Sinha A, Romesser PB, Costello CE. Identification of transcription complexes that contain the double bromodomain protein Brd2 and chromatin remodeling machines. J. Proteome Res. 2006;5:502–11.

[251] Cherrier T, Le Douce V, Eilebrecht S, Riclet R, Marban C, Dequiedt F, et al. CTIP2 is a negative regulator of P-TEFb. Proc. Natl. Acad. Sci. U. S. A. 2013;110:12655–60.

[252] Mbonye UR, Wang B, Gokulrangan G, Chance MR, Karn J. Phosphorylation of HEXIM1 at Tyr271 and Tyr274 Promotes Release of P-TEFb from the 7SK snRNP Complex and Enhances Proviral HIV Gene Expression. Proteomics. 2015;15:2078–86.

[253] Ji X, Zhou Y, Pandit S, Huang J, Li H, Lin CY, et al. SR proteins collaborate with 7SK and promoter-associated nascent RNA to release paused polymerase. Cell. 2013;153:855–68.

[254] Suñé C, Goldstrohm a C, Peng J, Price DH, Garcia-Blanco M a. An in vitro transcription system that recapitulates equine infectious anemia virus tat-mediated inhibition of human immunodeficiency virus type 1 Tat activity demonstrates a role for posi-

tive transcription elongation factor b and associated proteins in th. Virology. 2000;274:356–66.

[255] Lu H, Li Z, Xue Y, Schulze-Gahmen U, Johnson JR, Krogan NJ, et al. AFF1 is a ubiquitous P-TEFb partner to enable Tat extraction of P-TEFb from 7SK snRNP and formation of SECs for HIV transactivation. Proc. Natl. Acad. Sci.. 2014;111 :E15–24.

[256] He N, Liu M, Hsu J, Xue Y, Chou S, Burlingame A, et al. HIV-1 Tat and host AFF4 recruit two transcription elongation factors into a bifunctional complex for coordinated activation of HIV-1 transcription. Mol. Cell. 2010;38:428–38.

[257] Mueller D, García-Cuéllar MP, Bach C, Buhl S, Maethner E, Slany RK. Misguided transcriptional elongation causes mixed lineage leukemia. PLoS Biol. 2009;7.

[258] Lin C, Smith ER, Takahashi H, Lai K, Martin- S, Florens L, et al. AFF4, a component of the ELL/p-TEFb elongation complex and a shared subunit of MLL chimeras can link transcription elongation to leukemia. Mol. Cell. 2010;37:429–37.

[259] Yokoyama A, Lin M, Naresh A, Kitabayashi I, Cleary ML. A Higher-Order Complex Containing AF4 and ENL Family Proteins with P-TEFb Facilitates Oncogenic and Physiologic MLL-Dependent Transcription. Cancer Cell. 2010;17:198–212.

[260] Smith E, Lin C, Shilatifard A. The super elongation complex (SEC) and MLL in development and disease. Genes Dev. 2011;25:661–72.

[261] Chou S, Upton H, Bao K, Schulze-Gahmen U, Samelson AJ, He N, et al. HIV-1 Tat recruits transcription elongation factors dispersed along a flexible AFF4 scaffold. Proc. Natl. Acad. Sci. U. S. A. 2013;110:E123–31.

[262] Liu M, Hsu J, Chan C, Li Z, Zhou Q. The Ubiquitin Ligase Siah1 Controls ELL2 Stability and Formation of Super Elongation Complexes to Modulate Gene Transcription. Mol. Cell. Elsevier Inc.; 2012;46:325–34.

[263] Ammosova T, Berro R, Jerebtsova M, Jackson A, Charles S, Klase Z, et al. Phosphorylation of HIV-1 Tat by CDK2 in HIV-1 transcription. Retrovirology. 2006;3:78.

[264] Kiernan RE, Vanhulle C, Schiltz L, Adam E, Xiao H, Maudoux F, et al. HIV-1 tat transcriptional activity is regulated by acetylation. EMBO J. 1999;18:6106–18.

[265] Brès V, Tagami H, Péloponèse JM, Loret E, Jeang KT, Nakatani Y, et al. Differential acetylation of Tat coordinates its interaction with the co-activators cyclin T1 and PCAF. EMBO J. 2002;21:6811–9.

[266] Huo L, Li D, Sun X, Shi X, Karna P, Yang W, et al. Regulation of Tat acetylation and transactivation activity by the microtubule-associated deacetylase HDAC6. J. Biol. Chem. 2011;286:9280–6.

[267] Ott M, Schnölzer M, Garnica J, Fischle W, Emiliani S, Rackwitz HR, et al. Acetylation of the HIV-1 tat protein by p300 is important for its transcriptional activity. Curr. Biol. 1999;9:1489–92.

[268] Col E, Caron C, Seigneurin-Berny D, Gracia J, Favier A, Khochbin S. The Histone Acetyltransferase, hGCN5, Interacts with and Acetylates the HIV Transactivator, Tat. J. Biol. Chem. 2001;276:28179–84.

[269] Dorr A, Kiermer V, Pedal A, Rackwitz HR, Henklein P, Schubert U, et al. Transcriptional synergy between Tat and PCAF is dependent on the binding of acetylated Tat to the PCAF bromodomain. EMBO J. 2002;21:2715–23.

[270] Kaehlcke K, Dorr A, Hetzer-Egger C, Kiermer V, Henklein P, Schnoelzer M, et al. Acetylation of Tat defines a CyclinT1-independent step in HIV transactivation. Mol. Cell. 2003;12:167–76.

[271] Mahmoudi T, Parra M, Vries RGJ, Kauder SE, Verrijzer CP, Ott M, et al. The SWI/SNF chromatin-remodeling complex is a cofactor for Tat transactivation of the HIV promoter. J. Biol. Chem. 2006;281:19960–8.

[272] Tréand C, du Chéné I, Brès V, Kiernan R, Benarous R, Benkirane M, et al. Requirement for SWI/SNF chromatin-remodeling complex in Tat-mediated activation of the HIV-1 promoter. EMBO J. 2006;25:1690–9.

[273] Agbottah E, Deng L, Dannenberg LO, Pumfery A, Kashanchi F. Effect of SWI/SNF chromatin remodeling complex on HIV-1 Tat activated transcription. Retrovirology. 2006;3:48.

[274] Pagans S, Pedal A, North BJ, Kaehlcke K, Marshall BL, Dorr A, et al. SIRT1 regulates HIV transcription via Tat deacetylation. PLoS Biol. 2005;3:0210–20.

[275] Pagans S, Kauder SE, Kaehlcke K, Sakane N, Schroeder S, Dormeyer W, et al. The Cellular lysine methyltransferase Set7/9-KMT7 binds HIV-1 TAR RNA, monomethylates the viral transactivator Tat, and enhances HIV transcription. Cell Host Microbe. 2010;7:234–44.

[276] Sakane N, Kwon HS, Pagans S, Kaehlcke K, Mizusawa Y, Kamada M, et al. Activation of hiv transcription by the viral tat protein requires a demethylation step mediated by lysine-specific demethylase 1 (LSD1/KDM1). PLoS Pathog. 2011;7:1–12.

[277] Brès V, Kiernan RE, Linares LK, Chable-Bessia C, Plechakova O, Tréand C, et al. A non-proteolytic role for ubiquitin in Tat-mediated transactivation of the HIV-1 promoter. Nat. Cell Biol. 2003;5:754–61.

[278] Verdin E. DNase I-hypersensitive sites are associated with both long terminal repeats and with the intragenic enhancer of integrated human immunodeficiency virus type 1. J. Virol. 1991;65:6790–9.

[279] Verdin E, Paras P, Van Lint C. Chromatin disruption in the promoter of human immunodeficiency virus type 1 during transcriptional activation. EMBO J. 1993;12:3249–59.

[280] Van Lint C, Emiliani S, Ott M, Verdin E. Transcriptional activation and chromatin remodeling of the HIV-1 promoter in response to histone acetylation. EMBO J. 1996;15:1112–20.

[281] Rafati H, Parra M, Hakre S, Moshkin Y, Verdin E, Mahmoudi T. Repressive LTR nucleosome positioning by the BAF complex is required for HIV latency. PLoS Biol. 2011;9:e1001206.

[282] Cismasiu VB, Paskaleva E, Suman Daya S, Canki M, Duus K, Avram D. BCL11B is a general transcriptional repressor of the HIV-1 long terminal repeat in T lymphocytes through recruitment of the NuRD complex. Virology. 2008;380:173–81.

[283] Xue Y, Wong J, Moreno GT, Young MK, Côté J, Wang W. NURD, a novel complex with both ATP-dependent chromatin-remodeling and histone deacetylase activities. Mol. Cell. 1998;2:851–61.

[284] Zhang Y, Li Y. The Expanding Mi-2/NuRD Complexes: A Schematic Glance. Proteomics Insights. Libertas Academica; 2011;2010:79.

[285] Perkins ND, Felzien LK, Betts JC, Leung K, Beach DH, Nabel GJ. Regulation of NF-kappaB by cyclin-dependent kinases associated with the p300 coactivator. Science. 1997;275:523–7.

[286] Williams SA, Chen L-F, Kwon H, Ruiz-Jarabo CM, Verdin E, Greene WC. NF-kappaB p50 promotes HIV latency through HDAC recruitment and repression of transcriptional initiation. EMBO J. 2006;25:139–49.

[287] Keedy KS, Archin NM, Gates AT, Espeseth A, Hazuda DJ, Margolis DM. A limited group of class I histone deacetylases acts to repress human immunodeficiency virus type 1 expression. J. Virol. 2009;83:4749–56.

[288] Huber K, Doyon G, Plaks J, Fyne E, Mellors JW, Sluis-Cremer N. Inhibitors of histone deacetylases: Correlation between isoform specificity and reactivation of HIV type 1 (HIV-1) from latently infected cells. J. Biol. Chem. 2011;286:22211–8.

[289] Ying H, Zhang Y, Zhou X, Qu X, Wang P, Liu S, et al. Selective Histonedeacetylase Inhibitor M344 Intervenes in HIV-1 Latency through Increasing Histone Acetylation and Activation of NF-kappaB. PLoS One. 2012;7.

[290] Marban C, Suzanne S, Dequiedt F, de Walque S, Redel L, Van Lint C, et al. Recruitment of chromatin-modifying enzymes by CTIP2 promotes HIV-1 transcriptional silencing. EMBO J. 2007;26:412–23.

[291] Marban C, Redel L, Suzanne S, Van Lint C, Lecestre D, Chasserot-Golaz S, et al. COUP-TF interacting protein 2 represses the initial phase of HIV-1 gene transcription in human microglial cells. Nucleic Acids Res. 2005;33:2318–31.

[292] Bouchat S, Gatot J-S, Kabeya K, Cardona C, Colin L, Herbein G, et al. Histone methyltransferase inhibitors induce HIV-1 recovery in resting CD4(+) T cells from HIV-1-infected HAART-treated patients. AIDS. 2012;26:1473–82.

[293] Du Chéné I, Basyuk E, Lin Y-L, Triboulet R, Knezevich A, Chable-Bessia C, et al. Suv39H1 and HP1gamma are responsible for chromatin-mediated HIV-1 transcriptional silencing and post-integration latency. EMBO J. 2007;26:424–35.

[294] Friedman J, Cho W-K, Chu CK, Keedy KS, Archin NM, Margolis DM, et al. Epigenetic silencing of HIV-1 by the histone H3 lysine 27 methyltransferase enhancer of Zeste 2. J. Virol. 2011;85:9078–89.

[295] Imai K, Togami H, Okamoto T. Involvement of histone H3 lysine 9 (H3K9) methyltransferase G9a in the maintenance of HIV-1 latency and its reactivation by BIX01294. J. Biol. Chem. 2010;285:16538–45.

[296] Kauder SE, Bosque A, Lindqvist A, Planelles V, Verdin E. Epigenetic regulation of HIV-1 latency by cytosine methylation. PLoS Pathog. 2009;5:e1000495.

[297] Blazkova J, Trejbalova K, Gondois-Rey F, Halfon P, Philibert P, Guiguen A, et al. CpG methylation controls reactivation of HIV from latency. PLoS Pathog. 2009;5.

[298] Palacios JA, Pérez-Piñar T, Toro C, Sanz-Minguela B, Moreno V, Valencia E, et al. Long-term nonprogressor and elite controller patients who control viremia have a higher percentage of methylation in their HIV-1 proviral promoters than aviremic patients receiving highly active antiretroviral therapy. J. Virol. 2012;86:13081–4.

[299] Triboulet R, Mari B, Lin Y, Chable-bessia C, Bennasser Y, Lebrigand K, et al. Suppression of MicroRNA-Silencing Pathway by HIV-1 During Virus Replication. Science (80-.). 2007;1579–82.

[300] Nathans R, Chu C-Y, Serquina AK, Lu C-C, Cao H, Rana TM. Cellular microRNA and P bodies modulate host-HIV-1 interactions. Mol. Cell. Elsevier Ltd; 2009;34:696–709.

[301] Imam H, Shahr Bano A, Patel P, Holla P, Jameel S. The lncRNA NRON modulates HIV-1 replication in a NFAT-dependent manner and is differentially regulated by early and late viral proteins. Sci. Rep. 2015;5:8639.

[302] Wang X, Ye L, Hou W, Zhou Y, Wang Y-J, Metzger DS, et al. Cellular microRNA expression correlates with susceptibility of monocytes/macrophages to HIV-1 infection. Blood. 2009;113:671–4.

[303] Houzet L, Yeung ML, de Lame V, Desai D, Smith SM, Jeang K-T. MicroRNA profile changes in human immunodeficiency virus type 1 (HIV-1) seropositive individuals. Retrovirology. 2008;5:118.

[304] Witwer KW, Watson AK, Blankson JN, Clements JE. Relationships of PBMC micro-RNA expression, plasma viral load, and CD4+ T-cell count in HIV-1-infected elite suppressors and viremic patients. Retrovirology. 2012;9:5.

[305] Bignami F, Pilotti E, Bertoncelli L, Ronzi P, Gulli M, Marmiroli N, et al. Stable changes in CD4 + T lymphocyte miRNA expression after exposure to HIV-1 Stable changes in CD4 T lymphocyte miRNA expression after exposure to HIV-1. Blood. 2013;119:6259–67.

[306] Chiang K, Sung T-L, Rice a. P. Regulation of Cyclin T1 and HIV-1 Replication by MicroRNAs in Resting CD4+ T Lymphocytes. J. Virol. 2012;86:3244–52.

[307] Sung TL, Rice AP. miR-198 inhibits HIV-1 gene expression and replication in monocytes and its mechanism of action appears to involve repression of cyclin T1. PLoS Pathog. 2009;5.

[308] Chiang K, Rice AP. MicroRNA-mediated restriction of HIV-1 in resting CD4+ T cells and monocytes. Viruses. 2012;4:1390–409.

[309] Qian S, Zhong X, Yu L, Ding B, de Haan P, Boris-Lawrie K. HIV-1 Tat RNA silencing suppressor activity is conserved across kingdoms and counteracts translational repression of HIV-1. Proc. Natl. Acad. Sci. U. S. A. 2009;106:605–10.

[310] Hayes AM, Qian S, Yu L, Boris-Lawrie K. Tat RNA silencing suppressor activity contributes to perturbation of lymphocyte miRNA by HIV-1. Retrovirology. BioMed Central Ltd; 2011;8:36.

[311] Coley W, Van Duyne R, Carpio L, Guendel I, Kehn-Hall K, Chevalier S, et al. Absence of DICER in monocytes and its regulation by HIV-1. J. Biol. Chem. 2010;285:31930–43.

[312] Huang J, Wang F, Argyris E, Chen K, Liang Z, Tian H, et al. Cellular microRNAs contribute to HIV-1 latency in resting primary CD4+ T lymphocytes. Nat. Med. 2007;13:1241–7.

[313] Jiménez VC, Booiman T, de Taeye SW, van Dort K a., Rits M a. N, Hamann J, et al. Differential expression of HIV-1 interfering factors in monocyte-derived macrophages stimulated with polarizing cytokines or interferons. Sci. Rep. 2012;2:1–7.

[314] Mantri CK, Pandhare Dash J, Mantri JV, Dash CC V. Cocaine enhances HIV-1 replication in CD4+ T cells by down-regulating MiR-125b. PLoS One. Public Library of Science; 2012;7:e51387.

[315] Wang X, Ye L, Zhou Y, Liu MQ, Zhou DJ, Ho WZ. Inhibition of anti-HIV microRNA expression: A mechanism for opioid-mediated enhancement of HIV infection of monocytes. Am. J. Pathol. 2011;178:41–7.

[316] Swaminathan S, Murray DD, Kelleher AD. Mirnas and HIV: Unforeseen Determinants Of Host-Pathogen Interaction. Immunol. Rev. 2013;254:265–80.

[317] Sun G, Li H, Wu X, Covarrubias M, Scherer L, Meinking K, et al. Interplay between HIV-1 infection and host microRNAs. Nucleic Acids Res. 2012;40:2181–96.

[318] Betel D, Wilson M, Gabow A, Marks DS, Sander C. The microRNA.org resource: Targets and expression. Nucleic Acids Res. 2008;36:149–53.

[319] Landgraf P, Rusu M, Sheridan R, Sewer A, Iovino N, Aravin A, et al. A Mammalian microRNA Expression Atlas Based on Small RNA Library Sequencing. Cell. 2009;129:1401–14.

[320] Ahluwalia JK, Khan SZ, Soni K, Rawat P, Gupta A, Hariharan M, et al. Human cellular microRNA hsa-miR-29a interferes with viral nef protein expression and HIV-1 replication. Retrovirology. 2008;5:117.

[321] Ruelas DS, Chan JK, Oh E, Heidersbach AJ, Hebbeler AM, Chavez L, et al. MicroRNA-155 Reinforces HIV Latency. J. Biol. Chem. 2015;290:jbc.M115.641837.

[322] Zhang HS, Chen XY, Wu TC, Sang WW, Ruan Z. MiR-34a is involved in Tat-induced HIV-1 long terminal repeat (LTR) transactivation through the SIRT1/NF??B pathway. FEBS Lett. Federation of European Biochemical Societies; 2012;586:4203–7.

[323] Zhang H-S, Wu T-C, Sang W-W, Ruan Z. MiR-217 is involved in Tat-induced HIV-1 long terminal repeat (LTR) transactivation by down-regulation of SIRT1. Biochim. Biophys. Acta. 2012;1823:1017–23.

[324] Chen XY, Zhang HS, Wu TC, Sang WW, Ruan Z. Down-regulation of NAMPT expression by miR-182 is involved in Tat-induced HIV-1 long terminal repeat (LTR) transactivation. Int. J. Biochem. Cell Biol. Elsevier B.V.; 2013;45:292–8.

[325] Ma L, Shen CJ, Cohen É a., Xiong SD, Wang JH. MiRNA-1236 inhibits HIV-1 infection of monocytes by repressing translation of cellular factor VprBP. PLoS One. 2014;9:1–7.

[326] Bennasser Y, Le S-Y, Yeung ML, Jeang K-T. HIV-1 encoded candidate micro-RNAs and their cellular targets. Retrovirology. 2004;1:43.

[327] Yeung ML, Houzet L, Yedavalli VSRK, Jeang K-T. A genome-wide short hairpin RNA screening of jurkat T-cells for human proteins contributing to productive HIV-1 replication. J. Biol. Chem. 2009;284:19463–73.

[328] Schopman NCT, Willemsen M, Liu YP, Bradley T, Van Kampen A, Baas F, et al. Deep sequencing of virus-infected cells reveals HIV-encoded small RNAs. Nucleic Acids Res. 2012;40:414–27.

[329] Ouellet DL, Plante I, Landry P, Barat C, Janelle M-E, Flamand L, et al. Identification of functional microRNAs released through asymmetrical processing of HIV-1 TAR element. Nucleic Acids Res. 2008;36:2353–65.

[330] Omoto S, Fujii YR. Regulation of human immunodeficiency virus 1 transcription by nef microRNA. J. Gen. Virol. 2005;86:751–5.

[331] Whisnant AW, Bogerd HP, Flores O, Ho P, Powers JG, Sharova N, et al. In-depth analysis of the interaction of HIV-1 with cellular microRNA biogenesis and effector mechanisms. MBio. 2013;4:e000193.

[332] Raj A, van Oudenaarden A. Nature, Nurture, or Chance: Stochastic Gene Expression and Its Consequences. Cell. 2008;135:216–26.

[333] Weinberger LS, Burnett JC, Toettcher JE, Arkin AP, Schaffer D V. Stochastic gene expression in a lentiviral positive-feedback loop: HIV-1 Tat fluctuations drive phenotypic diversity. Cell. 2005;122:169–82.

[334] Singh A, Razooky B, Cox CD, Simpson ML, Weinberger LS. Transcriptional bursting from the HIV-1 promoter is a significant source of stochastic noise in HIV-1 gene expression. Biophys. J. Biophysical Society; 2010;98:L32–4.

[335] Weinberger AD, Weinberger LS. Stochastic fate selection in HIV-infected patients. Cell. Elsevier; 2013;155:497–9.

[336] Dar RD, Hosmane NN, Michelle RA, Siliciano RF, Weinberger LS. Screening for noise in gene expression identifies drug synergies. Science (80-.). 2014;344:1392–6.

[337] Lipinski CA. Lead- and drug-like compounds: the rule-of-five revolution. Drug Discov. Today. Technol. 2004;1:337–41.

[338] Leeson PD, Springthorpe B. The influence of drug-like concepts on decision-making in medicinal chemistry. Nat. Rev. Drug Discov. 2007;6:881–90.

[339] Chun TW, Engel D, Mizell SB, Ehler LA, Fauci AS. Induction of HIV-1 replication in latently infected CD4+ T cells using a combination of cytokines. J. Exp. Med. 1998;188:83–91.

[340] Prins JM, Jurriaans S, van Praag RM, Blaak H, van Rij R, Schellekens PT, et al. Immuno-activation with anti-CD3 and recombinant human IL-2 in HIV-1-infected patients on potent antiretroviral therapy. AIDS. 1999;13:2405–10.

[341] Brooks DG, Hamer DH, Arlen PA, Gao L, Bristol G, Kitchen CMR, et al. Molecular characterization, reactivation, and depletion of latent HIV. Immunity. 2003;19:413–23.

[342] Rasmussen TA, Schmeltz Søgaard O, Brinkmann C, Wightman F, Lewin SR, Melchjorsen J, et al. Comparison of HDAC inhibitors in clinical development: effect on HIV production in latently infected cells and T-cell activation. Hum. Vaccin. Immunother. 2013;9:993–1001.

[343] Wightman F, Lu HK, Solomon AE, Saleh S, Harman AN, Cunningham AL, et al. Entinostat is a histone deacetylase inhibitor selective for class 1 histone deacetylases and activates HIV production from latently infected primary T cells. AIDS. 2013;27:2853–62.

[344] Wei DG, Chiang V, Fyne E, Balakrishnan M, Barnes T, Graupe M, et al. Histone deacetylase inhibitor romidepsin induces HIV expression in CD4 T cells from patients

on suppressive antiretroviral therapy at concentrations achieved by clinical dosing. PLoS Pathog. Public Library of Science; 2014;10:e1004071.

[345] Archin NM, Liberty AL, Kashuba AD, Choudhary SK, Kuruc JD, Crooks AM, et al. Administration of vorinostat disrupts HIV-1 latency in patients on antiretroviral therapy. Nature. 2012;487:482–5.

[346] Archin NM, Bateson R, Tripathy MK, Crooks AM, Yang K-H, Dahl NP, et al. HIV-1 expression within resting CD4+ T cells after multiple doses of vorinostat. J. Infect. Dis. 2014;210:728–35.

[347] Bartholomeeusen K, Xiang Y, Fujinaga K, Peterlin BM. Bromodomain and extra-terminal (BET) bromodomain inhibition activate transcription via transient release of positive transcription elongation factor b (P-TEFb) from 7SK small nuclear ribonucleoprotein. J. Biol. Chem. 2012;287:36609–16.

[348] Falkenberg KJ, Johnstone RW. Histone deacetylases and their inhibitors in cancer, neurological diseases and immune disorders. Nat. Rev. Drug Discov. Nature Publishing Group, a division of Macmillan Publishers Limited. All Rights Reserved.; 2014;13:673–91.

[349] Xing S, Siliciano RF. Targeting HIV latency: pharmacologic strategies toward eradication. Drug Discov. Today. Elsevier Ltd; 2013;18:541–51.

[350] Bernhard W, Barreto K, Saunders A, Dahabieh MS, Johnson P, Sadowski I. The Suv39H1 methyltransferase inhibitor chaetocin causes induction of integrated HIV-1 without producing a T cell response. FEBS Lett. 2011;585:3549–54.

[351] Fernandez G, Zeichner SL. Cell line-dependent variability in HIV activation employing DNMT inhibitors. Virol. J. 2010;7:266.

[352] Blazkova J, Murray D, Justement JS, Funk EK, Nelson AK, Moir S, et al. Paucity of HIV DNA methylation in latently infected, resting CD4+ T cells from infected individuals receiving antiretroviral therapy. J. Virol. 2012;86:5390–2.

[353] Thibault S, Imbeault M, Tardif MR, Tremblay MJ. TLR5 stimulation is sufficient to trigger reactivation of latent HIV-1 provirus in T lymphoid cells and activate virus gene expression in central memory CD4+ T cells. Virology. 2009;389:20–5.

[354] Chang JJ, Altfeld M. Immune activation and the role of TLRs and TLR agonists in the pathogenesis of HIV-1 infection in the humanized mouse model. J. Infect. Dis. 2013;208 Suppl :S145–9.

[355] De Jong MAWP, de Witte L, Oudhoff MJ, Gringhuis SI, Gallay P, Geijtenbeek TBH. TNF-alpha and TLR agonists increase susceptibility to HIV-1 transmission by human Langerhans cells ex vivo. J. Clin. Invest. American Society for Clinical Investigation; 2008;118:3440–52.

[356] Moody MA, Santra S, Vandergrift NA, Sutherland LL, Gurley TC, Drinker MS, et al. Toll-like receptor 7/8 (TLR7/8) and TLR9 agonists cooperate to enhance HIV-1 envelope antibody responses in rhesus macaques. J. Virol. 2014;88:3329–39.

[357] Sanders CM, Cruse JM, Lewis RE. Toll-like receptors, cytokines and HIV-1. Exp. Mol. Pathol. 2008;84:31–6.

[358] De Nardo D. Toll-like receptors: Activation, signalling and transcriptional modulation. Cytokine. 2015;74:181–9.

[359] Winckelmann AA, Munk-Petersen L V, Rasmussen TA, Melchjorsen J, Hjelholt TJ, Montefiori D, et al. Administration of a Toll-like receptor 9 agonist decreases the proviral reservoir in virologically suppressed HIV-infected patients. PLoS One. Public Library of Science; 2013;8:e62074.

[360] Offersen R, Melchjorsen J, Paludan SR, Østergaard L, Tolstrup M, Søgaard OS. TLR9-adjuvanted pneumococcal conjugate vaccine induces antibody-independent memory responses in HIV-infected adults. Hum. Vaccin. Immunother. 2012;8:1042–7.

[361] Scheller C, Ullrich A, Lamla S, Dittmer U, Rethwilm A, Koutsilieri E. Dual activity of phosphorothioate CpG oligodeoxynucleotides on HIV: reactivation of latent provirus and inhibition of productive infection in human T cells. Ann. N. Y. Acad. Sci. 2006;1091:540–7.

[362] Scheller C, Ullrich A, McPherson K, Hefele B, Knöferle J, Lamla S, et al. CpG oligodeoxynucleotides activate HIV replication in latently infected human T cells. J. Biol. Chem. 2004;279:21897–902.

[363] Søgaard OS, Lohse N, Harboe ZB, Offersen R, Bukh AR, Davis HL, et al. Improving the immunogenicity of pneumococcal conjugate vaccine in HIV-infected adults with a toll-like receptor 9 agonist adjuvant: a randomized, controlled trial. Clin. Infect. Dis. 2010;51:42–50.

[364] Sloan DD, Irrinki A, Tsai A, Kaur J, Lalezari J, Murry J, et al. TLR7 Agonist GS-9620 Activates HIV-1 in PBMCs From HIV-Infected Patients on cART. Conf. Retroviruses Opportunistic Infect. 2015, Bost. Massachusetts. 2015. p. P – F7, 417.

[365] Whitney JB, Lim S-Y, Osuna CE, Sanisetty S, Barnes TL, Hraber PT, et al. Treatment With a TLR7 Agonist Induces Transient Viremia in SIV-Infected ART-Suppressed Monkeys. Conf. Retroviruses Opportunistic Infect. 2015, Bost. Massachusetts. 2015. p. O – 9, 108.

[366] Antoni BA, Rabson AB, Kinter A, Bodkin M, Poli G. NF-kappa B-dependent and -independent pathways of HIV activation in a chronically infected T cell line. Virology. 1994;202:684–94.

[367] Contreras X, Barboric M, Lenasi T, Peterlin BM. HMBA releases P-TEFb from HEXIM1 and 7SK snRNA via PI3K/Akt and activates HIV transcription. PLoS Pathog. 2007;3:1459–69.

[368] Choudhary SK, Archin NM, Margolis DM. Hexamethylbisacetamide and disruption of human immunodeficiency virus type 1 latency in CD4(+) T cells. J. Infect. Dis. 2008;197:1162–70.

[369] Klichko V, Archin N, Kaur R, Lehrman G, Margolis D. Hexamethylbisacetamide remodels the human immunodeficiency virus type 1 (HIV-1) promoter and induces Tat-independent HIV-1 expression but blunts cell activation. J. Virol. 2006;80:4570–9.

[370] Pérez M, de Vinuesa AG, Sanchez-Duffhues G, Marquez N, Bellido ML, Muñoz-Fernandez MA, et al. Bryostatin-1 synergizes with histone deacetylase inhibitors to reactivate HIV-1 from latency. Curr. HIV Res. 2010;8:418–29.

[371] Biancotto A, Grivel J-C, Gondois-Rey F, Bettendroffer L, Vigne R, Brown S, et al. Dual role of prostratin in inhibition of infection and reactivation of human immunodeficiency virus from latency in primary blood lymphocytes and lymphoid tissue. J. Virol. 2004;78:10507–15.

[372] Wang P, Qu X, Wang X, Liu L, Zhu X, Zeng H, et al. As2O3 synergistically reactivate latent HIV-1 by induction of NF-κB. Antiviral Res. 2013. p. 688–97.

[373] Mochly-Rosen D, Khaner H, Lopez J. Identification of intracellular receptor proteins for activated protein kinase C. Proc. Natl. Acad. Sci. 1991;88:3997–4000.

[374] Trushin SA, Bren GD, Asin S, Pennington KN, Paya C V, Badley AD. Human immunodeficiency virus reactivation by phorbol esters or T-cell receptor ligation requires both PKCalpha and PKCtheta. J. Virol. 2005;79:9821–30.

[375] Colin L, Vandenhoudt N, de Walque S, Van Driessche B, Bergamaschi A, Martinelli V, et al. The AP-1 binding sites located in the pol gene intragenic regulatory region of HIV-1 are important for viral replication. PLoS One. Public Library of Science; 2011;6:e19084.

[376] Hirsch I, Caux C, Hasan U, Bendriss-Vermare N, Olive D. Impaired Toll-like receptor 7 and 9 signaling: from chronic viral infections to cancer. Trends Immunol. 2010;31:391–7.

[377] Chang JJ, Altfeld M. Immune activation and the role of TLRs and TLR agonists in the pathogenesis of HIV-1 infection in the humanized mouse model. J. Infect. Dis. 2013;208 Suppl :S145–9.

[378] Gallastegui E, Marshall B, Vidal D, Sanchez-Duffhues G, Collado JA, Alvarez-Fernández C, et al. Combination of biological screening in a cellular model of viral latency and virtual screening identifies novel compounds that reactivate HIV-1. J. Virol. 2012;86:3795–808.

[379] Das B, Dobrowolski C, Mao H, Powell D, Miller M, Hazuda D, et al. Farnesyl transferase: A new target for eradication of latent HIV-1 provirus in jurkat T-cells. Proc. 6th Int. Work. HIV Persistence Dur. Ther. Miami, FL, USA, 3–6 December 2013. 2013.

[380] Hazuda D, Barnard R, Wolkenberg S, Powell D, Karn J, Das B, et al. HIV latency drug discovery: Optimizing drugs to induce latent HIV expression. Proc. 6th Int. Work. HIV Persistence Dur. Ther. Miami, FL, USA, 3–6 December 2013. 2013;

[381] Rafati H, LeMasters E, Lennert VDD, El-Sayyed M, Boucher C, Vries R, et al. Activation of the Wnt pathway by natural ligands or small molecule inhibitors activates latent HIV. Proc. 7th IAS Conf. HIV Pathog. Treat. Prev. (IAS), Kuala Lumpur, Malaysia, 30 June–3 July 2013.

[382] Sheridan PL, Sheline CT, Cannon K, Voz ML, Pazin MJ, Kadonaga JT, et al. Activation of the HIV-1 enhancer by the LEF-1 HMG protein on nucleosome-assembled DNA in vitro. Genes Dev. 1995;9:2090–104.

[383] McKnight RF, Adida M, Budge K, Stockton S, Goodwin GM, Geddes JR. Lithium toxicity profile: a systematic review and meta-analysis. Lancet. 2012;379:721–8.

[384] Oruch R, Elderbi MA, Khattab HA, Pryme IF, Lund A. Lithium: a review of pharmacology, clinical uses, and toxicity. Eur. J. Pharmacol. 2014;740:464–73.

[385] Dykhuizen EC, Carmody LC, Tolliday N, Crabtree GR, Palmer M a J. Screening for inhibitors of an essential chromatin remodeler in mouse embryonic stem cells by monitoring transcriptional regulation. J. Biomol. Screen. 2012;17:1221–30.

[386] Stoszko; M, Crignis; E De, Khalid; MM, Lungu; C, Palstra; R-J, Dykhuizen; EC, et al. Small molecule inhibitors of BAF; a new family of compounds in HIV latency reversal Article. EBioMedicine. 2015, submitted.

[387] Roth MD, Tashkin DP, Choi R, Jamieson BD, Zack JA, Baldwin GC. Cocaine Enhances Human Immunodeficiency Virus Replication in a Model of Severe Combined Immunodeficient Mice Implanted with Human Peripheral Blood Leukocytes. J. Infect. Dis. 2002;185:701–5.

[388] Bagasra O, Pomerantz RJ. Human immunodeficiency virus type 1 replication in peripheral blood mononuclear cells in the presence of cocaine. J. Infect. Dis. 1993;168:1157–64.

[389] Peterson PK, Gekker G, Chao CC, Schut R, Molitor TW, Balfour HH. Cocaine potentiates HIV-1 replication in human peripheral blood mononuclear cell cocultures. Involvement of transforming growth factor-beta. J. Immunol. 1991;146:81–4.

[390] Sahu G, Farley K, El-Hage N, Aiamkitsumrit B, Fassnacht R, Kashanchi F, et al. Cocaine promotes both initiation and elongation phase of HIV-1 transcription by activating NF-κB and MSK1 and inducing selective epigenetic modifications at HIV-1 LTR. Virology. 2015;483:185–202.

[391] Imai K, Yamada K, Tamura M, Ochiai K, Okamoto T. Reactivation of latent HIV-1 by a wide variety of butyric acid-producing bacteria. Cell. Mol. Life Sci. 2012;69:2583–92.

[392] Ye F, Karn J. Bacterial Short Chain Fatty Acids Push All The Buttons Needed To Reactivate Latent Viruses. Stem cell epigenetics. 2015;2.

[393] Shirakawa K, Chavez L, Hakre S, Calvanese V, Verdin E. Reactivation of latent HIV by histone deacetylase inhibitors. Trends Microbiol. Elsevier Ltd; 2013;21:277–85.

[394] Wherry EJ, Blattman JN, Murali-Krishna K, van der Most R, Ahmed R. Viral persistence alters CD8 T-cell immunodominance and tissue distribution and results in distinct stages of functional impairment. J. Virol. 2003;77:4911–27.

[395] Khaitan A, Unutmaz D. Revisiting immune exhaustion during HIV infection. Curr. HIV/AIDS Rep. 2011;8:4–11.

[396] Palmer BE, Neff CP, Lecureux J, Ehler A, Dsouza M, Remling-Mulder L, et al. In vivo blockade of the PD-1 receptor suppresses HIV-1 viral loads and improves CD4+ T cell levels in humanized mice. J. Immunol. 2013;190:211–9.

[397] (NIAID) NI of A and ID. Safety and Immune Response of BMS-936559 in HIV-Infected People Taking Combination Antiretroviral Therapy. 2015.

[398] Migueles SA, Osborne CM, Royce C, Compton AA, Joshi RP, Weeks KA, et al. Lytic granule loading of CD8+ T cells is required for HIV-infected cell elimination associated with immune control. Immunity. 2008;29:1009–21.

[399] Shan L, Deng K, Shroff NS, Durand CM, Rabi SA, Yang H-C, et al. Stimulation of HIV-1-specific cytolytic T lymphocytes facilitates elimination of latent viral reservoir after virus reactivation. Immunity. 2012;36:491–501.

[400] Hansen SG, Ford JC, Lewis MS, Ventura AB, Hughes CM, Coyne-Johnson L, et al. Profound early control of highly pathogenic SIV by an effector memory T-cell vaccine. Nature. 2011;473:523–7.

[401] Hansen SG, Piatak M, Ventura AB, Hughes CM, Gilbride RM, Ford JC, et al. Immune clearance of highly pathogenic SIV infection. Nature. 2013;502:100–4.

[402] Hansen SG, Sacha JB, Hughes CM, Ford JC, Burwitz BJ, Scholz I, et al. Cytomegalovirus vectors violate CD8+ T cell epitope recognition paradigms. Science. 2013;340:1237874.

[403] Barouch DH, Whitney JB, Moldt B, Klein F, Oliveira TY, Liu J, et al. Therapeutic efficacy of potent neutralizing HIV-1-specific monoclonal antibodies in SHIV-infected rhesus monkeys. Nature.; 2013;503:224–8.

[404] Barouch DH, Stephenson KE, Borducchi EN, Smith K, Stanley K, McNally AG, et al. Protective efficacy of a global HIV-1 mosaic vaccine against heterologous SHIV challenges in rhesus monkeys. Cell. 2013;155:531–9.

[405] Klein F, Halper-Stromberg A, Horwitz JA, Gruell H, Scheid JF, Bournazos S, et al. HIV therapy by a combination of broadly neutralizing antibodies in humanized mice. Nature. 2012;492:118–22.

[406] Shingai M, Nishimura Y, Klein F, Mouquet H, Donau OK, Plishka R, et al. Antibody-mediated immunotherapy of macaques chronically infected with SHIV suppresses viraemia. Nature. 2013;503:277–80.

[407] Lehrman G, Hogue IB, Palmer S, Jennings C, Spina C a, Wiegand A, et al. Depletion of latent HIV-1 infection in vivo: a proof-of-concept study. Lancet. 2005;366:549–55.

[408] Steel A, Clark S, Teo I, Shaunak S, Nelson M, Gazzard B, et al. No change to HIV-1 latency with valproate therapy. AIDS. 2006;20:1681–2.

[409] Siliciano JD, Lai J, Callender M, Pitt E, Zhang H, Margolick JB, et al. Stability of the latent reservoir for HIV-1 in patients receiving valproic acid. J. Infect. Dis. 2007;195:833–6.

[410] Archin NM, Eron JJ, Palmer S, Hartmann-Duff A, Martinson JA, Wiegand A, et al. Valproic acid without intensified antiviral therapy has limited impact on persistent HIV infection of resting CD4+ T cells. AIDS. 2008;22:1131–5.

[411] Archin NM, Cheema M, Parker D, Wiegand A, Bosch RJ, Coffin JM, et al. Antiretroviral intensification and valproic acid lack sustained effect on residual HIV-1 viremia or resting CD4+ cell infection. PLoS One. 2010;5:e9390.

[412] Sagot-Lerolle N, Lamine A, Chaix M-L, Boufassa F, Aboulker J-P, Costagliola D, et al. Prolonged valproic acid treatment does not reduce the size of latent HIV reservoir. AIDS. 2008;22:1125–9.

Section 3

Retroviral Infection Prevention Strategies

4

Which Vaccination Strategies and Immune Responses are more Effective Against HIV Infections?

Azam Bolhassani

Abstract

Vaccination is one of the most successful approaches for controlling various viral diseases. Novel approaches will be needed to develop highly effective vaccines to prevent infectious diseases such as HIV. There are many aspects of HIV-1 biology that make the development of an HIV vaccine difficult, including viral diversity, effective type of immune response, and suitable experimental model for preclinical trials. In spite of these challenges, recent published results showed that a vaccine regimen could reduce HIV infection rates by 31% in Thailand. This vaccine named as RV144 is composed of a recombinant canarypox vector expressing three HIV-1 proteins as a prime and two different recombinant HIV-1 gp120 envelope glycoproteins with alum adjuvant as a boost. In addition, a subunit vaccine constructed from the viral envelope protein could be efficiently developed using new techniques available through genetic engineering. The current HIV-1 vaccine development focuses on antibody-based approaches. It was shown that immunization with the viral envelope glycoprotein, gp120, should generate neutralizing antibodies that would prevent infection, thereby yielding protective immunity. However, HIV could develop many pathways to escape from antibodies that bind to the different parts of the viral envelope molecules. Thus, the generation of neutralizing antibodies is very difficult after viral infection or immunization protocols. Indeed, the viral envelope molecules (Env) possess glycosylated residues that cover surface epitopes for binding and neutralizing antibodies, even if the antibodies are produced. Furthermore, the trimeric structures of envelope molecules show rapid conformational changes due to the interaction with viral cell surface receptors, CCR5/CXCR4 and CD4; thus the transition state is very poor to be recognized by the immune system. Currently, studies focus on generating stable trimeric envelope molecules (gp120/gp41) as immunogens that can induce neutralizing antibodies that can compete for binding to the cell surface receptors. Altogether, it is clear that the design of a vaccine to elicit HIV-neutralizing antibodies is not straightforward, and it causes major challenges in structural biology and immunology, several other studies strongly suggest cytotoxic T-lymphocyte (CTL)-based immune responses against HIV infections. Indeed, $CD8^+$ T cells play a major role in controlling viral replication during primary HIV infections and in maintaining a stable viral load during the chronic phase. In this line, live-attenuated vaccines could elicit more potent and durable pathogen-specific immune responses than inactivated or subunit vaccines. Generally, DNA vaccines are poorly immunogenic alone, and viral vector vaccines are ineffective due to vector-specific immune responses if used repeatedly; hence, the two approaches have often been tested in combination as prime-boost vaccination strategies. Indeed, the prime-boost vaccination has been considered as an efficient strategy against HIV infections. In this chapter, we will represent challenges to determine the best vaccine strategies against HIV infections.

Keywords: HIV infection, Immune responses, Vaccination, DNA vaccine, Prime-boost vaccine, Adjuvant, Challenges for HIV vaccine

1. Introduction

1.1. HIV infection and vaccination

According to recent reports, 35 million people were living with HIV-1 at the end of 2013, the considerable majority being in Sub-Saharan Africa, with dynamic epidemics in Asia [1]. HIV infection results in gradual loss of $CD4^+T$ lymphocytes, containing immune competence, and progression to AIDS. Effective treatment with combined antiretroviral drugs decreases viral load below detectable levels but cannot eliminate the virus from the body. Furthermore, the success of combined antiretroviral drugs is hindered by accumulating drug toxicities and chronic immune activation leading to increased risk of several non-AIDS disorders, even when viral replication is inhibited. Therefore, there is a major need for therapeutic strategies as an alternative to the combined antiretroviral drugs [2]. HIV vaccine strategies are expected to be a critical component for controlling the HIV epidemic [3, 4]. Immunotherapy, or therapeutic vaccination, aims to enhance existing immune responses against HIV or stimulate immune responses. These immune responses should provide an efficient cure by controlling viral replication and preventing disease progression in the absence of combined antiretroviral drugs [2]. The cost-effective and different HIV-1 vaccine approaches have recently attracted a special interest. Both antibodies and cell-mediated immune responses are considered to be important to prevent HIV-1 infection in the mucosal compartment, *i.e.*, the entry point for sexual transmission [1]. A great majority of HIV-1 infections occur at mucosa during sexual contact. Therefore, it is important to provide mucosal barrier protection against this entry by mucosal vaccination. A number of mucosal routes of vaccination such as enteric oral or intranasal vaccines have significant barriers that limit vaccine efficacy or cause safety risks. In contrast, the sublingual region of the mouth could provide a simple route for mucosal vaccination with immunogens, but this site does not always induce strong immune responses, especially when protein antigens are used [5].

Currently, antibody-inducing vaccines are a major focus in the preventive HIV vaccine field [6]. In addition, T-cell-based therapeutic vaccines have focused on three strategies: a) to increase the levels of vaccine-induced responses, b) to enhance the responses targeting only conserved regions of the virus, and c) to use replication-competent viral vectors [1]. Generally, antiviral T-cell and B-cell responses play a crucial role in suppressing HIV replication during chronic infection [7, 8]. Novel approaches of HIV treatment include both conventional therapeutic vaccines (*i.e.*, active immunization strategies using HIV-derived immunogens) and the use of checkpoint blockers such as anti-PD-1 antibodies. These complex therapeutic strategies appear as promising approaches against HIV infection [7].

The biggest barrier for many vaccines is the pathogen's variability. Thus, studies should be further focused on the functionally most conserved regions of proteins common to many variants, including escape mutants inducing both antibody and T-cell immune responses. For vector-based vaccines, the "universal" subunit immunogens are efficiently delivered using heterologous prime-boost regimens, which can be further improved using adjuvants and delivery approaches [9–11]. Several studies have described the development of vaccine strategies, including improved envelope proteins formulated with potent adjuvants, DNA and vectors expressing mosaics or conserved sequences, capable of inducing strong relevant immune responses, such as neutralizing antibodies (NAbs) and non-neutralizing antibodies, $CD4^+$ and $CD8^+$ cell-mediated immune responses, mucosal immune responses, and immunological memory. The type of immune response elicited by different immunogens can also correlate with the risk of HIV infections. For example, IgG antibodies against the V2 loop of gp120 are associated with a decreased risk of HIV infection, while Env-specific IgA antibody is directly related to increased risk [12]. Generally, a combination of two independent approaches, containing the induction of broadly neutralizing antibodies (bNAbs) to prevent or reduce acquisition of infection and stimulation of effective CTL responses, is the currently used technique to slow disease progression in advance infections [13].

Briefly, more than 20 years after the discovery of HIV, researchers are trying to design a protective AIDS vaccine. The problem is the lack of basic knowledge about the immunological requirements for the protection against HIV. Virus diversity and escape from immune responses are the most important challenges to the development of an effective HIV vaccine. In this chapter, we will represent the challenges to vaccine design against HIV biologically and immunologically. Moreover, different vaccine strategies will be described to determine the best strategies already focused on HIV infections. In this line, the relationship between HIV biology and immunity will be demonstrated for the first time.

2. Immunogen-induced Neutralizing Antibodies (NAbs)

The most common tests for HIV infections rely on detecting antibodies against virus. Thus, these tests can also detect antibodies induced by a candidate HIV vaccine. The detection of vaccine-induced antibodies to HIV by serological tests is referred to as vaccine-induced seroreactivity (VISR) [6]. Neutralizing antibodies are useful in identifying the neutralizing epitopes of vaccine and for understanding the mechanism of potent and broad cross-neutralization, thereby providing a modality of preventive and therapeutic value [14, 15]. It has been shown that some NAbs confer protection toward neonatal HIV-1 infection [16].

Various types of HIV-1 Env immunogens were developed that express epitopes for broadly neutralizing antibodies and their precursors. There are three new structures proposed for the HIV-1 Env trimer, which will be more immunogenic than previously used immunogens: a) minimal immunogens that are fragments of HIV-1 Env-neutralizing epitopes, b) intermediate Env immunogens (*e.g.*, monomeric Env gp120), and c) various forms of Env trimers [17]. To date, these structures have not been capable of inducing the immune system to generate bNAbs after vaccination. Thus, a successful vaccination for HIV-1 and induction of bNAbs will need repetitive immunizations for a long time [17].

The current studies of HIV-positive patients with strongly neutralizing sera indicated that the immune system is able to produce antibodies neutralizing up to 90% of HIV strains [18]. The neutralizing antibodies bind to conserved gp120 sites, and the identification of these sites can help to design effective vaccines. Glycosylated residues (or carbohydrates) have a key role because of binding broadly neutralizing antibodies to carbohydrates and combining carbohydrate and peptide elements on gp120. However, carbohydrates partially cover some peptides on envelope surface recognized by bNAbs. Thus, the use of engineered glycoproteins as vaccines for the stimulation of bNAbs is a subject of interest in HIV vaccine design [18].Antibody responses to the HIV-1 envelope glycoproteins can be classified into two: a) non-neutralizing responses directed to peptide epitopes expressed on isolated envelope glycoproteins but not on the native envelope trimers responsible for mediating the entry of virus into target cells; b) broadly neutralizing antibody responses targeting epitopes expressed on the native envelope trimers. Currently, many potent broadly neutralizing antibodies have been isolated to stimulate prophylactic and therapeutic activities in animal models. These antibodies help us to improve vaccine design and therapeutic strategies for HIV-1 [19]. The recent characterization of new epitopes for stimulating broadly neutralizing antibodies has encouraged studies in the synthesis of novel antigenic constructs for the development of HIV-envelope-directed vaccines [20]. Thus, an important step in vaccine design is the determination of antibodies and epitopes associated with broad HIV neutralization. Indeed, immunogens and/or immunization protocols should be designed to increase antibody affinity maturation [1].

Regarding the studies, HIV-1 envelope gp120 is the target for neutralizing antibodies against the virus. HIV-1 envelope gp120 exhibits a great degree of variability that causes a major challenge for the development of vaccines against HIV/AIDS. Different approaches have been used to improve immunogenicity of broadly neutralizing epitopes on HIV-1 gp120 with limited success [21]. For example, immunogenicity of gp120 and its V3 epitopes was enhanced when gp120 was co-injected with monoclonal antibodies (mAb) to the CD4-binding sites (CD4bs). Indeed, the gp120/CD4bs complex was a potent immunogen for eliciting cross-reactive functional NAbs against V3 epitopes [21]. In contrast, the membrane proximal external region (MPER) of the gp41 subunit of the HIV-1 envelope glycoprotein (Env) includes epitopes for recognizing bNAb as an important region in vaccine development. However, designing an immunogen for the induction of bNAbs to MPER is difficult because of the relative inaccessibility of the MPER in the native conformation of Env [22]. Therefore, a group of oligomeric gp41 immunogens was designed to further expose MPER in a suitable conformation. The immunogens comprised different gp41 N-heptad lengths and insertion of extra epitopes and flexible C-termini. These immunogens were used in two different immunization strategies, including gp41/gp140 proteins and gp41/gp160 DNA associated with various adjuvants and modalities. It was observed that the gp41 immunogens elicit higher levels of MPER than the gp140 immunogens. In prime-boost strategies, the best MPER responses were shown in the groups receiving gp41 DNA followed by gp41 protein. Several agents may influence MPER immunogenicity such as the immunization route, dose, or adjuvant. Generally, these data encourage the researchers for designing MPER immunogens with optimized immunization protocols [22]. Furthermore, the aggregation of HIV-1 virions was detected by antibodies (IgG) to the viral envelope glycoprotein (Env). Neutralizing antibodies directed to a V3-base- and glycan-dependent epitope on gp120 and to the apex of the Env trimer, as well as non-neutralizing antibodies to the epitope cluster I on the gp41-ectodomain, could aggregate virions, but the neutralizing antibody 2G12, which is specific for a glycan-dependent monovalent epitope on gp120, could not aggregate. These data can potentially open the ways for the development of HIV-1 vaccine [23].

3. Preventive HIV vaccines

The studies indicated that a successful vaccine candidate needs to elicit broad antibodies targeting the Env protein. Immunogens targeting gp120 have been developed, which block infection in monkeys. Attempts to induce antibody persistence were complicated by increasing the number of HIV-target cells [24]. RV144 consisting of canary pox vector vaccine ALVAC-HIV (vCP1521) prime and AIDSVAX®gp120 B/E boost was the first vaccine against HIV-1 infection achieved in clinical trial [1, 25]. The analysis of vaccine-induced immune responses in vaccinated-infected and vaccinated-uninfected volunteers indicated that IgG specific for the V1V2 region of gp120 was related to the decreased risk of HIV-1 infection, and plasma Env IgA was directly associated with infection risk. Thus, RV144 studies indicated that Env is essential and possibly sufficient to stimulate protective antibody responses against mucosally acquired HIV-1. Efficacy trials were planned in heterosexual populations in southern Africa and Thailand [1]. Generally, the studies of nonhuman primates suggest that Env is a necessary component for a successful protection against viral infections. Two approaches are being followed to induce Env-specific-antibody-mediated protection: a) vaccines that elicit potent and broadly reactive neutralizing antibodies against viruses which are common in human transmission, b) vaccines that induce antibody neutralizing only in less commonly transmitted HIV strains, but that block HIV-1 infection by non-neutralizing (nNAbs) mechanisms [26, 27]. Monomeric gp120 HIV-1 envelope proteins alone failed to protect high-risk individuals against infection. In fact, the level and breadth of elicited NAb were not sufficient for protection [1]. Furthermore, the results indicated that IgG to linear epitopes in the V2 and V3 regions of gp120 is part of a complex interaction of immune responses that contribute to the protection in RV144 [28]. In RV144, Env IgG3 was correlated with decreased risk of HIV infection, a response that declined rapidly compared to overall IgG responses. Indeed, the rates of Env-specific IgG3 and V1/V2 IgG3 responses were high, and conversely IgG4 responses were considerably low in recipients of the RV144 vaccine. These findings indicated that V2 IgG plays a role in protection against HIV-1 infection. Generally, an increase in magnitude, affinity, breadth, and importantly in frequency and durability of V2- and V3-specific antibodies of IgG3 and IgG1 subclasses may confer a higher and more durable rate of protection against HIV-1 infection. The induction of cross-reactive V1V2-specific IgG raises the hypothesis of cross-clade protection. Additional booster vaccinations may increase the antibody levels. Residual antibody responses against gp120 were detected 6–8 years post vaccination in RV144 vaccinees. Additional boosts increased plasma IgG gp120 and gp70 V1/V2 antibodies at titers higher than those in RV144, while weak gp120 IgA responses were induced. These HIV-specific IgG antibodies were also detected in rectal secretions while IgAs were undetectable [29–34].

Regarding the studies, HIV antibodies capable of preventing mucosal cell-free or cell-to-cell HIV transmission are critical for the development of effective prophylactic and therapeutic vaccines. The interactions between antigen-presenting cells (APCs) and HIV result in cell-to-cell transmission of HIV. In the experimental macaque model, data indicated that the broadly neutralizing antibodies are capable of neutralizing an extensive range of HIV strains, preventing cell-to-cell transfer, and protecting from infection[35]. In addition, IgG Fcg receptor (FcgR)-mediated inhibition of antibodies at the mucosal site may play a role in protection against HIV mucosal transmission. On the contrary, mucosal IgA antibodies may be effective in protection against HIV sexual transmission. Thus, the determination of inhibitory effects of antibodies is critical for evaluating protection in HIV vaccines [35]. Furthermore, the majority of the antibodies against different viral proteins described a marker (shared idiotope) that is recognized by the monoclonal antibody 1F7. This shared idiotype on antibodies induced by HIV-1 was involved in the immune memory mechanism linking the early and late antibodies, the so-called back-boost effect. This finding was supported by auto-antibodies that bind to the 1F7 idiotope in sera of HIV-1-infected individuals. The expression of a shared idiotope in antibodies could provide a strategy to stimulate B cells selected to produce antibodies against HIV-1 and HCV, suggesting their implications in vaccine design [36].

The use of potent adjuvants may also enhance antigen-specific antibody responses. Several adjuvants have been tested in nonhuman primates and humans indicating a significant benefit of HIV envelope proteins formulated with MF59 and AS01 adjuvants. A study indicated that alum protected macaques from simian immunodeficiency virus (SIV)mac251 infection, while MF59 did not protect despite its ability to elicit higher systemic T-cell and antibody responses. Adjuvant-associated differences in the homing of plasmablasts and induction of key cellular signaling pathways may explain these effects. The formulation of HIV-1 gp120 with MPLA and alum induced significantly higher levels of neutralizing antibodies and T-cell lymphoproliferation compared to alum, MF59, or MPLA alone. Importantly, antibodies to gp70 V1V2 (subtypes B, C, and CRF01 AE) were induced more rapidly, to a higher magnitude and with a greater durability than alum-adjuvanted gp120 [37–42]. Also, the formulation of antigens with solid nanoparticles may prolong the duration of antibody responses by increasing antigen retention locally in the tissues driving B-cell responses, enhancing dendritic cell (DC) antigen presentation, and the development of $CD4^+$ Th cells that provide cytokines and signals that are required to initiate somatic hypermutation and affinity maturation for effective B-cell memory [1]. Furthermore, the ability of two mucosal adjuvants, including α-galactosylceramide (α-GalCer) as a potent stimulator of natural killer (NK) T cells and CpG-oligodeoxynucleotide (CpG-ODN) as a TLR9 agonist, was evaluated to enhance immune responses against clade C gp140 HIV-1 envelope protein antigen. The results showed that $CD4^+$ and $CD8^+$ T-cell responses in systemic and mucosal tissues were significantly higher in mice immunized with gp140 in the presence of either α-GalCer or CpG-ODN and further enhanced when both adjuvants were used. Also, the use of two adjuvants and especially their combination effectively increased gp140-specific serum IgG and vaginal IgA antibody levels. Memory T-cell responses detected at 60 days after immunization revealed that α-GalCer is more potent than CpG-ODN, and the combination of α-GalCer and CpG-ODN adjuvants was more effective than either alone [5]. Another approach is called B-cell lineage vaccine design. In this line, the recombinant antibodies belonging to bNAb members were used to determine HIV-1 envelope constructs as immunogens in a prime-boost strategy. These envelope constructs utilize the naïve B-cell repertoire residing in bone marrow and secondary lymphoid tissues *in vivo* and subsequently stimulate B-cell evolution until bNAb-producing cells are elicited [43, 44].

The nNAb B-cell progenitors become activated and internalize Env compared with bNAb B-cell progenitors. The reports showed that rational immunogen modifications can reduce the activation of naïve B cells that lead to such nNAbs, while promoting the activation of naïve B cells that result in germline-reverted bNAbs [43, 44]. A number of potent broadly neutralizing antibodies have been identified from HIV-infected individuals although the generation of bNAbs using traditional vaccine approaches has been obscure. The researchers tested a single dose of 3BNC117 or 10-1074 (mAb specific for the CD4-binding site and the V3 region, respectively) and also a combination of both antibodies. The data showed a total decline in viral loads post infusion in simian-human immunodeficiency virus (SHIV)-infected macaques. Another approach to the generation of bNAbs is to circumvent "normal" immune responses and direct non-lymphoid cells to produce bNAbs *in vivo* using gene therapy. Vectored immunoprophylaxis (VIP) is a gene therapy method in which transgenes encoding bNAbs are delivered directly into muscle tissue where bNAbs are produced. Two recent animal studies demonstrated that VIP could generate modest titers of NAb that can effectively prevent *in vivo* HIV infection in a humanized bone marrow–thymus–liver (BLT) HIV infection model and a simian immunodeficiency macaque infection model [7]. Recent findings showed that adeno-associated virus (AAV)-delivered broadly neutralizing antibodies can inhibit HIV replication. Indeed, a single injection of AAV could generate long-term antibody responses as a therapeutic approach in the lack of antiretroviral drugs. Induction of vector-mediated antibodies could inhibit cell-to-cell transmission and replication of HIV. This result represented an alternative to immunogen-based vaccine design and a novel therapeutic intervention by enabling particular manipulation of humoral immunity [45, 46].

A challenge for HIV-1 immunogen design is the induction of neutralizing antibodies against neutralization-resistant (Tier-2) viruses that control human transmissions. A soluble recombinant HIV-1 envelope glycoprotein trimer possessing a native conformation (BG505 SOSIP.664) could induce NAbs potently against the sequence-matched Tier-2 virus in rabbits but weaker and similar responses in macaques. The trimer also stably stimulated cross-reactive NAbs against more sensitive (Tier-1) viruses. Tier-2 NAbs recognized conformational epitopes that differed between animals and in some cases overlapped with those recognized by broadly neutralizing antibodies, whereas Tier-1 responses targeted linear V3 epitopes. A second trimer,

B41 SOSIP.664, also induced a strong autologous Tier-2 NAb response in rabbits. Thus, native-like trimers may represent a promising starting point for developing HIV-1 vaccines directed at inducing bNAbs [47].

The goal of an HIV vaccine is to generate robust and durable protective antibody. Thus, it is important to induce $CD4^+$ T-follicular helper (TFH) cells. However, very little is known about the TFH response to HIV vaccination and its relative contribution to magnitude and the quality of vaccine-elicited antibody titers [48]. In this line, a DNA/modified *vaccinia* virus Ankara SIV vaccine with and without gp140 boost in aluminum hydroxide was administered in rhesus macaques. The studies indicated that booster immunization with modified *vaccinia* virus Ankara induces a distinct and transient accumulation of proliferating $CXCR5^+$ and $CXCR5^-$ CD4 T cells in blood at day 7 post-immunization, and the frequency of the former but not the latter correlated with TFH and B-cell responses in germinal centers of the lymph node. Furthermore, gp140 boost elicited a skewing toward CXCR3 expression on germinal center TFH cells, which was strongly associated with longevity, avidity, and neutralization potential of vaccine-elicited antibody response. However, $CXCR3^+$ cells preferentially expressed the HIV co-receptor CCR5, and vaccine-induced $CXCR3^+CXCR5^+$ cells showed a moderate positive association with peak viremia following SIV251 infection. These data demonstrated that vaccine regimens eliciting CXCR3-biased TFH cell responses favor antibody persistence and avidity but may prompt higher acute viremia in advance infections [48, 49].

Generally, broadly neutralizing antibodies specific for conserved epitopes on the HIV-1 envelope (Env) are believed to be essential for the protection against multiple HIV-1 clades. Recently, an HIV vaccine incorporating the molecular adjuvants B-cell activating factor (BAFF) and a proliferation-inducing ligand (APRIL) was designed with the potential to facilitate the maturation of polyreactive and autoreactive B cells as well as to enhance the affinity and/or avidity of Env-specific antibodies. The results indicated that mice immunization with a DNA vaccine encoding BAFF or APRIL multitrimers, together with IL-12 and membrane-bound HIV-1 Env gp140, induced neutralizing antibodies against Tier-1 and Tier-2 viruses. The APRIL-containing vaccine was especially effective at generating Tier-2 neutralizing antibodies following a protein boost. Notably, BAFF and APRIL did not cause B-cell expansion or an increase in total IgG. Thus, BAFF and APRIL multitrimers were proposed as promising molecular adjuvants for inducing bNAbs against HIV-1 infection [50].

4. Stimulation of cell-mediated immune responses

A successful HIV vaccine must either completely prevent infection or eliminate the first round of infected CD4 T cells before the latent pool of HIV-infected cells is established. Thus, an effective HIV vaccine requires high levels of protective immunity at the time of virus contact with the host, and it cannot rely on memory immune responses to occur. CD8 T cells can effectively kill HIV-infected T cells, but in most cases of acute HIV infection, the virus rapidly escapes. For example, anti-HIV CD8 CTL activity is capable of eliminating virus-infected T cells in the setting of vaccination with an attenuated rhesus cytomegalovirus (rhCMV) containing simian immunodeficiency virus genes, but in the setting of acute HIV infection, the transmitted/founder virus usually escapes from CD8 T-cell control [51]. CTL responses targeting specific HIV proteins (*e.g.*, Gag) have been associated with relative control of viral replication *in vivo*. In a SIVmac251 intravenous challenge model, breadth of Gag CTL epitope recognition correlated with control of viremia peak [1]. These data have demonstrated that $CD8^+$ T cells are associated with the control and eradication of early retrovirus infections [17].

The vaccine trials were focused on whether cell-mediated immune-response-inducing vaccines can prevent infection or reduce post-infection plasma viral load [1]. Vaccinees with HLA alleles associated with HIV-1 control had a significantly lower mean viral load over time. Interestingly, the most highly conserved epitopes were detected at a lower frequency, suggesting that stronger responses to conserved sequences may be as important as breadth for protection [52, 53]. Interestingly, heterologous vector prime-boost regimens enhanced immunity by increasing the magnitude, onset, and multi-functionality of the insert-specific cell-mediated immune responses compared to homologous regimens [54]. New progress has been made in overcoming HIV-1 diversity through induction of cross-reactive T-cell responses to HIV-1 by vaccines designed *in silico* (called conserved and mosaic vaccines) [17]. Polyvalent mosaic immunogens derived by the recombination of natural HIV-1 strains were designed to induce cellular immune responses that recognize genetically diverse circulating virus isolates. Increasing the breadth and depth of epitope recognition may contribute to the protection against infection by genetically diverse viruses and also to the control of variant viruses that emerge as they mutate away from recognition by CTLs. For example, mosaic HIV-1 Gag, Pol, and Env antigens expressed by Ad26 vectors markedly augmented both the breadth and depth without compromising the magnitude of antigen-specific T-lymphocyte responses as compared with consensus or natural sequence HIV-1 antigens in rhesus monkeys [55, 56]. An alternative to multivalent wild-type or mosaic vaccines is the use of conserved element immunogens as a novel and effective strategy to broaden responses against highly diverse pathogens by avoiding decoy epitopes, while focusing responses to critical viral elements for which few escape pathways exist. Priming with conserved elements boosted with the complete immunogen induced broad cellular and humoral immunity focused on the conserved regions of the virus. In contrast, full-length HIV-1 immunogens elicited greater magnitude and comparable breadth of T-lymphocyte responses to conserved HIV-1 regions compared with conserved-region only HIV-1 immunogens in rhesus monkeys [57, 58].

An important point is the use of replicating vector. A replication-competent rhesus cytomegalovirus vaccine expressing SIV proteins induced and maintained high frequency of SIV-specific $CD4^+$ and $CD8^+$ T-cell effector memory responses at extra-lymphoid sites without measurable antibody responses to SIV. Half of vaccinated monkeys showed a severe control of three routes of SIVmac239 transmission including intrarectal, intravaginal, and intravenous. The conservation of particular cytotoxic epitopes does make them good candidates for a global HIV-1 vaccine [59, 60].

5. Therapeutic HIV vaccines

Recent studies have focused on the improvement of effective prophylactic and therapeutic approaches to combat persistent viral infections. Therapeutic vaccines for HIV infection should aim to elicit antiviral CD8 T cells (CTLs), CD4 T cells, and neutralizing antibody since these immune responses control viral replication [7, 61]. It is critical to generate broad cellular responses as HIV mutates very rapidly to escape immune system. In addition, recent studies determined that T-follicular helper cells constitute a significant source of virus production and contribute to the total viral reservoir. Since these cells reside in B-cell follicles/germinal centers, it may be critical to generate CD8 T cells that can home to B-cell follicles and exert immune response on these cells. The HIV-specific CD4 T-cell response is also important for maintaining the functional CD8 T-cell and B-cell responses. However, these HIV-specific CD4 T cells could also serve as potential targets for virus replication [7]. Interestingly, CD4 T cells with cytolytic function have been shown to be associated with enhanced viral control, although it is demonstrated whether these responses can be primed by vaccination. The function of dendritic cells may be also critical for generating a protective cellular and humoral immune response, as chronic HIV infections are associated with impaired DC function. Thus, therapeutic vaccines may also need to use strategies such as adjuvants to enhance the function of innate immunity [7].

Several therapeutic vaccine strategies have already been used such as live-attenuated microbes, viral vectors, and dendritic cell-based vaccines that led to suppress and/or clear infections. Among them, improved DNA vaccines have emerged as a promising candidate for the treatment of infectious diseases especially HIV infections [61–63]. Some strategies have been considered to improve immune responses stimulated by DNA vaccines such as *in vivo* efficient DNA delivery systems, co-delivery with molecular adjuvants as well as the development of potent heterologous prime-boost regimens [61, 62, 64]. DNA vaccines have been utilized as candidate HIV vaccines because of their ability to generate cellular and humoral immune responses, the lack of anti-vector response allowing for repeat administration, and their ability to prime the response to viral-vectored vaccines. Because the HIV epidemic has unreasonably affected the developing world, the favorable thermostability profile and relative ease and low cost of manufacture of DNA vaccines offer additional advantages. *In vivo* electroporation (EP) has been utilized to improve immune responses to DNA vaccines as candidate HIV-1 vaccines alone or prime-boost regimens with both proteins and viral-vectored vaccines in several animal models, and recently, in human clinical trials [65]. In addition, intradermal electroporation of HIV DNA was well tolerated. Strong cell- and antibody-mediated immune responses were elicited by the HIV-DNA prime and HIV-Modified Vaccinia Ankara (MVA) boosting regimen, with or without intradermal electroporation use [66]. DNA vaccines have an intrinsic bias toward generating cellular immunity against intracellular pathogens. By manipulating the DNA formulation and delivery, effective antibody responses can also be induced. For instance, studies showed that the immunized monkeys with DNA vaccine developed HIV-specific T-cell immune responses that persisted for months [62].

The use of live-attenuated invasive bacteria as a carrier for DNA-based vaccines has previously been reported [67]. Immunization with recombinant invasive bacteria including *Shigella, Salmonella,* and *Listeria* carrying plasmid DNA (pDNA) vaccines has been shown to induce protective immune responses in mice. The use of human enteric bacteria is especially useful due to their ability to infect human colonic mucosa, and their tropism for the activation of dendritic cells and macrophage of internal mucosa. Thus, they are very efficient for the delivery of DNA vaccines to APCs in the mucosa resulting in stimulation of potent systemic and local immune responses. Such responses may be critical for the development of an effective prophylactic HIV vaccine, because a large number of HIV transfer through human mucosal routes [67]. For instance, a live-attenuated strain of *Salmonella typhimurium* was used to deliver plasmid orally and showed an adjuvant role through the release of various cytokines [68]. In addition, intranasal immunization of mice with live recombinant *Shigella* cells induced an HIV Gag-specific cellular immune response similar to that observed by intramuscular injection of naked DNA. Importantly, a strong boosting effect was obtained in mice primed with DNA, suggesting the efficacy of bacterial vectors in prime-boost vaccination regimens [67]. The studies indicated that a novel vaccine delivery system using bacterial ghosts (BGs) can be considered as an efficient and nontoxic delivery system for DNA vaccines *in vitro* and *in vivo*. In this line, a new strategy of HIV vaccine delivery was designed using *Salmonella typhi* Ty21a bacterial ghosts. The data showed that Ty21a BGs loaded with an HIV gp140 DNA vaccine (Ty21a BG-DNA) are easily taken up by murine macrophage cells (RAW264.7), and gp140 is efficiently expressed in these cells. Peripheral and intestinal mucosal anti-gp120 antibody responses in mice vaccinated with BGs–DNA vaccine were significantly higher than those in mice immunized with naked DNA vaccine. The enhancement of antibody responses was associated with BG-induced production of IL-10 through TLR4 pathway [69].

Attenuated virus vaccines have traditionally been potent and relatively easy to produce and deliver. Vaccination with a live virus results in high intracellular synthesis of viral proteins. This high-level expression stimulates strong cellular and humoral immune responses and results in the production of long-lasting memory B and T cells. However, attenuated HIV vaccines replicate strongly in animal models to retain residual virulence. Recent studies indicated that priming with a DNA vaccine induces a Th1 response that can be boosted by the subsequent administration of a viral vector encoding the same gene. This prime-boost strategy elicited strong protective immunity in several primate models [70]. Currently, immunogenicity of a highly attenuated *vaccinia* virus with low neuro-virulence, LC16m8 strain, was studied as an HIV vaccine vector. The data showed that the recombinant *vaccinia* virus-based vaccine (vLC-Env) combined with DNA vaccine expressing the HIV *env* gene (pCAG-Env) produces a protective immune response against HIV infection in BALB/c mice. Vaccination of vLC-Env alone induced much higher HIV-specific humoral and cellular immune responses than that of pCAG-Env. Priming with pCAG-Env further enhanced vLC-Env-induced immune responses, especially cell-mediated immune response. In addition, administration of vLC-Env-infected dendritic cells to mice generated a high cellular immune response. These results demonstrated that priming with pCAG-Env and boosting with vLC-Env represents a potential candidate for vaccination against HIV infection [70]. A few studies have used DC presenting either the autologous virus or virus-derived peptides as therapeutic vaccines (DC-based vaccines) in macaques and humans. They also showed that a similar approach could successfully control HIV replication in humans. Similarly, a recent study showed that an efficient HIV-1-specific immune response could be generated using an autologous monocyte-derived DC (MDDC) transfer. Thus, HIV-specific immune responses could be elicited by DC-based therapeutic vaccinations. In general, therapeutic vaccinations should be explored as a combination therapy with other immune modulators to achieve a functional cure (*i.e.*, long-term control of virus replication in the absence of antiretroviral therapy) [7].

There is growing interest in the role of anti-HIV antibody-dependent cellular cytotoxicity (ADCC) antibodies in the prevention and control of HIV infection. Passive transfer studies in macaques supported a role for the Fc region of antibodies in the prevention of simian–human immunodeficiency virus infection. The Thai RV144 HIV-1 vaccine trial induced anti-HIV ADCC antibodies that may play a role in the partial protection observed. Several studies showed a role for ADCC antibodies in slowing HIV disease progression. However, HIV evolves to escape ADCC antibodies, and chronic HIV infections cause the dysfunction of effector cells such as natural killer cells that mediate the ADCC functions. Furthermore, four recent studies showed that the HIV-1 *Vpu* protein, by promoting release of virions, reduces the capacity of ADCC antibodies to recognize HIV-infected cells [71]. On the contrary, The HIV-1 transactivator of transcription (Tat) is a key HIV virulence factor, which plays critical roles in virus gene expression, replication, transmission, and disease progression. The results indicated that Tat-induced immune responses are necessary to restore immune homeostasis, to block the replenishment and to reduce the size of the viral reservoir. Anti-Tat antibodies are uncommon in natural infection and, when present, correlate with the asymptomatic state and lead to lower or no disease progression. Hence, targeting Tat represents a pathogenesis-driven intervention [72].

6. Adjuvants

Although the importance of DNA vaccines, especially as a priming immunization, has been well proved in different HIV vaccine studies, the immunogenicity of DNA vaccines is generally moderate. Novel adjuvant is necessary for improving the immunogenicity of DNA vaccine [73]. Multiple groups have demonstrated the potential of co-administering plasmid DNA that express cytokines, chemokines, or co-stimulatory molecules together with plasmids encoding target viral antigens [74]. The potency of DNA vaccines can be developed by the co-delivery of plasmid-encoded molecular adjuvants. The pDNAs encoding granulocyte-macrophage colony-stimulating factor (GM-CSF), hematopoietic factor fms-like tyrosine kinase 3 ligand (Flt-3L), and interleukin-12 (IL-12) could markedly enhance cell-mediated immune responses elicited by an HIV-1 *env* pDNA vaccine in BALB/c mice [74]. Plasmid GM-CSF also increased the immune responses elicited by DNA vaccines expressing HIV-1 Gag and Nef-Tat-Vif. In addition, the use of pGM-CSF as a vaccine adjuvant appeared to increase antigen-specific proliferative responses and the percentage of polyfunctional memory $CD8^+$ T cells. Co-delivery of pFlt-3L with pGM-CSF did not result in a further increase in adjuvant activity. However, the co-administration of pGM-CSF with pIL-12 significantly enhanced Env-specific proliferative responses and vaccine efficacy in the murine *vaccinia* virus challenge model relative to mice immunized with the *env* pDNA vaccine adjuvanted with either pGM-CSF or pIL-12 alone [74]. In another study, co-administration of the HIV-1 DNA vaccine with pIL-12 and pGM-CSF by topical application to the skin enhanced the levels of both the HIV-specific cytotoxic T-lymphocyte response and delayed-type hypersensitivity (DTH). Indeed, the skin is accessible for generating immune responses by both intradermal injection and topical use of gene delivery vectors [75]. In addition, co-administration of plasmids encoding the codon-optimized GM-CSF sequence with the HIV-1 Gag DNA vaccine resulted in a strong antibody and CTL response and a protective immune response against infection with recombinant *vaccinia* virus expressing HIV-1 Gag [76]. On the contrary, researchers strongly support the use of IL-6 or IL-15 as a cytokine adjuvant in HIV DNA vaccination. The data indicated that intranasal administration of DNA vaccine and pIL-15 can enhance Th1-dependent HIV-1-specific cell-mediated immunity. However, co-injection of pIL-15 with pIL-2 or pIL-12 did not show any synergistic effect on the immune responses induced by DNA vaccine *in vivo* [77]. Furthermore, the immunogenicity of HIV-1 DNA vaccine expressing the chimeric gene Gag-gp120 (pVAX1-Gag-gp120) was increased by co-inoculating pVAX1-IL6 in BALB/c mice [78]. The studies demonstrated that in-frame fusion of tumor necrosis factor alpha (TNF-α), DNA to DNA, encoding a large fragment of HIV gp120 could enhance Th1 immune responses against gp120 antigen. Also, in-frame fusion of IFNγ-encoding DNA at the 5′ end of the chimeric molecule, to create a tripartite fusion, had no additional effect on immunogenicity [79]. A number of studies have shown that α-galactosylceramide, a natural killer T-cell (NKT) ligand, was applied as an adjuvant for various vaccines, including viral, parasite, and protein-based vaccines. The α-GalCer was able to enhance HIV-specific antibody responses. Furthermore, co-administration of α-GalCer with suboptimal doses of DNA vaccines greatly increased antigen-specific $CD4^+$ and $CD8^+$ T-cell responses. The level of cell-mediated immune responses in mice vaccinated with 5 μg of DNA in the presence of α-GalCer was similar to that of mice vaccinated with 50 μg of DNA in the absence of α-GalCer [80].

Recombinant adjuvants composed of a fusion between surfactant protein-D (SP-D) and either CD40 ligand (CD40L) or GITR ligand (GITRL) were previously shown to enhance HIV-1 Gag DNA vaccines. It was demonstrated that similar fusion constructs composed of the TNF superfamily ligands (TNFSFL) including 4-1BBL, OX40L, RANKL, LIGHT, CD70, and BAFF can also enhance immune responses to an HIV-1 Gag DNA vaccine. Importantly, the SP-D-4-1BBL, SP-D-OX40L, and SP-D-LIGHT constructs enhanced $CD8^+$ T-cell avidity and $CD8^+$/$CD4^+$ T-cell proliferation 7 weeks after vaccination [81]. Also, the SP-D-OX40L, SP-D-LIGHT, and SP-D-BAFF constructs increased Gag-specific IL-2 secretion in memory T cells, suggesting their potency to elevate the number of self-renewing Gag-specific $CD8^+$ and $CD4^+$ T cells. Finally, the SP-D-OX40L and SP-D-CD70 adjuvants augmented IgG2a but not IgG1 antibody responses in the immunized animals. Interestingly, the B-cell-activating protein BAFF did not enhance anti-Gag antibody responses when administered as an SP-D fusion adjuvant, but augmented $CD4^+$ and $CD8^+$ T-cell responses. Indeed, various SP-D-TNFSFL fusion constructs can enhance immune responses following DNA vaccination with HIV-1 Gag expression plasmid [81]. Several studies indicated that 4-1BB and 4-1BB ligand (4-1BBL) interactions are important for inducing robust CTL responses and also long-lived memory T cells. Recently, plasmid DNAs expressing either the membrane bound or soluble form of 4-1BBL were designed to enhance the Gag DNA vaccine as an adjuvant. The data showed that 4-1BBL DNA increased the Gag-specific IgG and cellular immune responses. Importantly, the expression of Gag and 4-1BBL from the same plasmid was critical for the adjuvant activity [82].

To improve the immunogenicity of DNA vaccines, some studies were focused on the immunoglobulin (Ig) fusion antigen. These reports showed that cytokine-coding plasmids fused with Ig have higher expression efficiency and better adjuvanticity. Furthermore, these plasmids have features that make them useful such as augmentation of half-life *in vivo*, formation of a multivalent antigen, and solubilization of hydrophobic proteins [83]. The possibility of increasing HIV gp120-specific cellular immune responses was determined in mice using a DNA vaccine encoding a mouse Ig fragment fused with gp120 in two directions (gp120-Ig or Ig-gp120). *In vitro* expression analysis revealed that the efficiency of HIV gp120 protein expression was higher in cells transfected with the gp120-Ig-coding plasmid (pGp120Ig) than in those transfected with the gp120 and Ig-gp120 expression plasmids (pGp120 and pIgGp120, respectively). The gp120-Ig-coding plasmid elicited more HIV-specific $CD8^+$ T cells and effector memory $CD8^+$ T cells than pGp120 in immunized mice. Furthermore, pGp120Ig significantly reduced the viral load after challenge with an HIV Env gp160-expressing *vaccinia* virus. These results represented that covalent antigen modification with an Ig sequence can modulate antigen-specific cellular immune responses [83].

Conversely, polysaccharide and nucleic acid fraction extracted from *Mycobacterium bovis bacillus Calmette–Guérin* (BCG-PSN) could be used as a novel adjuvant of DNA vaccine to elicit potent cellular and humoral immune responses against the HIV-1 Env antigen in BALB/c mouse model. In this experiment, the BCG-PSN was mixed with 10 μg or 100 μg of DNA vaccine and injected intramuscularly two or three times. BCG-PSN co-immunization with 10 μg DNA vaccine could elicit cellular and humoral immune responses which were comparable to that induced by 100 μg DNA vaccine alone. Moreover, BCG-PSN could activate TLR signaling pathways and induce Th1-type cytokine secretion. These findings suggested that BCG-PSN can be applied as a new and effective adjuvant for DNA vaccination [73].

Chemokines are largely bioactive inflammatory molecules which play a major role in a variety of immune and inflammatory responses, acting primarily as chemoattractants and activators of various leukocytes. In addition, some chemokines play a critical role in the transmission and progression of HIV-1 and HIV-2 viruses responsible for AIDS. Recent studies have indicated that chemokines and their receptors may play an important role in the differentiation and expansion of T cells in response to immune activation. These regulatory properties of chemokines make them suitable as molecular adjuvants [84]. For example, the modulation and regulation of immune responses were evaluated from the co-delivery of two β-chemokines as gene expression cassettes, MIP-1α or RANTES, along with HIV-1 DNA immunogen constructs. The data showed that MIP-1α had the greatest effect on antibody responses. In addition, co-expression of MIP-1α also modulated the shift of immune responses to Th2-type (*i.e.*, the increase of IgG1/IgG2a ratio). RANTES co-immunization also enhanced the levels of antigen-specific Th1 and CTL responses. The use of chemokine adjuvanted vaccines as HIV vaccine modulators may be important due to the interesting relationship between HIV cell entry and the receptors for β-chemokines. Indeed, β-chemokines as vaccine adjuvants increased β-chemokine production in an antigen-specific manner [84].

7. Heterologous prime-boost strategies

Most of the current DNA vaccines utilize CMV, β-actin, or muscle-specific desmin promoters to potentiate expression of one or two fused genes of HIV-1 including the Env, Gag, Pol, and Tat. DNA vaccines comprising multiple plasmids encoding different HIV-1 proteins have been used to obtain a broader spectrum of immunity than individual plasmids expressing single proteins. The use of these plasmid DNA vaccines proved to be safe and immunogenic in macaques; however, these constructs needed to be boosted with viral proteins expressed by various vector systems including recombinant *pox virus*, modified *vaccinia virus Ankara*, and *adenovirus* for enhancing their efficiency in preventing AIDS [85].

The heterologous prime-boost regimen uses the ability of the immune system to generate large numbers of secondary antigen-specific T cells following an initial priming step. The same antigen is delivered subsequently using different vectors. Following a priming immunization, the antigen-specific T-cell populations develop to modest levels and then reduce. Indeed, a percentage of these cells transform into antigen-specific memory T cells. In a heterologous boost, because the priming and boosting vectors are different, T cells that specifically target the viral vector are not boosted and do not activate cell number control mechanisms, therefore allowing for greater development of the disease antigen-specific T-cell populations [82]. Several groups have now established that heterologous prime-boost regimens are the most potent strategies to induce cellular immune responses [86, 87]. In a plasmid DNA vaccine priming and viral vector-boosting regimen, the order of DNA followed by recombinant virus is important, as the reverse order did not induce higher levels of antigen-specific $CD8^+$ T cells. It seems that the cytokine microenvironment created by a local virus infection during boosting is responsible for the efficient expansion of effector T cells [86]. In 2004, a consecutive immunization strategy involving priming with DNA and boosting with recombinant fowlpoxvirus (rFPV) vaccines encoding multiple common HIV-1 antigens was evaluated in 30 macaques. The vaccines were well tolerated, and a significant enhancement of DNA-vaccine primed HIV-1-specific T-lymphocyte responses was observed following rFPV boosting. Co-expression of IFNγ or IL-12 by the rFPV vaccines did not further enhance immune responses [88]. In addition, a subtype A or B HIV gp160 plasmid DNA and Env gp140 trimeric glycoprotein co-immunization was superior to immunization with glycoprotein alone by enhancing neutralizing antibodies. These data showed that co-delivering DNA and protein can increase antibody responses to Env. Hence, this approach has the potential to simplify vaccine regimens by inducing higher antibody responses using fewer vaccinations, an advantage for a successful HIV vaccine design [89].

The reports showed that co-immunization of a DNA vaccine encoding HIV-1P24-Nef with GM-CSF in DNA priming and peptide boost strategy increases the immunogenicity of the candidate vaccine. Cytokine profile studies showed that both IL-4 and IFN-γ levels were increased. Also, co-immunization with GM-CSF resulted in a higher level of total IgG, comprising approximately equal levels of both specific IgG1 and IgG2a subtypes. Taken together, the results suggested that GM-CSF is able to induce long-term memory for the HIV-1 P24-Nef vaccine candidate [90]. Recent studies have used DNA/protein or DNA/adeno-vector regimens for HIV immunization. The essential mechanisms of heterogeneous prime/boost regimens are not well understood, but DNA priming results in much lower antigen expression compared to protein vaccines, and this may prime T-helper cell responses with the humoral response subsequently being boosted by the high-dose protein or viral vector (*e.g.*, RV 144 tested in clinical trials) [87].

8. Clinical trials

Several vaccine candidates were used in different phases of clinical trials. DNA prime-viral vector boost regimens have become the primary choice for stimulation of T-cell immune responses. For example, *Poxvirus* vector-based vaccines including the *Modified Vaccinia Ankara* and the genetically modified NYVAC-based vaccines appeared to be efficient in inducing the immune responses and could be evaluated in combination with DNA priming in clinical trials [91]. In addition, the safety and immunogenicity of several *Canarypox*-based vaccines with multiple HIV-1 gene inserts have been studied in humans. A phase III trial, RV144, using *Canarypox* (vCP1521) prime and AIDSVAX B/E boost has demonstrated modest protective efficacy in Thailand. The protection in RV144 trial was short-time and needed to use the additional boosters in participants for improving recall responses and continuing protection among them. AIDSVAX was also a component of the prime-boost (ALVAC/AIDSVAX) RV 144 vaccine in Thailand that showed successful results. In both cases, the vaccines targeted gp120 and were specific for the geographical regions. Among the adenoviral vector vaccine candidates, replication-defective Ad5 candidate indicated high immunogenicity in phase I clinical trials and reduced viral load in the SHIV/NHP model. However, this strategy failed to prevent new infections as well as reduce post-infection viral RNA levels in the vaccinated individuals in phase IIb. Furthermore, participants with preexisting antibodies against Ad5 vector showed increased HIV infection rates [91]. The heterologous prime-boost strategy using DNA prime and Ad5 boost was considered to avoid the problem of preexisting immunity. It has been shown that the preexisting Ad5-neutralizing antibodies did not affect the levels of cell-mediated responses in the DNA/rAd5 prime-boost recipients, as compared to participants who received rAd5 alone. However, in spite of robust immune responses induced by DNA/Ad5 strategy in phase I and phase II trials, the strategy failed to show protection from new infections in phase IIb [91]. Generally, prime-boost vaccination is an efficient approach compared to other strategies, but it still needs to develop against HIV infections in future.

9. Challenges for HIV vaccines

There are some major challenges against HIV vaccine design as described below. Figure 1 shows these challenges briefly.

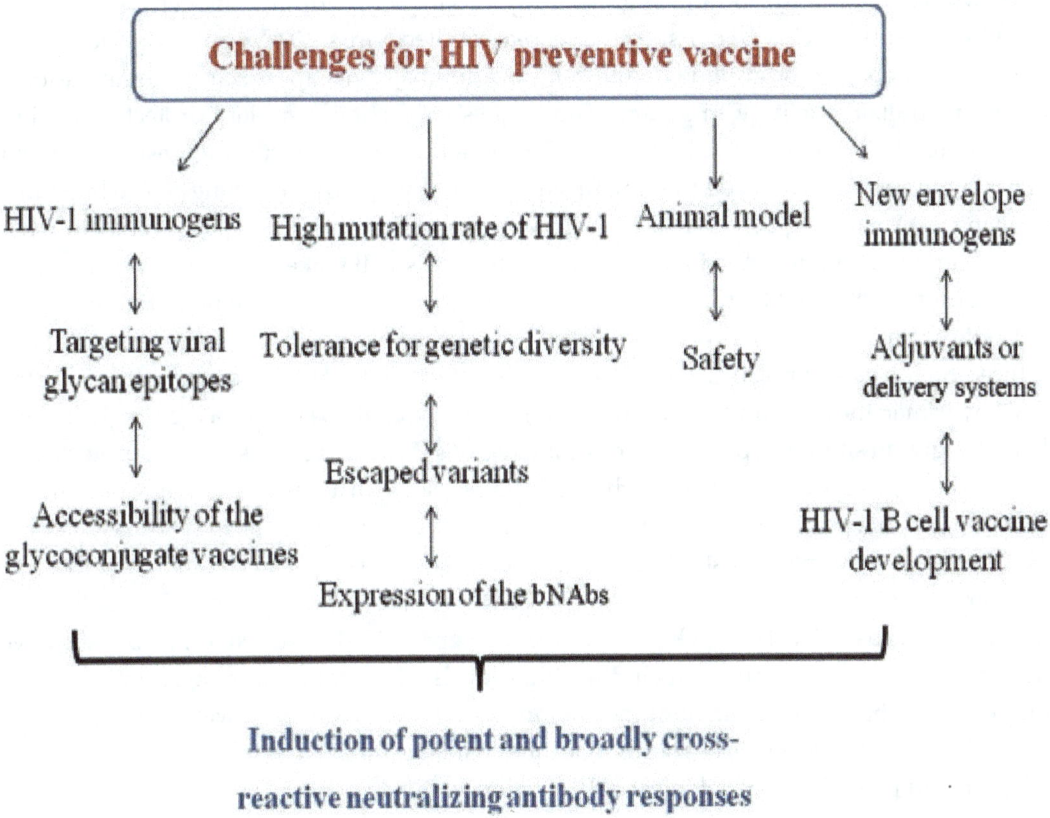

Figure 1. Challenges for HIV-preventive vaccines

9.1. Designing trimeric HIV-1 envelopes

The challenge remains to develop HIV-1 immunogens that will elicit protective immunity [13]. The challenge is to design, engineer, and produce a pure stable envelope immunogen that mimics the antigenic profile of the functional envelope spike. The engineered trimeric envelope was unable to induce bNAb in animals. Modification of the trimers, including removal of individual glycans proximal to CD4-binding region, elimination of the glycosylation site near the gp41 loop, linker-stabilized gp140 trimeric envelopes, have resulted in improved immunogenicity but have not yielded the desired bNAb. A combination of mosaic envelopes increased the magnitude of NAbs but not the breadth of the response in macaques. Therefore, no trimeric envelope induces bNAb in humans [1]. Some studies showed that HIV-1 bNAbs identifies four conserved Env targets for HIV neutralization. To date, more than 30 bNAbs specific for conserved neutralizing Env epitopes have been characterized [17].

9.2. Why does vaccination with HIV envelope not induce bNAbs?

Whether bNAb will effectively confer protection against HIV infection in humans remains unknown. An alternative to inducing bNAb by vaccination with immunogens is to deliver these bnMAbs with viral vectors (*e.g.*, adeno-associated virus (AAV) gene transfer vector expressing antibodies or antibody-like immunoadhesins). This approach generated a long-lasting neutralizing activity in serum of macaques conferring complete protection against intravenous challenge with virulent SIVmac316. Similarly, full protection against intravenous HIV-1 challenge was observed in humanized mice receiving AAV carrying b12, while those receiving AAV carrying 2G12, 4E10, and 2F5 were partially protected [1]. A recent study has demonstrated that up to 50% of HIV-infected individuals will make cross-reactive antibodies that neutralize 50% of HIV primary strains. However, when bNAbs develop in HIV infection, they only occur after 2–4 years of infection. In contrast, no vaccine immunizations to date have induced high levels of bNAbs. The bNAbs are targeted to one of five conserved sites on the HIV Env trimer: the CD4 binding site, the membrane proximal gp41 region, the V3-glycan site, the V1V2-glycan site, and gp41-gp120 bridging regions [51]. Each of these sites is protected by surrounding glycans, and each one of these sites is restricted in access, such that relatively few antibody variable heavy (VHDJH) and variable light (VL) combinations may be used to bind these Env sites. Example of restricted VHDJH/VL usage is the use of VH1-2 paired with a 5 aa VL complementarity-determining region 3 (LCDR3) for the VRC01 type of CD4 binding site bNAb, and the use of VH1-69, Vk3-20 for 4E10-like gp41 bNAbs. Moreover, all bNAbs have one or more unusual features, including high levels of somatic mutations, and poly- or autoreactivity that can result in immune tolerance control for bNAbs. However, in the simian-human immunodeficiency virus rhesus macaque challenge model, passive infusion of the new bNAbs could potently protect against SHIV challenge [51].

9.3. Effective adjuvants

Adjuvants are important for the use of recombinant envelope immunogens, since these proteins by themselves generate only weak immune responses. For potent vaccine formulations delivered by mucosal routes, incorporation of adjuvants that controls the potential of innate immune modulators is important for overcoming immune tolerance and enhancing the immunogenicity of co-administered antigens [25]. The RV144 trial used alum as an adjuvant, which was then the only licensed vaccine adjuvant. However, alum is not believed to support robust cellular immune responses. Also, bacterial toxins are the most potent mucosal adjuvant candidates but concerns remain regarding their safety even when mutated to reduce toxicity. In contrast, ligands for TLRs 7/8 and 9 serve as potent adjuvants for parenteral and mucosal vaccines based on plasmid DNA, viral vectors, and recombinant proteins [25]. In particular, CpG-containing synthetic oligodeoxynucleotides (CpG-ODN) that activate TLR9 on dendritic cells appear potent in stimulating antigen presentation and induction of antigen-specific immune responses. The synthetic glycolipid α-galactosylceramide has been tested primarily in cancer immunotherapy studies because of its capacity to serve as a ligand and potent activator of invariant natural killer T cells. In addition, the repeated mucosal delivery of α-GalCer adjuvant was done in primary and booster immunizations that resulted in repeated activation of NKT cells and DC to progressively increase adaptive immune responses [25]. The parallel development of adjuvants along with better HIV-1 immunogens will be needed for a successful AIDS vaccine. Additional comparative testing will be required to determine the optimal adjuvant and immunogen regimen that can elicit antibody responses capable of blocking HIV-1 transmission [92].

Both flagellin (fliC) and IL-18 (interferon-γ-inducing factor) have been developed as adjuvants to improve immunogenicity in DNA-vaccinated hosts. An HIV-1 Gag plasmid encodes a protein harboring broad epitopes for CTLs [93]. The immunogenicity of BALB/c mice immunized with an HIV-1 Gag plasmid (pVAX/Gag), combined with a chimeric plasmid encoding IL-18 fused to flagellin (pcDNA3/IL-18_fliC) or a single plasmid encoding IL-18 (pcDNA3/IL-18) and/or flagellin (pcDNA3/fliC), was studied. The IL-18 and flagellin fusion protein effectively induced IFN-γ by lymphocytes. During a 12-week immunization, both Gag-specific IgG in sera and spleen cell proliferation were elevated in all murine groups. However, the IgG2a/IgG1 ratio, Th1 cytokine (IL-2 and IFN-γ) production, and the proportion of Gag-specific $CD3^+$ $CD8^+$ IFN-γ-secreting cells were significantly increased in the murine group co-immunized with the pVAX/Gag plasmid and pcDNA3/IL-18_fliC compared with the mice immunized with the pVAX/Gag plasmid combined with either the pcDNA3/fliC or pcDNA3/IL-18 plasmid or both of them. The data suggested that the chimeric plasmid encoding IL-18 fused to flagellin can be used as an adjuvant-like plasmid to improve the Th1 immune response, particularly for the induction of $CD3^+$ $CD8^+$ IFN-γ-secreting cells in Gag plasmid-vaccinated mice [93].

9.4. High mutation rate of HIV-1

The high mutation rate of HIV-1 and tolerance for genetic diversity represented central challenges for vaccine design. Because the immune response is itself adaptive, the optimal HIV-1 sequence within an individual also differs over time. HIV-1 develops specific mutations within its genome that allow it to escape detection by human leukocyte antigen (HLA) class I-restricted immune responses, notably those of $CD8^+$ CTLs. HLA thus represents a major force driving the evolution and diversity of HIV-1 within individuals [94].

9.5. Escaped variants

A major challenge is how to induce effective immune responses against escaped variants. It is important that a CTL-based vaccine stimulate effective cellular responses across the range of HLA class I alleles expressed in a host population. These observations have led to the idea that immune-mediated control of HIV-1 replication to levels that slow disease progression might be feasible through the design of vaccines that focus CTL responses against viral regions where escape cannot occur [94]. To date, adenovirus vector prime and pox vector boost vaccines have been one among the most immunogenic vaccines for inducing HIV CD8 T-cell responses in humans. Efforts continue to overcome HIV diversity for T-cell epitope recognition by the *in silico* design of centralized consensus or mosaic HIV gene inserts based on optimizing the coverage of T-cell epitopes in HIV strains in the Los Alamos HIV Sequence Database, or based on conserved epitopes in the vaccine [51].

9.6. Expression of the bNAbs is limited by host tolerance mechanisms

The studies showed that two human recombinant bNAbs, called 2F5 and 4E10, that bind near the virion membrane to Env gp41 were reactive in human autoantibody assays. In a subsequent study, 2F5 was shown to avidly bind the human protein kynureninase (KYNU), and 4E10 was shown to react with the mammalian RNA splicing factor 3B3. The nominal gp41 epitope of the 2F5 bNAb is the linear peptide ELDKWAS and an identical 6-residue sequence is present in KYNU (ELDKWA). This ELDKWA motif in KYNU is conserved in nearly all mammalian species and absent in all proteins other than the HIV Env. Thus, the autoantigens for these two bNAbs, 2F5 and 4E10, have been identified, suggesting that expression of these bNAbs is limited by host tolerance mechanisms [17].

9.7. Challenges for developing vaccines targeting viral glycan epitopes

Generation of antibodies to glycans has several challenges: a) due to the inherent weakness of carbohydrate–protein interactions, binding affinities must be enhanced through avidity effects. For example, lectins are able to overcome this problem using interaction of multiple carbohydrate binding domains with arrays of glycan ligands; b) glycoproteins usually always exist as a number of different glycoforms where the same protein backbone is glycosylated with different glycan structures. This microheterogeneity weakens the antigenic response to the individual glycan structures. These multiple conformations may be presented to the immune system further weakening the response; c) glycosylation is ubiquitous to all mammalian cells, thus the host may display tolerance toward these sugars. Generally, these effects result in glycans are poorly immunogenic. The major concern of generating antibodies against self-glycan structures is their potential autoreactivity *in vivo* [95].

9.8. Animal model for preclinical studies

There are few nonhuman primate models of enhanced HIV susceptibility [96]. Animal model research during the past years has focused on the development of models in order to explore key questions about HIV entry, immune control, and persistence and also their use for testing therapeutic vaccines [97].

9.9. Design of new envelope immunogens

A major challenge for HIV-1 B-cell vaccine development is the design of new envelope immunogens that can trigger the selection and expansion of germline precursor and intermediate memory B cells associated with the maturation of a broadly neutralizing antibody response. The identification of delivery systems, prime-boost strategies, and synergistic adjuvant combinations is important to induce the magnitude and quality of antigen-specific T-follicular helper cell responses needed to induce somatic hypermutation (SHM) and B-cell maturation against heterologous primary virus envelopes [98].

9.10. Safety of vaccines

Safety of vaccines is one of the most important subjects for the design of vaccines that should be determined in clinical trials [99].

9.11. Accessibility of the glycoconjugate vaccines

Accessibility of these glycoconjugate vaccines in resource-poor regions which bear the highest disease burden from these pathogens remains challenging largely due to high vaccine pricing [100].

9.12. Induction of potent and broadly cross-reactive neutralizing antibody responses

Induction of potent and broadly cross-reactive neutralizing antibody responses remains a major challenge for the development of HIV vaccines because of the high diversity of gp120. The high glycosylation, large conformational changes, and steric restriction of the epitopes in gp120 during receptor binding and membrane fusion processes prevent the access of antibodies to these sites [101].

9.13. Challenges associated with antigen immunogenicity

The failure to date of Env-based antigens to stimulate bNAb is likely to result from several specific reasons that influence BCR recognition of unusual structural antigenic elements:

a. **Incorrect presentation of the vaccine antigen**: Immunization with linear peptides of MPER failed to re-elicit neutralizing responses because the MPER peptide mimics adopted an inappropriate conformation in solution and failed to present the correct surface for B-cell recognition [5].

b. **Cross-reactivity with self**: The 4E10 mAb, and to a lesser extent the 2F5 mAb, binds lipid as part of their epitope by using an array of hydrophobic residues. This binding appears to make them autoreactive, resulting in B-cell tolerance mechanisms [5].

c. **Epitopes with steric constraints for BCR recognition**: The CD4bs is an apparent target for eliciting NAbs as it requires conservation for function, and needs to be exposed for CD4 binding. In contrast, most infected individuals do not make CD4bs-specific bNmAbs. The main reason is the intrinsic immunorecessive nature of the conserved segments of CD4bs [5].

d. **Unique antigenic features for BCR recognition**: The 2G12 bNmAb has an epitope composed entirely of oligomannose groups. The PG and PGT series of bNmAbs have complex glycan–peptide binding surfaces in which the glycans are heterogeneous.

Preparation of such epitopes will require powerful synthetic chemistry related to scaffolded peptide design. Both MPER bNmAbs 2F5 and 4E10 require a lipid component to their epitopes and up to now this has not been incorporated into a successful immunogen [5].

e. **Germline BCR recognition and requirement for extensive antibody affinity maturation**: There are two possible consequences of the steric constraints imposed on BCRs during the recognition of structurally unusual antigens: a) the frequency of germline BCRs available to recognize such complex antigens will be low, thus an extensive degree of affinity maturation will be required to generate a high-affinity bNAbs recognizing structurally "difficult" epitopes; b) the germline BCR affinity for a bNmAb epitope may be undetectable. A feasible outcome of these constraints is that the host will require long-term antigen exposure to select and clonally expand the rare B cells with appropriate BCRs and to affinity mature them into bNAbs, because most bNmAbs arise in individuals after chronic HIV-1 infection [5].

f. **Conceptual concerns relating to epitope recognition by BCRs:** There are concerns that isolating an epitope from its antigenic context will not lead to re-elicitation of the same type of antibody against the epitope, which is a reasonable concern. However, although an epitope mimic may not re-stimulate an antibody identical to the template bNmAb, it may be sufficiently balance between elicited Ab and epitope mimic to allow specific binding to trimeric Env. If this is achieved, trimeric Env may be used to boost and affinity mature those B cells reactive with the epitope mimetic [5].

g. **Responders and non-responders**: The finding that among large cohorts of HIV-1-infected individuals only a minor percentage makes a bNmAb response suggests that this may apply also to responses to vaccination [5].

Finally, the key difficulty in the development of an HIV vaccine is our ignorance of the immune responses that control viral replication, how these responses can be elicited, and how they can be monitored [2]. The question of whether to focus on induction of antibody or CTLs continues to be discussed in the HIV-1 field. However, evidence from many other vaccine-preventable infectious diseases indicates that antibody titers correlate with protection from infection, but CTL-mediated immune responses are required for protection against disease. This suggests that a dual approach is still necessary. Aspects of CTL vaccine technology such as replicating or persistent vectors may need to express Env-based antigens to allow long-term antigenic exposure for the induction of bNAb. In contrast, approaches to elicit bNmAbs may need to be immunologically compatible with the generation of a parallel CTL response.

9.14. Conclusions

The development of a safe and effective vaccine for HIV is a major global priority. To date, efforts to design an HIV vaccine have not been successful due to HIV diversity, HIV integration into the host genome, and ability of HIV to consistently evade antiviral immune responses. While the RV144 immunization strategy remains a priority for future efficacy trials, newer prime-boost mosaic and conserved sequence immunization strategies inducing efficient and long-time immune responses as well as the development of immunogens inducing broadly neutralizing antibodies should be followed and tested in humans. Recent success in isolation of potent broadly neutralizing antibodies, in discovery of mechanisms of bNAb induction and atypical mechanisms of CD8 T-cell killing of HIV infected cells, has opened new ways for HIV vaccine design. Indeed, the most protective HIV vaccine will require the combination of T-cell-inducing and antibody-inducing vaccine candidates with appropriate adjuvant formulations, since the innate and adaptive arms of the immune system cooperate for virus neutralization and pathogen-infected cell elimination. In general, acceleration of vaccine discovery depends on basic research and new technologies. Novel strategies should be safe, but rapidly tested in humans.

10. Key points

- Development of a safe and effective vaccine for HIV is a major global priority
- Efforts to design an efficient HIV vaccine have not succeeded due to HIV diversity, HIV integration into the host genome, and ability of HIV to consistently evade antiviral immune responses
- The cost-effective and different HIV-1 vaccine approaches have recently attracted a special interest
- Both antibodies and cell-mediated immune responses are considered to be important to prevent HIV-1 infection in the mucosal compartment
- Novel prime-boost mosaic and conserved sequence immunization strategies as well as the development of immunogens inducing broadly neutralizing antibodies have attracted a special interest
- Recent success in isolation of potent broadly neutralizing antibodies, in discovery of mechanisms of bNAb induction and atypical mechanisms of CD8 T-cell killing of HIV infected cells, has opened new ways for HIV vaccine design
- Several therapeutic vaccine strategies have already been used such as live-attenuated microbes, viral vectors, and dendritic cell-based vaccines that led to suppress and/or clear HIV infections
- Among different vaccines, improved DNA vaccines have emerged as a promising candidate for the treatment of infectious diseases especially HIV infections
- Some strategies have been considered to improve immune responses stimulated by DNA vaccines such as *in vivo* efficient DNA delivery systems, co-delivery with molecular adjuvants as well as the development of potent heterologous prime-boost regimens

Author details

Azam Bolhassani*

Address all correspondence to: A_bolhasani@pasteur.ac.ir; azam.bolhassani@yahoo.com

Department of Hepatitis and AIDS, Pasteur Institute of Iran, Tehran, Iran

References

[1] Excler, J.L., Robb, M.L., Kim, J.H. (2015). Prospects for a globally effective HIV-1 vaccine. Vaccine, 1–9

[2] de Goede, A.L., Vulto, A.G., Osterhaus, A.D., Gruters, R.A. (2015). Understanding HIV infection for the design of a therapeutic vaccine; Part I: Epidemiology and pathogenesis of HIV infection. Ann. Pharm. Fr., 73(2), 87–99

[3] Richert, L., Lhomme, E., Fagard, C., Lévy, Y., Chêne, G., Thiébaut, R. (2015). Recent developments in clinical trial designs for HIV vaccine research. Hum. Vaccin. Immunother., 11(4),1022–1029

[4] Wilson, C.B., Karp, C.L. (2015). Can immunological principles and cross-disciplinary science illuminate the path to vaccines for HIV and other global health challenges? Philos. Trans. R. Soc. Lond. B. Biol. Sci.,19, 370

[5] Schiffner, T., Sattentau, Q.J., Dorrell, L. (2013). Development of prophylactic vaccines against HIV-1. Retrovirology, 10(72), 1–16

[6] VISR Working Group of the Global HIV Vaccine Enterprise. (2015). HIV vaccine-induced sero-reactivity: A challenge for trial participants, researchers, and physicians. Vaccine, 33, 1243–1249

[7] Mylvaganam, G.H., Silvestri, G., Amara, R.R. (2015). HIV therapeutic vaccines: moving towards a functional cure. Curr. Opin. Immunol., 35, 1–8

[8] Goepfert, P., Bansal, A. (2014). Human immunodeficiency virus vaccines. Infect. Dis. Clin. North. Am., 28(4), 615–631

[9] Hanke, T. (2014). Conserved immunogens in prime-boost strategies for the next-generation HIV-1 vaccines. Expert Opin. Biol. Ther.,14(5), 601–616

[10] Alchin, D.R. (2014). HIV vaccine development: an exploratory review of the trials and tribulations. Immunol. Res. 60(1), 35–37

[11] Pavot, V., Rochereau, N., Lawrence, P., Girard, M.P., Genin, C., Verrier, B., Paul, S. (2014). Recent progress in HIV vaccines inducing mucosal immune responses. AIDS, 28(12), 1701–1718

[12] Excler, J.L., Robb, M.L., Kim, J.H. (2014). HIV-1 vaccines: challenges and new perspectives. Hum. Vaccin. Immunother. 10(6), 1734–1746

[13] Mann, J.K., Ndungu, T. (2015). HIV-1 vaccine immunogen design strategies. Mann and Ndungu Virol. J., 12 (3), 1–11

[14] Ringe, R., Bhattacharya J. (2013). Preventive and therapeutic applications of neutralizing antibodies to human immunodeficiency virus type 1 (HIV-1). Ther. Adv. Vaccines, 1(2), 67–80

[15] Mouquet, H. (2014). Antibody B cell responses in HIV-1 infection. Trends Immunol., 35(11), 549–561

[16] Braibant, M., Barin, F. (2013). The role of neutralizing antibodies in prevention of HIV-1 infection: what can we learn from the mother-to-child transmission context? Braibant and Barin Retrovirology, 10 (103), 1–14

[17] Haynes, B.F., Moody, M.A., Alam, M., Bonsignori, M., Verkoczy, L., Ferrari, G., et al. (2014). Progress in HIV-1 vaccine development. J. Allergy Clin. Immunol., 134, 3–10

[18] Horiya, S., MacPherson, I.S., Krauss, I.J. (2014). Recent strategies targeting HIV glycans in vaccine design. Nat. Chem. Biol., 10(12), 990–999

[19] Burton, D.R., Mascola, J.R. (2015). Antibody responses to envelope glycoproteins in HIV-1 infection. Nat. Immunol., 16(6), 571–576

[20] Fernández-Tejada, A., Haynes, B.F., Danishefsky, S.J. (2015). Designing synthetic vaccines for HIV. Expert Rev. Vaccines, 14(6), 815–831

[21] Kumar, R., Tuen, M., Liu, J., Nàdas, A., Pan, R., Kong, X., Hioe, C.E. (2013). Elicitation of broadly reactive antibodies against glycan-modulated neutralizing V3 epitopes of HIV-1 by immune complex vaccines. Vaccine, 31, 5413–5421

[22] Vassell, R., He, Y., Vennakalanti, P., Dey, A.K., Zhuang, M., Wang, W., et al. (2015). Immunogens modeling a fusion-intermediate conformation of gp41 elicit antibodies to the membrane proximal external region of the HIV envelope glycoprotein. PLoS One, 10(6), e0128562

[23] Alexander, M.R., Sanders, R., Moore, J., Klasse, P.J. (2015). Virion aggregation by neutralizing and non-neutralizing antibodies to the HIV-1 envelope glycoprotein. AIDS Res. Hum. Retroviruses

[24] Gallo, R.C. (2015). Developing a successful HIV vaccine. J. Infect. Dis., 212(1), S40–S41

[25] Singh, S., Yang, G., Byrareddy, S.N., Barry, M.A., Sastry, J. (2014). Natural killer T cell and TLR9 agonists as mucosal adjuvants forsublingual vaccination with clade C HIV-1 envelope protein. Vaccine, 32, 6934–6940

[26] Zolla-Pazner, S. (2014). A critical question for HIV vaccine development: which antibodies to induce? Science, 345, 167–168

[27] Mascola, J.R. (2007). HIV/AIDS: allied responses. Nature, 449, 29–30

[28] Gottardo, R., Bailer, R.T., Korber, B.T., Gnanakaran, S., Phillips, J., Shen, X., et al. (2013). Plasma IgG to linear epitopes in the V2 and V3 regions of HIV-1 gp120 correlate with a reduced risk of infection in the RV144 vaccine efficacy trial. PLoS One, 8, e75665

[29] Yates, N.L., Liao, H.X., Fong, Y, Decamp, A., Vandergrift, N.A., Williams, W.T., et al. (2014). Vaccine-induced Env V1–V2 IgG3 correlates with lower HIV-1 infection risk and declines soon after vaccination. Sci. Transl. Med., 6, 228–239

[30] Chung, A.W., Ghebremichael, M., Robinson, H., Brown, E., Choi, I., Lane, S., et al. (2014). Poly-functional Fc-effector profiles mediated by IgG subclass selection distinguish RV144 and VAX003 vaccines. Sci. Transl. Med., 6(228), 228–238

[31] O'Connell, R.J., Kim, J.H., Excler, J.L. (2014). The HIV-1 gp120 V1V2 loop: structure, function and importance for vaccine development. Expert Rev. Vaccines, 1–12

[32] Plotkin, S.A., Gilbert, P.B. (2012). Nomenclature for immune correlates of protection after vaccination. Clin. Infect. Dis., 54, 1615–1617

[33] Karasavvas, N., Karnasuta, C., Ngauy, V., Vasan, S., Tricharavoj, R., de Souza, M.S., et al. (2013). Investigation of antibody responses induced in RV305 a late boost vaccination of HIV-1 uninfected volunteers that participated in RV144, a Thai trial. In: P03.68LB, AIDS Vaccine

[34] Akapirat, S., Karnasuta, C., Madnote, S., Savadsuk, H., Puangkaew, J., Rittiroongrad, S., et al. (2014). HIV-specific antibody in rectal secretions following late boosts in RV144 participants (RV305). In: OA11.05, HIV R4P.

[35] Su, B., Moog, C. (2014). Which antibody functions are important for an HIV vaccine? Frontiers in Immunology, HIV and AIDS, 5, 1–12

[36] Kohler, H. (2015). Novel vaccine concept based on back-boost effect in viral infection. Vaccine, 33, 3274–3275

[37] O'Hagan, D.T., Ott, G.S., De Gregorio, E., Seubert, A. (2012). The mechanism of action of MF59-an innately attractive adjuvant formulation. Vaccine, 30, 4341–4348

[38] Leroux-Roels, I., Koutsoukos, M., Clement, F., Steyaert, S., Janssens, M., Bourguignon, P., et al. (2010). Strong and persistent CD4+ T-cell response in healthy adults immunized with a candidate HIV-1 vaccine containing gp120, Nef and Tat antigens formulated in three adjuvant systems. Vaccine, 28, 7016–7024

[39] Garcon, N., Van Mechelen, M. (2011). Recent clinical experience with vaccines using MPL- and QS-21-containing adjuvant systems. Expert Rev. Vaccines, 10, 471–486

[40] Vaccari, M., Gordon, S.N., Fourati, S., Schifanella, L., Cameron, M., Keele, B.F., et al. (2014). Adjuvant dependent mucosal V2 responses and RAS activation in vaccine induced protection from SIVmac251 acquisition. In: OA25.01. HIV R4P.

[41] McElrath, M.J. (1995). Selection of potent immunological adjuvants for vaccine construction. Semin Cancer Biol., 6, 375–385

[42] Rao, M., Onkar, S., Peachman, K., Padilla-Sanchez, V., Yamini, G., Jobe, O., et al. (2014). Potent V2-specific antibodies induced in humans using liposome-encapsulated HIV-1 gp120 recognize a well-exposed V2 epitope on envelope trimer. In: Abstract2049. Keystone, HIV Vaccines: Adaptive Immunity and Beyond.

[43] Bonsignori, M., Alam, S.M., Liao, H.X., Verkoczy, L., Tomaras, G.D., Haynes, B.F., et al. (2012). HIV-1 antibodies from infection and vaccination: insights for guiding vaccine design. Trends Microbiol., 20, 532–539

[44] McGuire, A.T., Dreyer, A.M., Carbonetti, S., Lippy, A., Glenn, J., Scheid, J.F., et al. (2014). HIV antibodies, antigen modification regulates competition of broad and narrow neutralizing HIV antibodies. Science, 346, 1380–1383

[45] Deal, C.E., Balazs, A.B. (2015). Vectored antibody gene delivery for the prevention or treatment of HIV infection. Curr. Opin. HIV AIDS, 10(3), 190–197

[46] Ensoli, B., Cafaro, A., Monini, P., Marcotullio, S., Ensoli, F. (2014). Challenges in HIV vaccine research for treatment and prevention. Frontiers in Immunology, HIV and AIDS, 5, 1–11

[47] Sanders, R.W., van Gils, M.J., Derking, R., Sok, D., Ketas, T.J., Burger, J.A., et al. (2015). HIV-1 neutralizing antibodies induced by native-like envelope trimers. Science, doi: 10.1126/science.aac4223

[48] Iyer, S.S., Gangadhara, S., Victor, B., Gomez, R., Basu, R., Hong, J.J., et al. (2015). Co-delivery of envelope protein in alum with MVA vaccine induces CXCR3-Biased CXCR5+ and CXCR5- CD4 T Cell responses in Rhesus Macaques. J. Immunol., doi: 10.4049/jimmunol.1500083

[49] Klasse, P.J. (2012). The molecular basis of HIV entry. Cell Microbiol., 14(8), 1183–1192

[50] Gupta, S., Clark, E.S., Termini, J.M., Boucher, J., Kanagavelu, S., LeBranche, C.C., et al. (2015). DNA vaccine molecular adjuvants SP-D-BAFF and SP-D-APRIL enhance anti-gp120 immune response and increase HIV-1 neutralizing antibody titers. J. Virol., 89(8), 4158–4169

[51] Haynes, B.F. (2015). New approaches to HIV vaccine development. Current Opinion in Immunology, 35, 39–47

[52] Fitzgerald, D.W., Janes, H., Robertson, M., Coombs, R., Frank, I., Gilbert, P., et al. (2011). An Ad5-vectored HIV-1 vaccine elicits cell-mediated immunity but does not

affect disease progression in HIV-1-infected male subjects: results from a randomized placebo-controlled trial (the step study). J. Infect. Dis., 203,765–772

[53] Li, F., Finnefrock, A.C., Dubey, S.A., Korber, B.T., Szinger, J., Cole, S., et al. (2011). Mapping HIV-1 vaccine induced T-cell responses: bias towards less-conserved regions and potential impact on vaccine efficacy in the step study. PLOS ONE, 6, e20479

[54] Ratto-Kim, S., Currier, J.R., Cox, J.H., Excler, J.L., Valencia-Micolta, A., Thelian, D., et al. (2012). Heterologous prime-boost regimens using rAd35 and rMVA vectors elicit stronger cellular immune responses to HIV proteins than homologous regimens. PLOS ONE, 7, e45840

[55] Santra, S., Liao, H.X., Zhang, R., Muldoon, M., Watson, S., Fischer, W., et al. (2010). Mosaic vaccines elicit CD8+ T lymphocyte responses that confer enhanced immune coverage of diverse HIV strains in monkeys. Nat. Med., 16, 324–328

[56] Barouch, D.H., O'Brien, K.L., Simmons, N.L., King, S.L., Abbink, P., Maxfield, L.F., et al. (2010). Mosaic HIV-1 vaccines expand the breadth and depth of cellular immune responses in rhesus monkeys. Nat. Med., 16, 319–323

[57] Kulkarni, V., Valentin, A., Rosati, M., Rolland, M., Mullins, J.I., Pavlakis, G.N., et al. (2014). HIV-1 conserved elements p24CE DNA vaccine induces humoral immune responses with broad epitope recognition in macaques. PLOS ONE, 9, e111085

[58] Stephenson, K.E., SanMiguel, A., Simmons, N.L., Smith, K., Lewis, M.G., Szinger, J.J., et al. (2012). Full-length HIV-1 immunogens induce greater magnitude and comparable breadth of T lymphocyte responses to conserved HIV-1 regions compared with conserved-region-only HIV-1 immunogens in rhesus monkeys. J. Virol. 86, 11434–11440

[59] Hansen, S.G., Ford, J.C., Lewis, M.S., Ventura, A.B., Hughes, C.M., Coyne-Johnson, L., et al. (2011). Profound early control of highly pathogenic SIV by an effector memory T-cell vaccine. Nature, 473, 523–527

[60] Hansen, S.G., Piatak, J. M., Ventura, A.B., Hughes, C.M., Gilbride, R.M., Ford, J.C., et al. (2013). Immune clearance of highly pathogenic SIV infection. Nature, 50, 100–104

[61] Bolhassani, A., Yazdi, S.R. (2009). DNA immunization as an efficient strategy for vaccination. J. Med. Biotechnol., 1, 71–88

[62] Habibzadeh, N., Bolhassani, A., Vahabpour, R., Sadat, S.M. (2015). How can Improve DNA Vaccine Modalities as a Therapeutic Approach against HIV Infections? J. AIDS Clin. Res., 6(4), 1–8

[63] Daemi, A., Bolhassani, A., Rafati, S., Zahedifard, F., Hosseinzadeh, S., et al. (2012). Different domains of glycoprotein 96 influence HPV16 E7 DNA vaccine potency via electroporation mediated delivery in tumor mice model. Immunol. Lett., 148, 117–125

[64] Bolhassani, A., Rafati, S. (2013). Mini-chaperones: potential immuno-stimulators in vaccine design. Hum. Vaccin. Immunother., 9, 153–161

[65] Vasan, S. (2014). Electroporation-mediated administration of candidate DNA vaccines against HIV-1. Methods Mol. Biol., 1121, 291–307

[66] Nilsson, C., Hejdeman, B., Godoy-Ramirez, K., Tecleab, T., Scarlatti, G., Brave, A., et al. (2015). HIV-DNA given with or without intradermal electroporation is safe and highly immunogenic in healthy swedish HIV-1 DNA/MVA vaccines: A phase I randomized trial. PLoS One,10(6), e0131748

[67] Xu, F., Hong, M., Ulmer, J.B. (2003). Immunogenicity of an HIV-1 gag DNA vaccine carried by attenuated Shigella. Vaccine, 21, 644–648

[68] Wolchok, J.D., Houghton, A.N. (2002). DNA-based cancer vaccines. Encyclopedia of Cancer, 2, 73–78

[69] Wen, J., Yang, Y., Zhao, G., Tong, S., Yu, H., et al. (2012). *Salmonella typhi* Ty21a bacterial ghost vector augments HIV-1 gp140 DNA vaccine-induced peripheral and mucosal antibody responses via TLR4 pathway. Vaccine, 30, 5733–5739

[70] Shinoda, K., Xin, K.Q., Kojima, Y., Saha, S., Okuda, K., et al. (2006). Robust HIV-specific immune responses were induced by DNA vaccine prime followed by attenuated recombinant vaccinia virus (LC16m8 strain) boost. Clin. Immunol., 119, 32–37

[71] Kramski, M., Stratov, I., Kent, S.J. (2015). The role of HIV-specific antibody-dependent cellular cytotoxicity in HIV prevention and the influence of the HIV-1 *Vpu* protein. AIDS, 29(2), 137–144

[72] Cafaro, A., Tripiciano, A., Sgadari, C., Bellino, S., Picconi, O., Longo, O., et al. (2015). Development of a novel AIDS vaccine: the HIV-1 transactivator of transcription protein vaccine. Expert Opin. Biol. Ther., 1–17

[73] Sun, J., Hou, J., Li, D., Liu, Y., Hu, N., et al. (2013). Enhancement of HIV-1 DNA vaccine immunogenicity by BCG-PSN, a novel adjuvant. Vaccine, 31, 472–479

[74] Xu, R., Megati, S., Roopchand, V., Luckay, A., Masood, A., et al. (2008). Comparative ability of various plasmid-based cytokines and chemokines to adjuvant the activity of HIV plasmid DNA vaccines. Vaccine, 26, 4819–4829.

[75] Liu, L.J., Watabe, S., Yang, J., Hamajima, K., Ishii, N., et al. (2001). Topical application of HIV DNA vaccine with cytokine-expression plasmids induces strong antigen-specific immune responses. Vaccine, 20, 42–48

[76] Qiu, J.T., Chang, T.C., Lin, C.T., Chen, Y.M., Li, F.Q., et al. (2007). Novel codon-optimized GM-CSF gene as an adjuvant to enhance the immunity of a DNA vaccine against HIV-1 Gag. Vaccine, 25, 253–263

[77] Xin, K.Q., Hamajima, K., Sasaki, S., Tsuji, T., Watabe, S., et al. (1999). IL-15 expression plasmid enhances cell-mediated immunity induced by an HIV-1 DNA vaccine. Vaccine, 17, 858–866

[78] Jiang, W., Jin, N., Cui, S., Li, Z., Zhang, L., et al. (2006). Enhancing immune responses against HIV-1 DNA vaccine by coinoculating IL-6 expression vector. J. Virol. Methods, 136, 1–7

[79] Nimal, S., Heath, A.W., Thomas, M.S. (2006). Enhancement of immune responses to an HIV gp120 DNA vaccine by fusion to TNF alpha cDNA. Vaccine, 24, 3298–3308

[80] Huang, Y., Chen, A., Li, X., Chen, Z., Zhang, W., et al. (2008). Enhancement of HIV DNA vaccine immunogenicity by the NKT cell ligand, alpha-galactosylceramide. Vaccine, 26, 1807–1816

[81] Kanagavelu, S.K., Snarsky, V., Termini, J.M., Gupta, S., Barzee, S., et al. (2012). Soluble multi-trimeric TNF superfamily ligand adjuvants enhance immune responses to a HIV-1 Gag DNA vaccine. Vaccine, 30, 691–702

[82] Scott, J.K., Gulzar, N., Klaric, K., Ross, T.M., Lu, S., et al. (2012). DNA vaccines that express the MPER of HIV-1gp41 elicit different antibodies depending upon their transmembrane and cytoplasmic domains. Retrovirology, 9, P348

[83] Shimada, M., Yoshizaki, S., Jounai, N., Kondo, A., Ichino, M., et al. (2010). DNA vaccine expressing HIV-1 gp120/immunoglobulin fusion protein enhances cellular immunity. Vaccine 28, 4920–4927

[84] Boyer, J.D., Kim, J., Ugen, K., Cohen, A.D., Ahn, L., et al. (1999). HIV-1 DNA vaccines and chemokines. Vaccine, 17, S53–S64

[85] Liu, Z., Singh, D.K., Sheffer, D., Smith, M.S., Dhillon, S., et al. (2006). Immunoprophylaxis against AIDS in macaques with a lentiviral DNA vaccine. Virology, 351, 444–454

[86] Ivanoff, L.A., Dubay, J.W., Morris, J.F., Roberts, S.J., Gutshall, L., et al. (1992). V3 loop region of the HIV-1 gp120 envelope protein is essential for virus infectivity. Virology, 187, 423–432

[87] Li, L., Saade, F., Petrovsky, N. (2012). The future of human DNA vaccines. J. Biotechnol., 162, 171–182

[88] Shinoda, K., Xin, K.Q., Kojima, Y., Saha, S., Okuda, K., et al. (2006). Robust HIV-specific immune responses were induced by DNA vaccine prime followed by attenuated recombinant vaccinia virus (LC16m8 strain) boost. Clin. Immunol., 119, 32–37

[89] Pissani, F., Malherbe, D.C., Schuman, J.T., Robins, H., Park, B.S., et al. (2014). Improvement of antibody responses by HIV envelope DNA and protein co-immunization. Vaccine, 32, 507–513

[90] Mahdavi, M., Ebtekar, M., Khorram Khorshid, H.R., Azadmanesh, K., Hartoonian, C., Hassan, Z.M., (2011). ELISPOT analysis of a new CTL based DNA vaccine for

HIV-1 using GM-CSF in DNA prime/peptide boost strategy: GM-CSF induced long-lived memory responses. Immunol. Lett., 140, 14–20

[91] Shete, A., Thakar, M., Mehendale, S., Paranjape, R. (2014). Is prime boost strategy a promising approach in HIV vaccine development? J. AIDS Clin. Res., 5, 293

[92] Moody, M.A. (2014). Modulation of HIV-1 immunity by adjuvants. Curr. Opin. HIV AIDS, 9, 242–249

[93] Chen, Y.L., Chen, Y.S., Hung, Y.C., Liu, P.J., Tasi, H.Y., Ni, W.F., et al. (2015). Improvement of the Th1-related immune response in BALB/c mice immunized with an HIV-1 gag plasmid combined with a chimeric plasmid encoding IL-18 and flagellin. Microbiol. Immunol., doi: 10.1111/1348-0421.12274

[94] Carlson, J.M., Le, A.Q., Shahid, A., Brumme, Z.L. (2015). HIV-1 adaptation to HLA: a window into virus–host immune interactions. Trends in Microbiology, 23 (4), 212–224

[95] Crispin, M., Doores, K.J. (2015). Targeting host-derived glycans on enveloped viruses for antibody-based vaccine design. Curr. Opin. Virol., 2015, 11, 63–69

[96] Henning, T.R., McNicholl, J.M., Vishwanathan, S.A., Kersh, E.N. (2015). Macaque models of enhanced susceptibility to HIV. Virol. J., 12 (90), 1–9

[97] Hessell, A.J., Haigwood, N.L. (2015). Animal models in HIV-1 protection and therapy. Curr. Opin. HIV AIDS, 10(3), 170–176

[98] Ahlers, J.D. (2014). All eyes on the next generation of HIV vaccines: strategies for inducing a broadly neutralizing antibody response. Discov. Med., 17(94), 187–199

[99] Kagina, B.M., Wiysonge, C.S., Lesosky, M., Madhi, S.A., Hussey, G.D. (2014). Safety of licensed vaccines in HIV-infected persons: a systematic review protocol. Syst. Rev., 11(3), 101

[100] Vella, M., Pace, D. (2015). Glycoconjugate vaccines: an update. Expert Opin. Biol. Ther., 15(4), 529–546

[101] Liu, H., Bi, W., Wang, Q., Lu, L., Jiang, S. (2015). Receptor binding domain based HIV vaccines. Hindawi Publishing Corporation, BioMed Research International, 1–9

Section 4

Retroviruses as Vectors in Gene Therapy

5

Retroviral Vectors in Gene Therapy

Miroslava Matuskova and Erika Durinikova

Abstract

Several decades ago, the first retroviral vectors were constructed. They have been proved as delivery vehicles in basic and translational research; many of them were used in clinical trials in the treatment of genetic and immunologic disorders or malignancies to deliver therapeutic genes into target tissue. Gammaretroviral and lentiviral vectors are popular viral delivery vehicles; their ability to integrate into genome of the host cell enables permanent genetic modification of the target cell and long-term expression of the transgene. Besides classical cancer gene therapy, they are used in cell-mediated cancer gene therapy in combination with mesenchymal stromal cells (MSC) or neural progenitors. Based on the promising preclinical studies, clinical trials with genetically engineered cell vehicles were initiated.

Keywords: Retroviral vector, lentiviral vector, gene therapy

1. Introduction

Besides negative and pathogenic attributes, viruses can also be beneficial when used as delivery vehicles in gene therapy. The advocates of viral vectors even claim that just viruses are the right tools for delivery of foreign genetic information into the cell because they have been evolving for this purpose for millions of years. Gene therapy can be defined as the delivery of nucleic acid into the cell for the purpose of acquiring new features or restoration of physiologic status. The idea that disorders can be treated by genes arose in the 1960s, when the mechanism of cell transformation by SV40 virus and papovaviruses was described [1]. Gene therapy enables modification of cell by the replacement of non-functional or missing gene, suppression of another gene, or induction of cell death as in the case of oncologic diseases. Monogenic diseases and age-related disorders can be treated by retrovirus-mediated gene therapy, but (retro)viral vectors are most frequently used in cancer gene therapy.

1.1. Basic terminology

- Viral vector – a synthetic construct containing given viral sequences determined to transfer genetic information into the target cell.
- Transfection – transfer of genetic information by a non-viral system.
- Transduction – transfer of genetic information by a viral vector (genetic information is packaged in the viral particle).
    - Transient – DNA is not integrated into the genome of host cell, genetic modification is temporary.
    - Stable – transgene is an integral part of the genome of the host cell, and it is transferred into daughter cells by cell division.
- Provirus – viral genome integrated into chromosome of the host cell.
- Episome – foreign genetic information in the cytoplasm or nucleus, which replicates independently from the genome of the host cell.
- Replication-competent vector – the genetic information of the virus is complete, and the realisation of the whole life cycle of the virus is facilitated in the target cells. Viral progeny infects surrounding cells.
- Replication-defective vector – some viral sequences are removed from the genome; helper cell lines are necessary for the production of virion. Replication-defective vectors are not able to replicate in target transduced cells.

2. History of gene therapy and retroviral vectors

The transfer of genetic information among bacteria by a bacteriophage was first described by Joshua Lederberg and Norton Zinder. They named this phenomenon as 'transduction' [2].

The work of Howard M. Temin performed on Rous sarcoma virus (RSV) is the fundamental part in the research of retroviruses and retroviral vectors. He discovered that specific genetic mutations could be inherited as a result of viral infection. Moreover, his study showed that RSV infection required cellular DNA for replication, and genetic information can also flow in the direction RNA → DNA, and he postulated the provirus hypothesis [3-6].

In the 1970s, specific viral genes involved in the transformation were discovered. The SRC and other (proto-)oncogenes with cellular origin were described. The insertional mutagenesis was revealed as another mechanism of transformation [7].

While pioneer work was performed on avian-infectious alpharetroviral Rous sarcoma virus, Moloney murine leukaemia virus (MoMLV) belonging to gammaretroviruses was initially used for the preparation of therapeutic vector [8, 9], and until now, MoMLV-derived constructs along with human immunodeficiency virus (HIV)-derived vectors are most frequently used. The construction of mutant Moloney murine leukaemia virus defective in the packaging of genomic RNA into virions represents an important step towards the development of retroviral vectors [7]. The first gene delivery systems based on HIV-1 were prepared in the early 1990s [10].

The first officially approved clinical study was conducted by Rosenberg. In this initial study, the gene for neomycin phosphotransferase was introduced into the tumour-infiltrating lymphocytes (TIL) of patients with advanced cancer. Subsequently, he performed clinical trials in which the gene for tumour necrosis factor (TNF) was inserted by retroviral vector into TIL in an effort to increase their therapeutic effectiveness [11]. During the following years, the gene therapy became a very promising approach for the treatment of genetic and oncologic diseases. But serious complication halted the progress of this therapeutic approach. In 1999, Jesse Gelsinger, suffering from a partial deficiency of ornithine transcarbamylase, took part in a gene therapy clinical trial at the University of Pennsylvania. He died due to excessive immune response to a high dose of adenoviral vector [12]. There was a study conducted in Paris, in which 20 children suffering from severe combined immunodeficiency (SCID) took part. They were treated by *ex vivo*–transduced autologous $CD34^+$ haematopoietic progenitor cells. Five of these patients developed T-cell leukaemia [13], and the disease was fatal for one out of these patients [14].

Since then, many steps were taken to improve the safety and efficacy of the gene therapy, and until now, many clinical studies have been conducted.

3. Comparison between retroviral and other viral and non-viral systems

There are two systems for the delivery of transgene into the cell – viral and non-viral. The non-viral approaches are represented by polymer nanoparticles, lipids, calcium phosphate, electroporation/nucleofection or biolistic delivery of DNA-coated microparticles. The safety is mentioned as the major advantage of non-viral approaches. In general, non-viral delivery of transgene is less effective in comparison to viral systems.

There are several categories of viral vectors. We distinguish two main types of vectors depending on whether the DNA is integrated into chromatin of the host cell or not. Retroviral vectors derived from gammaretroviruses or lentiviruses persist in the nucleus as integrated provirus and reproduce with cell division. Other types of vectors (e.g. those derived from herpesviruses or adenoviruses) remain in cell in the episomal form.

The overview of viral vectors is depicted in Figure 1. Adenoviral and retro/lentiviral vectors are most frequently used in research and gene therapy clinical trials.

As stated above, adenoviral vectors are very popular. They have been used for several decades. Since adenoviruses are non-enveloped dsDNA viruses, they are relatively resistant to chemical and physical agents, which enable them to persist out of host cells and make the work in laboratory easier in comparison to enveloped RNA viruses. They are often used in cancer gene therapy as replication-defective or replication-competent vectors. They infect proliferating as

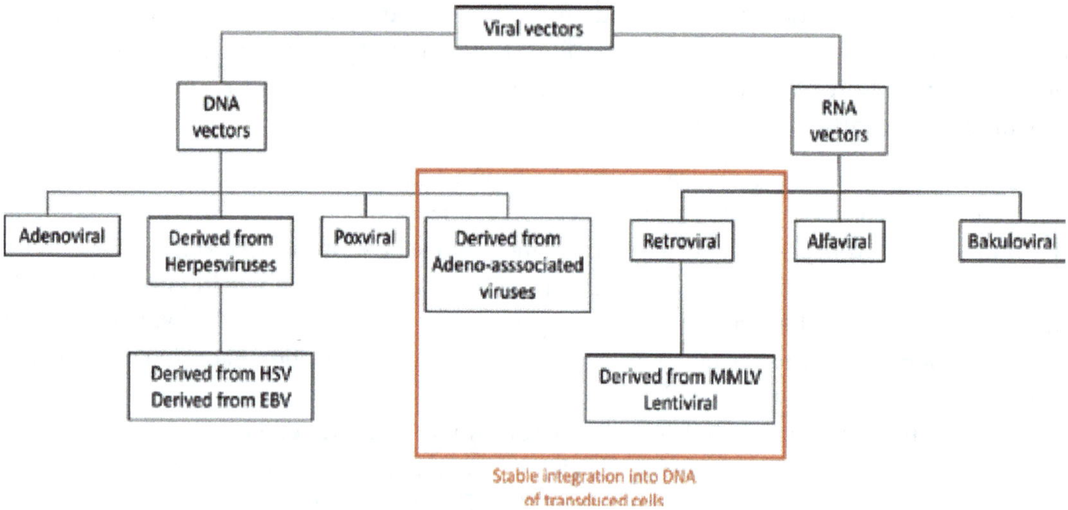

Figure 1. Overview of viral vectors

well as non-dividing cells. In general, adenoviral vectors are considered safe. Since they do not integrate into host DNA, the transduction is transient. The drawback is their immunogenicity [15-17].

Adeno-associated vectors share with retroviral vectors the ability to integrate into host DNA. Wild type adeno-associated virus integrates into specific site of the chromosome 19 (19q13.3-qter). Recombinant vectors lack this characteristic and the risk of insertional mutagenesis exists. These vectors transduce dividing and non-dividing cells, and the transgene expression is long term. Transduced cells are minimally immunogenic [18].

Herpetic viruses are relatively complex enveloped dsDNA viruses. The vectors have been prepared from Herpes simplex type 1 virus, Epstein–Barr virus or cytomegalovirus. They are less immunogenic in comparison to adenoviruses. The transduction is transient; the drawback of HSV 1-derived vector is the short-term expression of the transgene. Herpes virus-derived vectors are preferentially used in vaccination [19, 20].

Poxviruses are the most complex viruses. Their major advantage lies in the huge cloning capacity. Up to 25 kbp of foreign DNA can be cloned into vaccinia-derived vectors. Similarly to herpes virus-derived vectors, they are popular in the preparation of vaccines including cancer immunotherapy [21, 22].

Baculoviruses, the viruses specific for invertebrates, are not competitors of retroviruses in gene therapy. They have been used for more than 30 years for transduction of insect cells for expression of recombinant proteins. Pseudotyping enables the transduction of mammalian cells [23].

Vectors derived from alfaviruses (ssRNA viruses) are also used in cancer gene therapy and immunotherapy [24, 25].

4. Retroviral vectors

Retroviruses are relatively complex enveloped RNA viruses with diploid ssRNA genome. Typical feature of retroviruses and retroviral vectors is their ability to integrate into host DNA. Viral RNA is reversibly transcribed and integrated in the form of provirus. They very effectively cooperate with enzymes of the host cell, and they use it for their own replication and long-term expression of viral proteins. The entry of virus into the host cell is receptor-dependent [26].

Many types of retroviruses (bovine leukaemia virus, Rous sarcoma virus, lentiviruses and spumaviruses) were used for preparation of vectors. The most popular vectors are constructs based on MoMLV and HIV.

4.1. Gammaretroviral vectors

The first MoMLV-based vectors were prepared more than 30 years ago [8, 9], and they are still very popular. The construct is relatively small, and it is possible to achieve high titres in inoculum. The diagram of MoMLV provirus and MoMLV-derived vector is depicted in Figure 2.

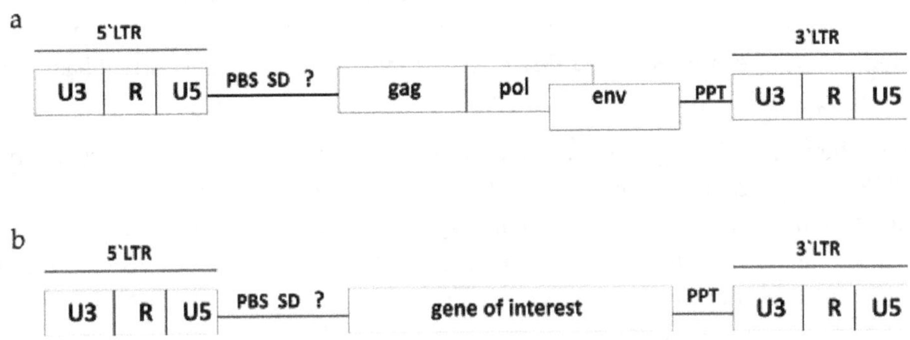

Figure 2. Genome structure of integrated MoMLV and MoMLV-derived vector. (a) Diagram of MoMLV provirus; (b) Diagram of integrated MoMLV-derived retroviral vector; LTR = long terminal repeats; U3 = unique sequence derived from 3′ end of the viral RNA; R = repeated sequence; U5 = unique sequence derived from 5′ end of the viral RNA; PBS = primer binding site; SD = splice donor; Ψ = packaging signal; gag = genes for structural proteins; pol = region coding for genes needed for replication of retrovirus; env = genes for envelope proteins; PPT = polypurine tract (according to [27, 28], adjusted).

Contrary to lentiviruses, the packaging system of gammaretroviruses does not require incorporation of sequences overlapping coding region gag, pol or additional genes.

Since MoMLV lacks elements necessary for active transport of genetic information through nuclear membrane, integration of viral DNA is possible only during the mitosis. MoMLV-derived vectors transduce only dividing cells. Integration of viral genome is mediated by the pre-integration complex (PIC) consisting of integrase, capsid, p12, proviral DNA and host cell proteins. Proviral DNA is surrounded by two long terminal repeats (LTR), which are composed of U3, R and U5 regions. Transcription of proviral DNA starts from enhancer/promotor in 5′ U3 region [27].

MoMLV constructs can be prepared as replication-competent or replication-deficient.

Gag and env genes are removed in replication-deficient vectors. Gene of interest is cloned in the free space. A typical replication-deficient vector contains packaging signal (ψ), primer binding site (PBS) and LTRs. Viral genes gag, pol and env are cloned on separate expression helper plasmids, and helper cell lines are co-transfected with more plasmids. Stably transfected helper cell lines expressing gag pol (GP) and env were prepared.

Packaging cell lines GP+E-86 and GP-envAM-12 were derived from NIH-3T3 cell line by electroporation of two plasmids. One plasmid contained the gag and pol regions of MoMLV, and the other contained the env region. GP+E-86 cells are used for the production of ecotropic viral particles, and line GP+envAM-12 is used for amphotropic viral particles (explained later) [29]. Expression cassette with gene of interest should not contain introns, internal polyadenylation signals and large secondary structures, which could interfere with reverse transcriptase. The cloning capacity of retroviral vectors is up to 10 kbps, but the size of transgene significantly influences its expression and viral titre [27].

Envelope protein coded by gene env is responsible for the tropism of MoMLV. Three natural variants of MoMLV were described. They differ in their envelope proteins. Ecotropic MoMLV infects only mouse and rats cells; amphotropic is also able to infect the cells derived from other mammals including human cells. Xenotropic MoMLV was supposed to infect many types of cells excluding murine, but recent studies indicate that it also infects mice and it has the widest tropism among MoMLV [27, 30].

Tropism of the viral vector can be modulated by pseudotyping using a particular envelope protein. The glycoprotein of vesicular stomatitis virus (VSVg) is very popular. It enables broad host range (mammalian and non-mammalian cells). It also has stabilisation properties, and viral particles can be purified by ultracentrifugation. On the other hand, VSVg is recognised by complement, and this fact can decrease the transduction efficacy *in vivo* [31].

4.2. Lentiviral vectors

In comparison to MoMLV-derived vectors, lentiviral vectors are more complex. Three generations of lentiviral vectors were prepared in order to increase the safety and efficacy of the gene transfer. The viral genome was split into packaging and transfer vectors. The first and second generation are composed of three plasmids; the third generation consists of four plasmids [31]. The main difference between lentiviral vectors and vectors derived from other retroviruses is their ability to infect/transduce quiescent non-dividing cells [32]. They are able to pass through nucleopores into the intact nucleus. The mechanism of this phenomenon has not been completely clarified; yet, it is known that both viral and cellular proteins participate in this process. In addition to HIV vectors, vectors based on feline immunodeficiency virus (FIV), simian immunodeficiency virus (SIV) or equine infectious anaemia virus (EIAV) have also been prepared. Pseudotyping (VSVg is the most common) is typical for lentiviral vectors [31].

4.3. Self-inactivating vectors

The risk of insertional mutagenesis is a drawback of retroviral vectors. With the purpose of increasing the safety of gene therapy, self-inactivating (SIN) vectors were prepared in the 1980s. SIN vectors have a deletion in the 3′ U3 region, where promotor and enhancer sequences occur. This deletion is copied into 5′ LTR during the reverse transcription, and virus becomes free of LTR-bound promotor activity. Transcription control is under the chosen cloned promotor [33].

SIN vectors are characterised by decreased risk of insertional mutagenesis; the vector is not activated via infection by another retrovirus, and the internal promotor is autonomous. Controlled/inducible or cell-specific expression of transgene can be achieved based on the chosen promotor [33]. The use of tetracycline (Tet)-inducible system was published in the 1990s [34], and until now it is being used for enhanced expression of an exogenous gene in a cell-type-specific manner.

4.4. Replication-competent retroviral vectors

Retroviral vectors can be constructed as replication-defective to transduce target cells and enable long-term expression of transgene (immunology disorders, genetic diseases), or they carry transgenes inducing cell death (cancer gene therapy). On the other hand, the replication-competent vectors (RCV) are prepared in order to replicate in the target (tumour) cells. Their progeny infects surrounding malignant cells. Since the targeting is an inevitable characteristic of RCV, they can be engineered to express ligands to tumour cell-specific markers. The advantage of MoMLV-derived vectors is their natural preference to tumour cells [35]. MoMLV is unable to infect quiescent cells, making them suitable vehicles for the treatment of brain tumours. RCV with suicide gene mediate synchronised cell killing after prodrug administration, and due to their stable integration into DNA of infected cells, residual cancer cells serve as a reservoir for long-term viral persistence even when they migrate to new sites. Multiple cycles of prodrug administration to achieve improved treatment efficacy are possible [36].

Although gammaretroviral and lentiviral vectors are derived from the same viral family, they differ in some characteristics. Advantages of gammaretroviral vectors reside in the complete absence of viral gene remnants in the transfer vector, efficient pseudotyping and the lack of mobilisation by human-infectious viruses. In comparison to HIV-derived vectors, there are also minor concerns related to potential seroconversion *in vivo*. Lentiviral vectors are clearly superior for the ability to transduce non-dividing cells. Both types of vectors are equally potent in terms of expression properties when containing similar internal expression cassettes [37]. Compared to vectors derived from non-integrating viruses, retroviral vectors possess the risk of insertional mutagenesis. Gammaretroviral vectors preferentially integrate close to transcription start sites and CpG islands, which are enriched in gene-regulatory elements. Lentiviral vectors prefer integration inside of the transcription units of actively transcribed genes [38].

Despite the promise for success in the clinic, one major drawback of the retrovirus-based vector is that any unintended insertion events from the therapy can potentially lead to deleterious effects or it can cause an abnormal expression of nearby host genes driven by the enhancer of the inserted viral DNA in patients, as demonstrated by the development of malignancies in both animal and human studies. The better prediction of the integration sites by elucidation of this mechanism might lead to the development of retroviral vectors capable of selective integration. This understanding could provide the ultimate solution to the problems of insertional mutagenesis [39]. The definition of the precise mechanism of the retroviral pre-integration complex is required. Many efforts have been made in designing modified integrases with sequence-specific integration capability, which can be accomplished by rational modification of the protein or by using the directed evolution approach [40]. One approach to directing integration into predetermined DNA sites is fusing integrase to a sequence-specific DNA-binding protein, which results in a bias of integration near the recognition site of the fusion partner [41]. Efforts to engineer integrase to recognise specific target DNA sequences within the host genome may lead to development of effective retroviral vectors that can safely deliver gene-based therapeutics in a clinical setting. Insertion of a lentiviral vector via virion-associated Cre protein, capable of directing site-specific insertion of a gene in the vector, into a defined loxP site in the host genome was described [42].

A detailed study of the vector integration sites performed on haematopoietic stem cells by Aiuti *et al.* [43] concluded that lentiviral gene therapy was safer than retroviral gene therapy, and lentiviral gene therapy did not induce selection of integrations near oncogenes, and no aberrant clonal expansion was observed after 20 to 32 months follow-up. Also, the so-called integration-deficient lentiviral vectors (IDLVs) can be produced through the use of integrase mutations that specifically prevent proviral integration. These lentiviral episomes lack replication signals and are gradually lost by dilution in dividing cells, but are stable in quiescent cells. Compared to integrating lentivectors, IDLVs have a greatly reduced risk of causing insertional mutagenesis [44].

4.5. Overview of preparation of the gammaretroviral vector

Recently, many viral cloning systems are available. Transgene is cloned either directly or via a bacterial intermediate. Circular dsDNA is found at the beginning of the process. It contains viral sequences including LTR necessary for the integration, genes coding for antibiotic resistance and target sequences for different restriction endonucleases (multicloning sites). The first part of vector construction takes place in bacteria; therefore, the bacterial origin of replication (ORI) is a necessary component. The gene of interest is cloned into the vector, then propagated in bacteria and verified by sequencing. Subsequently, the packaging cell line is transfected, and the sequences between LTRs are integrated into host-cell DNA. Vector-containing cells are selected via antibiotic resistance. In order to increase the titre of the viral vector, ecotropic and amphotropic packaging cell lines can be used. First, the ecotropic cell line is transfected, and then the amphotropic packaging cells are transduced by virus-containing cultivation supernatant from ecotropic cells. The 'ping-pong' method – mutual exchange of virus-containing medium between ecotropic and amphotropic cells – is performed to further increase the viral titre. The viral titre is determined and cultivation supernatant from transduced cells is collected for transduction of target cells. Transgene-containing cells are selected via antibiotic resistance (Figure 3).

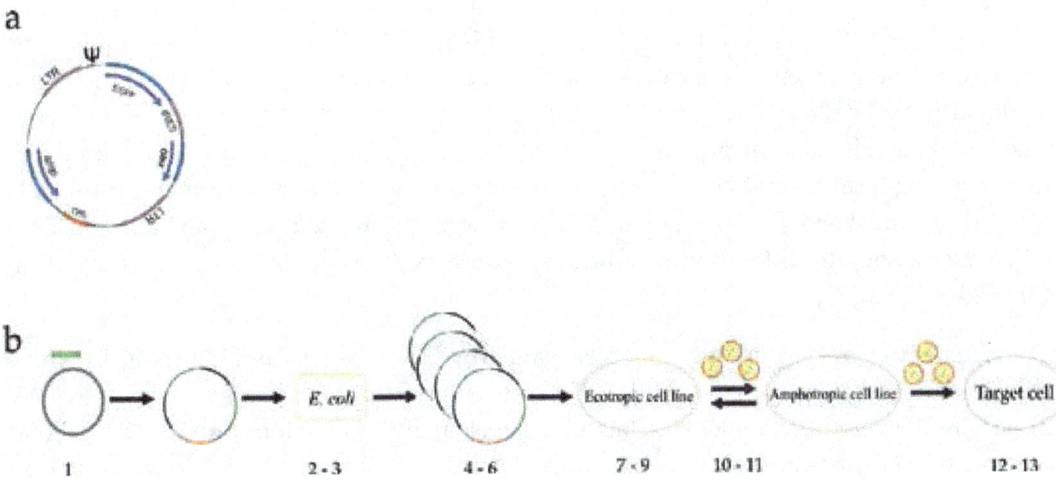

Figure 3. The preparation of replication-defective retrovirus vector. (a) Circular dsDNA construct, which is transfected to bacteria; (b) particular steps of vector preparation: 1. cloning of desired gene into the vector; 2. transformation of bacteria; 3. selection of bacterial clones with desired transgene (NeoR); 4. verification of cloned gene by sequencing; 5. multiplication of desirable clone; 6. purification of plasmid DNA; 7. transfection via packaging cell line; 8. selection of transfected cells (NeoR); 9. virus titration; 10. ping-pong; 11. virus titration after ping-pong; 12. transduction of target cells; 13. selection of transduced target cells [NeoR/G418R – resistance of bacteria to neomycin and mammalian cells' resistance to geneticin (G418) coding by *NeoR* gene].

5. Gene therapy

Treatment of genetic diseases and cancer gene therapy are the main targets of recent gene therapy. They belong to serious diseases, which are difficult to treat or are incurable using conventional treatment, or the treatment is accompanied by severe adverse effects.

Over 60% of ongoing gene therapy clinical trials represent cancer treatment followed by monogenetic and cardiovascular diseases [11]. Two approaches of gene therapy are defined: (i) The *ex vivo* method is characterised by the collection of target cells from the organism, genetic modification and subsequent administration to the patient. (ii) The *in vivo* method is characterised by the direct administration of therapeutic gene to the patient.

5.1. Treatment of genetic, immunologic and other non-oncologic diseases

Genetic as well as age-related diseases can be treated by gene therapy. They are caused by deficiency or aberrant expression of one or more gene(s). Patients suffering from severe combined immunodeficiency (SCID) – devastating disorder of adapted immunity – are not able to defend against infections. The term SCID covers several genetic defects. Adenosine deaminase (ADA) deficiency was the first SCID condition for which a genetic and molecular cause was identified [45]. The patients suffering from ADA-SCID are ideal candidates for gene therapy, when haematopoietic cells are transduced by a gene encoding adenosine deaminase. A retroviral vector carrying ADA was the first construct to contain a therapeutic gene (Rosenberg transduced cells with neomycin resistance gene to track them *in vivo*) used in an FDA-approved clinical trial. Two children were treated [46]. SCID-X1 is characterised by various mutations in the gene encoding interleukin 2 receptor-γ (IL2RG) [47, 48]. Transduction of functional IL2RG restores expression of functional interleukin 2 receptor-γ. It was demonstrated that gene therapy for primary immunodeficiencies is an effective treatment modality providing long-term clinical benefit for patients. Lentiviral vectors contributed significantly to this achievement [49]. Haematopoietic stem cells were engineered *ex vivo* and administered to the patient. Long-term T-cell reconstitution was achieved in patients suffering from ADA-SCID and SCID-X [50].

Age-related macular degeneration is accompanied with excessive vascularisation in which vascular endothelial growth factor (VEGF) takes place. Its function can be inhibited by retrovirally delivered antiangiogenic factors such as angiostatin, endostatin or extracellular domain of VEGF receptor. Vectors are administered into vitreous body or under retina. Antiangiogenic genes are also used in cancer gene therapy [51-53].

Epidermolysis bullosa is a group of devastating skin disorders. Mutations in the COL7A1 gene result in the absence or dysfunction of type VII collagen protein and cause recessive dystrophic epidermolysis bullosa (RDEB). Collagen VII expression has been restored by retroviral and lentiviral vectors carrying COL7A1 gene. Long-term expression of transgene in keratinocytes, fibroblasts or epidermal stem cells was achieved [54]. Mutations in genes encoding the basement membrane component laminin 5 (LAM5) are the cause of junctional epidermolysis bullosa. Retrovirally transduced epidermal stem cells have been used for preparation of epidermal grafts. Analysis revealed that synthesis and proper assembly of normal levels of functional LAM5 were observed, together with the development of a firmly adherent epidermis that remained stable at least for 1 year [55].

The treatment of congenital disorders of liver metabolism is currently limited, and prognosis of patients suffering from Crigler–Najjar syndrome, urea cycle disorders, familial hypercholesterolemia and primary hyperoxaluria type 1 is unfavourable. In comparison to the liver transplantation, *ex vivo* gene therapy offers a less invasive method without the need for lifelong immunosuppression [56].

Homozygous mutation in LDL receptor causes familial hypercholesterolemia. The gene therapy approach for treatment of this lethal disease was developed. Grossman *et al.* [57] described the treatment of a 29-year-old woman by *ex vivo* gene therapy. Autologous hepatocytes isolated from the patient were genetically corrected with recombinant retroviruses carrying the LDL receptor and subsequently reimplanted. The patient's LDL/HDL ratio improvements have remained stable for the duration of the treatment (18 months). *Ex vivo* gene therapy using modified haematopoietic stem cells has generated encouraging results for treatment of multisystemic lysosomal storage disorders [58]. Retroviral and lentiviral vector-transduced bone marrow-derived cells overexpressing lysosomal enzymes can migrate into the central nervous system (CNS) and mediate cross-correction of the neighbouring brain cells. This approach has resulted in excellent outcomes, preventing the development of clinical manifestations in metachromatic leukodystrophy. According to a study by Cartier *et al.* [59], patients with peroxisomal disorder X-linked adrenoleukodystrophy treated with *ex vivo* lentiviral vector-mediated gene therapy have also exhibited clinical benefits.

Based on the significant progress made to date, in spite of the expected setbacks of all drug development efforts, gene therapy for liver metabolic disorders is becoming a viable option for treatment in future clinical trials.

5.2. Cancer gene therapy

Cancer is a complex disease accompanied by progressive accumulation of genetic and epigenetic alterations, which enables the cell to escape from cell and environmental control. Conventional treatment consists of surgery, chemotherapy and radiotherapy. Despite progress in the treatment, it is ineffective or accompanied by severe adverse effects in many cases; therefore, novel approaches are needed. Recently, research has been focusing on targeted therapy. Gene therapy – classical or mediated by cellular vehicles – is the promising approach, primarily for patients suffering from glioblastoma multiforme (GBM), neuroblastoma, metastatic melanoma and other metastatic cancers. As stated above, cancer treatment is recently the most elaborated area of gene therapy.

The treatment is focused on suppression of activated oncogenes, restoration of expression of tumour-suppressor genes, activation of anti-tumour immunity and inhibition of angiogenesis- or metastatic potential-suppressing genes. The separate group is represented by genes inducing autodestruction of tumour – 'toxic' genes or genes coding for enzymes converting non-toxic prodrug to toxic product [60].

Prodrug-converting genes are also known as 'suicide genes'. They are of viral, bacterial or yeast origin, and they do not have the equivalent in mammalian cells. Mammalian cells obtain the ability to utilise a new substrate after transduction – an inactive compound is converted to chemotherapeutics. The apoptosis is induced in transduced cells; therefore, this approach is called suicide gene therapy. Bystander effect is the phenomenon of suicide gene therapy. Toxic metabolites released from transduced cells are received by the surrounding bystander cells and mediate toxic effect in them [61]. This targeted chemotherapy is considerate to the organism because chemotherapeutic agent is present only in the tumour vicinity – the site where (systemically administered) prodrug and therapeutic enzyme meet together.

Herpes simplex virus thymidine kinase (HSVtk) is one of the most frequently used therapeutic genes. Its affinity to nucleotide analogue ganciclovir (GCV) is approximately 1000 times higher than mammalian thymidine kinase. HSVtk phosphorylates GCV to GCV-monophosphate (GCV-P), which is subsequently phosphorylated by cellular kinases and incorporated into replicating DNA instead of guanosine triphosphate. GCV lacks deoxyribose at 3'OH and bound between carbons 2' and 3', which are necessary for the elongation of DNA chain. Incorporation of GCV-3P yields into termination of DNA synthesis and subsequent cell death – preferentially apoptosis, advisable cell death in cancer treatment. GCV becomes a charged molecule after phosphorylation, and it is not able to diffuse across the membrane. As stated above, the bystander effect is an important phenomenon in contributing to the efficacy of gene therapy. Gap junctional intercellular communication (GJIC) is necessary for transport of phosphorylated GCV. GJIC is often a limiting factor of HSVtk/GCV system because many cells are defective in expression of the connexins, which are the major components of gap junctions. Transduction of connexin gene can improve the efficacy of the treatment [62].

The second most frequently used system in suicide gene therapy is cytosine deaminase (CD) derived from bacteria (bCD) or yeasts (yCD) combined with prodrug 5-fluorycytosine (5-FC), which is used in conventional antimycotic therapy. Yeast CD is 15 times more efficient in comparison to its bacterial counterpart [63].

Transduced cell is able to convert non-toxic 5-FC to conventionally used chemotherapeutic 5-fluorouracil (5-FU). More active molecules arise by the metabolism of 5-FU, and the synthesis of both DNA and RNA is impaired. Thymidylate synthase (TS) is the key target enzyme in 5-FC/5-FU treatment. 5-fluorouridine monophosphate, one of the active metabolites of 5-FU, binds irreversibly to TS, and starvation for thymine leads to inhibition of DNA synthesis. 5-fluorothymidine triphosphate, another metabolite, impairs RNA by incorporation instead of UTP [64].

The efficacy of the CD/5-FC approach can be increased by the addition of another enzyme, bacterial or yeast-derived uracil phosphoribosyl transferase (UPRT), which supplements low expression of mammalian orotate phosphoribosyl transferase important for the activation of 5-FU. It also continuously utilises 5-FU supporting its synthesis [65]. CD and UPRT can be cloned separately or as a synthetic fusion gene CD::UPRT.

Purine nucleoside phosphatase (PNP) is an *E. coli*-derived enzyme, which activates fludarabine. PNP-transduced cells convert fludarabine into metabolites, which are highly toxic for proliferating as well as for quiescent cells, because they inhibit ATP-dependent reactions. The metabolism of nucleic acid and proteins is impaired.

When the therapeutic gene is inserted into a replication-defective vector, it is expressed only in the transduced cell, and surrounding cells are affected only via the bystander effect. On the other hand, the replication-competent vectors replicate in target cells, and despite the bystander killing, they are able to infect their neighbours and spread the transgene. RCV have been constructed for treatment of aggressive tumours.

Promising results by MoMLV-derived replication-competent vector carrying cytosine deaminase were achieved on orthotopic glioblastoma model. Considerable infection of target cells, virus spread and significant bystander effect were demonstrated [66]. Based on promising preclinical results, a phase I/II clinical study is ongoing using the replication-competent cytosine deaminase-expressing retroviral vector to patients with recurrent or progressive grade III or grade IV gliomas (NCT01985256) [67].

It is important to note that despite of very promising preclinical data, many clinical studies failed because of low efficiency of the transfer of genetic information into target cells, insufficient infiltration of target tissue by the vector or low expression of transgene in tumour. The identification of mesenchymal stromal cells (MSC) and the discovery of their high affinity to tumour tissue facilitate important improvement in the cancer gene therapy.

MSC were originally isolated from bone marrow and characterised as rare non-haematopoietic population with clonogenic capacity and plastic adherence [68]. They have the self-renewal potential and are able to differentiate into specialised progeny [69-72]. Bone marrow is the most popular source of MSC, but adipose tissue is also very suitable because of its accessibility. Moreover, the frequency of MSC in the adipose tissue is 500 times higher in comparison to bone marrow. MSC can also be isolated from umbilical cord, dental pulp and different connective tissues. They serve as a source of regenerative cells in fractures, inflammation and necrosis. The injured tissue produces chemotactic signals which attract MSC [73]. Tumour can be compared to a wound that never heals [74]. Many tumours produce chemotactic signals which attract MSC [75-77]. It was demonstrated that MSC-derived cells are a component of tumour stroma, and they can support proliferation and vascularisation of malignant tissue [78]. The natural affinity of MSC to malignant tissue can be used for targeted therapy. They can be used as delivery vehicles in cancer gene therapy [28, 79]. The approach can be compared to the Trojan horse. MSC are able to pass across the endothelium, enter the blood stream and engraft in the tumour. Therefore, genetically engineered MSC can be administered intravenously, and they reach the target site. This enables to treat the disseminated tumours and metastases. MSC even cross the blood-brain barrier [80, 81]. In this regard, the MSC-mediated cancer gene therapy is superior to the 'classic' cancer gene therapy. On the other hand, it is also necessary to note that the therapy by genetically engineered MSC is limited by the fact that tumours differ in the attractiveness for MSC. Many paracrine factors are involved in MSC – tumour cell signalling. The SDF-1α and CXCR4 (CXCL12) signalling seem to play an important role in homing of stromal cells [76, 77].

Besides MSC, other cellular vehicles can be transduced and used in cancer gene therapy. Neural progenitors or neural stem cells isolated from brain tissue are used as delivery vehicles in targeted therapy of aggressive tumours of central nervous system [82].

The ideal candidates of cellular therapy for clinical use are the cells harvested without difficulty, which can be processed *ex vivo* very efficiently and afterwards transplanted. The unique biological features of MSC predetermine them as valuable gene carriers for therapeutic approaches. MSC can be easily transduced with retroviral and lentiviral vectors, which is a key prerequisite for the introduction and durable expression of marker and/or therapeutic genes within the tumour environment after homing to target tissues [83, 84].

The immunophenotype and the ability to differentiate are not affected by transduction. In order to address the safety of retrovirally transduced MSC, many studies have been performed. Particular transgene can give a proliferative advantage, but it does not preclude the entering of cells to senescence and has no impact on the safety of cancer gene therapy mediated by MSC [85].

The combination of cellular and gene therapy provides a unique opportunity to bypass the obstacles connected with direct viral delivery of the transgene. Cell vehicle protects the vector from the immune surveillance and supports targeting of a therapeutic molecule to the tumour [86].

Cell-mediated gene therapy is based on the bystander effect. The suicide effect is not the main goal; neighbouring bystander cells should be impaired at first. Therefore, it is more appropriate to use the term 'prodrug-converting gene' instead of 'the suicide gene'.

The simple retroviral plasmid pJZ308 derived from Moloney murine leukaemia virus [87] was used for delivery of yeast *cytosine deaminase::uracil phosphoribosyltransferase* (CD::UPRT) and *Herpes simplex* virus-thymidine kinase (HSVtk) prodrug-converting genes into adipose tissue-derived MSC. The retroviral transduction of AT-MSC by CD gene was published for the first time in 2007. In this pilot study, the capability of AT-MSC expressing fusion yeast CD::UPRT gene in combination with prodrug 5-fluorocytosine (5-FC) to eradicate human colon carcinoma cells HT-29 *in vitro*, and their significant role in inhibition of tumour growth in a therapeutic paradigm *in vivo* were demonstrated [88]. A number of published papers reported the cytotoxic efficiency of CD::UPRT-MSC/5-FC enzyme/prodrug therapeutic system, both *in vitro* and *in vivo*, in the treatment of experimental prostate tumour [89, 90], melanoma [91, 92] and medullary thyroid carcinoma [93, 94]. The 3D multicellular culture conditions for better prediction of the therapeutic outcome in mouse xenograft models are suggested to be used according to the study performed on melanoma model [95]. Contrary to 5-FU, 5-FC is able to cross the blood-brain barrier, thus making this enzyme/prodrug approach suitable for the treatment of CNS tumours [96], which was proved on malignant glioma model [97]. The complete regression of glioblastoma simulating clinical therapeutic scenario was demonstrated by Altaner *et al.* [98].

AT-MSC were shown to form gap junctional intercellular communication with glioblastoma cell lines, thus rendering them suitable vehicles for the enzyme/prodrug therapy system HSVtk/GCV relying on transport of polar metabolites [99]. AT-MSC transduced by this suicide gene HSVtk also via lentiviral vector proved strong candidates of gene therapy for U-87-derived model of glioblastoma multiforme [100].

Efficacy of gene-directed enzyme/prodrug therapy can be improved by the combination of individual systems. Matuskova *et al.* [101] demonstrated various levels of synergy depending on tested cell line and experimental set-up. Systemic administration of CD::UPRT-MSC and HSVtk-MSC in combination with both prodrugs, 5-FC and GCV, inhibited growth of experimental lung metastases derived from human breast adenocarcinoma cells.

MSC were also retrovirally transduced to stably express an exogenous gene encoding the therapeutic agent hTNFα whose effect was tested on tumour cell lines of different origins. Co-injection of such therapeutic cells with melanoma cells inhibited the tumour mass growth up to 97.5% *in vivo* [102].

MSC isolated from the Wharton's jelly of the human umbilical cord were lentivirally transduced by gene carrying the soluble human tumour necrosis factor-related apoptosis-inducing ligand (sTRAIL). The specific expression of the transgene in the tumour was ensured by alpha-fetoprotein promoter. Significant therapeutic effect was observed on orthotopic hepatocarcinoma model established on athymic mice, and the treatment was even more efficient in combination with 5-fluorouracil [103].

The effectiveness of therapeutic system using TRAIL expression from bone marrow-derived MSC with significantly increased survival of nude mice was noted as suitable for use in the prevention of the recurrence of hepatocellular carcinoma after radiofrequency ablation [104].

A different therapeutic strategy comprises carboxyl esterase (CE), an enzyme hydrolysing prodrug Irinotecan. Hong *et al*. [105] transduced neural stem cells by this gene, which led to the development of a novel strategy for delivering therapeutic genes to brain tumours. The significant inhibition of the growth of human non-small-cell lung adenocarcinoma cells was achieved for these lung cancer brain metastases also *in vivo*.

PNP-transduced AT-MSC were tested for treatment of ovarian cancer in immunodeficient mice model (unpublished data). Cell vehicles were also retrovirally transduced by interleukin-coding genes (IL-2, IL-4, IL-12, IL-23) and interferon-β in order to treat primary or metastatic brain tumours [96].

As stated above, MSC are the long-term reservoir for tissue regeneration. They are naturally radio- and chemo-resistant [73, 106, 107]. Despite being equipped by enzymes and efflux mechanisms enabling resistance to chemotherapeutics, their resistance is not absolute. The transgene or metabolite activated by prodrug-converting gene also affects MSC, which after a certain time undergo cell death [108]. In the context of tumour-promoting potential of MSC [76], this fact should be considered as the advantage improving the safety of MSC-mediated gene therapy. It was shown that expression of yCD::UPRT transgene sensitises MSC to 5-FC, and its expression as well as the expression of HSVtk lead to suicide effect of therapeutic cells in the presence of GCV [108]. As shown on neural stem cells, if they are co-expressed together, the effect is even stronger [109].

Promising preclinical data enabled the approval of clinical trials mediated by engineered cellular vehicles. Patients suffering from aggressive, by conventional approaches incurable tumours can be included. The protocol for the first clinical study utilising genetically engineered MSC was published in 2015. Patients suffering from advanced, recurrent or metastatic gastrointestinal or hepatopancreatobiliary adenocarcinoma will be treated by autologous retrovirally transduced bone marrow-derived MSC. The gammaretroviral self-inactivating vector carrying HSVtk will be used [110].

To conclude, it is important to note that despite many clinical studies, the gene therapy is in the early stage of clinical use. For now, it presents an experimental approach. Besides clinical efficacy, safety is the crucial criterion of gene therapy. It is undisputed that retroviral vectors are indispensable tools for genetic modification, and they have the potential to significantly contribute to the improvement in targeted treatment of immunologic, oncologic and genetic disorders.

Acknowledgements

Studies mentioned in this chapter were also supported by the Slovak Research and Development Agency under contract nos. APVV-0052-12 and APVV-0230-11, by VEGA grant nos. 2/0171/13, 2/0087/15, d2/0130/13 and by the Slovak Cancer Research Foundation.

Author details

Miroslava Matuskova* and Erika Durinikova

*Address all correspondence to: exonmigu@savba.sk

Cancer Research Institute, Slovak Academy of Sciences, Bratislava, Slovakia

References

[1] Sambrook J, Westphal H, Srinivasan PR, Dulbecco R: The integrated state of viral DNA in SV40-transformed cells. Proceedings of the National Academy of Sciences of the United States of America. 1968;60(4):1288-95.

[2] Zinder ND, Lederberg J: Genetic exchange in Salmonella. Journal of bacteriology. 1952;64(5):679-99.

[3] Temin HM, Rubin H: Characteristics of an assay for Rous sarcoma virus and Rous sarcoma cells in tissue culture. Virology. 1958;6(3):669-88.

[4] Rubin H, Temin HM: A radiological study of cell-virus interaction in the Rous sarcoma. Virology. 1959;7(1):75-91.

[5] Temin HM: Mixed infection with two types of Rous sarcoma virus. Virology. 1961;13(2):158-63.

[6] Temin HM: The Participation of DNA in Rous sarcoma virus production. Virology. 1964;23(4):486-94.

[7] Suerth JD, Labenski V, Schambach A: Alpharetroviral vectors: from a cancer-causing agent to a useful tool for human gene therapy. Viruses. 2014;6(12):4811-38. DOI: 10.3390/v6124811.

[8] Cone RD, Mulligan RC: High-efficiency gene transfer into mammalian cells: generation of helper-free recombinant retrovirus with broad mammalian host range. Proceedings of the National Academy of Sciences of the United States of America. 1984;81(20):6349-53.

[9] Mann R, Mulligan RC, Baltimore D: Construction of a retrovirus packaging mutant and its use to produce helper-free defective retrovirus. Cell. 1983;33(1):153-59.

[10] Page KA, Landau NR, Littman DR: Construction and use of a human immunodeficiency virus vector for analysis of virus infectivity. Journal of virology. 1990;64(11): 5270-76.

[11] Wirth T, Parker N, Yla-Herttuala S: History of gene therapy. Gene. 2013;525(2): 162-69. DOI: 10.1016/j.gene.2013.03.137.

[12] Stolberg SG: The biotech death of Jesse Gelsinger. The New York Times Magazine. 1999:136-40, 49-50.

[13] Fischer A, Hacein-Bey-Abina S, Cavazzana-Calvo M: 20 years of gene therapy for SCID. Nature Immunology. 2010;11(6):457-60. DOI: 10.1038/ni0610-457.

[14] Check E: Gene therapy: a tragic setback. Nature. 2002;420(6912):116-18. DOI: 10.1038/420116a.

[15] Goncalves MAFV, de Vries AAF: Adenovirus: from foe to friend. Reviews in Medical Virology. 2006;16(3):167-86. DOI: 10.1002/Rmv.494.

[16] Ahmed AU, Thaci B, Alexiades NG, Han Y, Qian SO, Liu FF, et al.: Neural stem cell-based cell carriers enhance therapeutic efficacy of an oncolytic adenovirus in an orthotopic mouse model of human glioblastoma. Molecular Therapy. 2011;19(9): 1714-26. DOI: 10.1038/Mt.2011.100.

[17] Bischoff JR, Kirn DH, Williams A, Heise C, Horn S, Muna M, et al.: An adenovirus mutant that replicates selectively in p53-deficient human tumor cells. Science. 1996;274(5286):373-76.

[18] Daya S, Berns KI: Gene therapy using adeno-associated virus vectors. Clinical Microbiology Reviews. 2008;21(4):583-93. DOI: 10.1128/Cmr.00008-08.

[19] Hellebrand E, Mautner J, Reisbach G, Nimmerjahn F, Hallek M, Mocikat R, et al.: Epstein-Barr virus vector-mediated gene transfer into human B cells: potential for anti-tumor vaccination. Gene therapy. 2006;13(2):150-62. DOI: 10.1038/sj.gt.3302602.

[20] Lentz TB, Gray SJ, Samulski RJ: Viral vectors for gene delivery to the central nervous system. Neurobiology of Disease. 2012;48(2):179-88. DOI: 10.1016/j.nbd.2011.09.014.

[21] Kim JW, Gulley JL: Poxviral vectors for cancer immunotherapy. Expert Opinion on Biological Therapy. 2012;12(4):463-78. DOI: 10.1517/14712598.2012.668516.

[22] Sanchez-Sampedro L, Perdiguero B, Mejias-Perez E, Garcia-Arriaza J, Di Pilato M, Esteban M: The evolution of poxvirus vaccines. Viruses. 2015;7(4):1726-803. DOI: 10.3390/v7041726.

[23] Li SF, Wang HL, Hu ZH, Deng F: Genetic modification of baculovirus expression vectors. Virologica Sinica. 2012;27(2):71-82. DOI: 10.1007/s12250-012-3236-y.

[24] Zajakina A, Vasilevska J, Zhulenkovs D, Skrastina D, Spaks A, Plotniece A, et al.: High efficiency of alphaviral gene transfer in combination with 5-fluorouracil in a mouse mammary tumor model. BMC Cancer. 2014;14(1):460. DOI: 10.1186/1471-2407-14-460.

[25] Riezebos-Brilman A, de Mare A, Bungener L, Huckriede A, Wischut J, Daemen T: Recombinant alphaviruses as vectors for anti-tumour and anti-microbial immunotherapy. Journal of Clinical Virology. 2006;35(3):233-43. DOI: 10.1016/j.jcv.2005.12.001.

[26] Baum C, Schambach A, Bohne J, Galla M: Retrovirus vectors: toward the plentivirus? Molecular Therapy. 2006;13(6):1050-63. DOI: 10.1016/j.ymthe.2006.03.007.

[27] Maetzig T, Galla M, Baum C, Schambach A: Gammaretroviral vectors: biology, technology and application. Viruses. 2011;3(6):677-713. DOI: 10.3390/v3060677.

[28] Durinikova E, Kucerova L, Matuskova M: Mesenchymal stromal cells retrovirally transduced with prodrug-converting genes are suitable vehicles for cancer gene therapy. Acta virologica. 2014;58(1):1-13.

[29] Bank A, Markowitz DG, Goff SP: Retroviral packaging cell lines and process of using same. Google Patents; 1994.

[30] Tailor CS, Nouri A, Lee CG, Kozak C, Kabat D: Cloning and characterization of a cell surface receptor for xenotropic and polytropic murine leukemia viruses. Proceedings of the National Academy of Sciences of the United States of America. 1999;96(3): 927-32.

[31] Durand S, Cimarelli A: The inside out of lentiviral vectors. Viruses. 2011;3(2):132-59. DOI: 10.3390/v3020132.

[32] Naldini L, Blomer U, Gallay P, Ory D, Mulligan R, Gage FH, et al.: In vivo gene delivery and stable transduction of nondividing cells by a lentiviral vector. Science. 1996;272(5259):263-67.

[33] Yu SF, von Ruden T, Kantoff PW, Garber C, Seiberg M, Ruther U, et al.: Self-inactivating retroviral vectors designed for transfer of whole genes into mammalian cells. Proceedings of the National Academy of Sciences of the United States of America. 1986;83(10):3194-208.

[34] Gossen M, Bujard H: Tight control of gene expression in mammalian cells by tetracycline-responsive promoters. Proceedings of the National Academy of Sciences of the United States of America. 1992;89(12):5547-51.

[35] Russell SJ, Peng KW, Bell JC: Oncolytic virotherapy. Nature Biotechnology. 2012;30(7):658-70. DOI: 10.1038/nbt.2287.

[36] Tai CK, Kasahara N: Replication-competent retrovirus vectors for cancer gene therapy. Frontiers in Bioscience : A Journal and Virtual Library. 2008;13:3083-95.

[37] Schambach A, Bohne J, Chandra S, Will E, Margison GP, Williams DA, et al.: Equal potency of gammaretroviral and lentiviral SIN vectors for expression of O6-methylguanine-DNA methyltransferase in hematopoietic cells. Molecular Therapy. 2006;13(2):391-400. DOI: 10.1016/j.ymthe.2005.08.012.

[38] Gabriel R, Schmidt M, von Kalle C: Integration of retroviral vectors. Current Opinion in Immunology. 2012;24(5):592-97. DOI: 10.1016/j.coi.2012.08.006.

[39] Yi Y, Noh MJ, Lee KH: Current advances in retroviral gene therapy. Current Gene Therapy. 2011;11(3):218-28.

[40] Chen R: Enzyme engineering: rational redesign versus directed evolution. Trends in Biotechnology. 2001;19(1):13-14.

[41] Su K, Wang D, Ye J, Kim YC, Chow SA: Site-specific integration of retroviral DNA in human cells using fusion proteins consisting of human immunodeficiency virus type 1 integrase and the designed polydactyl zinc-finger protein E2C. Methods. 2009;47(4):269-76. DOI: 10.1016/j.ymeth.2009.01.001.

[42] Michel G, Yu Y, Chang T, Yee JK: Site-specific gene insertion mediated by a Cre-loxP-carrying lentiviral vector. Molecular Therapy. 2010;18(10):1814-21. DOI: 10.1038/mt.2010.150.

[43] Aiuti A, Biasco L, Scaramuzza S, Ferrua F, Cicalese MP, Baricordi C, et al.: Lentiviral hematopoietic stem cell gene therapy in patients with Wiskott-Aldrich syndrome. Science. 2013;341(6148):1233151. DOI: 10.1126/science.1233151.

[44] Wanisch K, Yanez-Munoz RJ: Integration-deficient lentiviral vectors: a slow coming of age. Molecular Therapy. 2009;17(8):1316-32. DOI: 10.1038/mt.2009.122.

[45] Giblett ER, Anderson JE, Cohen F, Pollara B, Meuwissen HJ: Adenosine-deaminase deficiency in two patients with severely impaired cellular immunity. Lancet. 1972;2(7786):1067-69.

[46] Blaese RM, Culver KW, Miller AD, Carter CS, Fleisher T, Clerici M, et al.: T lymphocyte-directed gene therapy for ADA-SCID: initial trial results after 4 years. Science. 1995;270(5235):475-80.

[47] Noguchi M, Yi H, Rosenblatt HM, Filipovich AH, Adelstein S, Modi WS, et al.: Interleukin-2 receptor gamma chain mutation results in X-linked severe combined immunodeficiency in humans. Cell. 1993;73(1):147-57.

[48] Puck JM, Deschenes SM, Porter JC, Dutra AS, Brown CJ, Willard HF, et al.: The interleukin-2 receptor gamma chain maps to Xq13.1 and is mutated in X-linked severe combined immunodeficiency, SCIDX1. Human Molecular Genetics. 1993;2(8):1099-104.

[49] Sauer AV, Di Lorenzo B, Carriglio N, Aiuti A: Progress in gene therapy for primary immunodeficiencies using lentiviral vectors. Current Opinion in Allergy and Clinical Immunology. 2014;14(6):527-34. DOI: 10.1097/ACI.0000000000000114.

[50] Ott de Bruin LM, Volpi S, Musunuru K: Novel genome-editing tools to model and correct primary immunodeficiencies. Frontiers in Immunology. 2015;6:250. DOI: 10.3389/fimmu.2015.00250.

[51] Binley K, Widdowson PS, Kelleher M, de Belin J, Loader J, Ferrige G, et al.: Safety and biodistribution of an equine infectious anemia virus-based gene therapy, RetinoStat((R)), for age-related macular degeneration. Human Gene Therapy. 2012;23(9):980-91. DOI: 10.1089/hum.2012.008.

[52] Murakami Y, Ikeda Y, Yonemitsu Y, Miyazaki M, Inoue M, Hasegawa M, et al.: Inhibition of choroidal neovascularization via brief subretinal exposure to a newly developed lentiviral vector pseudotyped with Sendai viral envelope proteins. Human Gene Therapy. 2010;21(2):199-209. DOI: 10.1089/hum.2009.102.

[53] McFarland TJ, Zhang Y, Appukuttan B, Stout JT: Gene therapy for proliferative ocular diseases. Expert Opinion on Biological Therapy. 2004;4(7):1053-58. DOI: 10.1517/14712598.4.7.1053.

[54] Perdoni C, Osborn MJ, Tolar J: Gene editing toward the use of autologous therapies in recessive dystrophic epidermolysis bullosa. Translational Research: The Journal of Laboratory and Clinical Medicine. 2015;S1931-5244(15)00175-9. DOI: 10.1016/j.trsl.2015.05.008.

[55] Mavilio F, Pellegrini G, Ferrari S, Di Nunzio F, Di Iorio E, Recchia A, et al.: Correction of junctional epidermolysis bullosa by transplantation of genetically modified epidermal stem cells. Nature Medicine. 2006;12(12):1397-402. DOI: 10.1038/nm1504.

[56] Piccolo P, Brunetti-Pierri N: Gene therapy for inherited diseases of liver metabolism. Human gene therapy. 2015;26(4):186-92. DOI: 10.1089/hum.2015.029.

[57] Grossman M, Raper SE, Kozarsky K, Stein EA, Engelhardt JF, Muller D, et al.: Successful ex vivo gene therapy directed to liver in a patient with familial hypercholesterolaemia. Nature Genetics. 1994;6(4):335-41. DOI: 10.1038/ng0494-335.

[58] Biffi A, Montini E, Lorioli L, Cesani M, Fumagalli F, Plati T, et al.: Lentiviral hematopoietic stem cell gene therapy benefits metachromatic leukodystrophy. Science. 2013;341(6148):1233158. DOI: 10.1126/science.1233158.

[59] Cartier N, Hacein-Bey-Abina S, Bartholomae CC, Veres G, Schmidt M, Kutschera I, et al.: Hematopoietic stem cell gene therapy with a lentiviral vector in X-linked adrenoleukodystrophy. Science. 2009;326(5954):818-23. DOI: 10.1126/science.1171242.

[60] Portsmouth D, Hlavaty J, Renner M: Suicide genes for cancer therapy. Molecular Aspects of Medicine. 2007;28(1):4-41. DOI: 10.1016/j.mam.2006.12.001.

[61] Adachi M, Sampath J, Lan LB, Sun D, Hargrove P, Flatley R, et al.: Expression of MRP4 confers resistance to ganciclovir and compromises bystander cell killing. The Journal of Biological Chemistry. 2002;277(41):38998-9004. DOI: 10.1074/jbc.M203262200.

[62] Nicholas TW, Read SB, Burrows FJ, Kruse CA: Suicide gene therapy with Herpes simplex virus thymidine kinase and ganciclovir is enhanced with connexins to improve gap junctions and bystander effects. Histology and Histopathology. 2003;18(2):495-507.

[63] Kievit E, Nyati MK, Ng E, Stegman LD, Parsels J, Ross BD, et al.: Yeast cytosine deaminase improves radiosensitization and bystander effect by 5-fluorocytosine of human colorectal cancer xenografts. Cancer Research. 2000;60(23):6649-55.

[64] Scartozzi M, Maccaroni E, Giampieri R, Pistelli M, Bittoni A, Del Prete M, et al.: 5-Fluorouracil pharmacogenomics: still rocking after all these years? Pharmacogenomics. 2011;12(2):251-65. DOI: 10.2217/pgs.10.167.

[65] Ramnaraine M, Pan W, Goblirsch M, Lynch C, Lewis V, Orchard P, et al.: Direct and bystander killing of sarcomas by novel cytosine deaminase fusion gene. Cancer Research. 2003;63(20):6847-54.

[66] Hlavaty J, Jandl G, Liszt M, Petznek H, Konig-Schuster M, Sedlak J, et al.: Comparative evaluation of preclinical in vivo models for the assessment of replicating retroviral vectors for the treatment of glioblastoma. Journal of Neuro-Oncology. 2011;102(1): 59-69. DOI: 10.1007/s11060-010-0295-5.

[67] Perez OD, Logg CR, Hiraoka K, Diago O, Burnett R, Inagaki A, et al.: Design and selection of Toca 511 for clinical use: modified retroviral replicating vector with improved stability and gene expression. Molecular Therapy. 2012;20(9):1689-98. DOI: 10.1038/mt.2012.83.

[68] Friedenstein AJ, Chailakhjan RK, Lalykina KS: The development of fibroblast colonies in monolayer cultures of guinea-pig bone marrow and spleen cells. Cell and Tissue Kinetics. 1970;3(4):393-403.

[69] Caplan AI: Mesenchymal stem cells. Journal of Orthopaedic Research: Official Publication of the Orthopaedic Research Society. 1991;9(5):641-50. DOI: 10.1002/jor. 1100090504.

[70] Prockop DJ: Marrow stromal cells as stem cells for nonhematopoietic tissues. Science. 1997;276(5309):71-74.

[71] Morrison SJ, Wandycz AM, Hemmati HD, Wright DE, Weissman IL: Identification of a lineage of multipotent hematopoietic progenitors. Development. 1997;124(10): 1929-39.

[72] Pittenger MF, Mackay AM, Beck SC, Jaiswal RK, Douglas R, Mosca JD, et al.: Multilineage potential of adult human mesenchymal stem cells. Science. 1999;284(5411): 143-47.

[73] Strioga M, Viswanathan S, Darinskas A, Slaby O, Michalek J: Same or not the same? Comparison of adipose tissue-derived versus bone marrow-derived mesenchymal stem and stromal cells. Stem Cells and Development. 2012;21(14):2724-52. DOI: 10.1089/scd.2011.0722.

[74] Dvorak HF: Tumors: wounds that do not heal. Similarities between tumor stroma generation and wound healing. The New England Journal Of Medicine. 1986;315(26): 1650-59. DOI: 10.1056/NEJM198612253152606.

[75] Kidd S, Spaeth E, Klopp A, Andreeff M, Hall B, Marini FC: The (in) auspicious role of mesenchymal stromal cells in cancer: be it friend or foe. Cytotherapy. 2008;10(7): 657-67. DOI: 10.1080/14653240802486517.

[76] Klopp AH, Gupta A, Spaeth E, Andreeff M, Marini F, 3rd: Concise review: dissecting a discrepancy in the literature: do mesenchymal stem cells support or suppress tumor growth? Stem Cells. 2011;29(1):11-19. DOI: 10.1002/stem.559.

[77] Kucerova L, Matuskova M, Hlubinova K, Altanerova V, Altaner C: Tumor cell behaviour modulation by mesenchymal stromal cells. Molecular Cancer. 2010;9:129. DOI: 10.1186/1476-4598-9-129.

[78] Junttila MR, de Sauvage FJ: Influence of tumour micro-environment heterogeneity on therapeutic response. Nature. 2013;501(7467):346-54. DOI: 10.1038/nature12626.

[79] Cihova M, Altanerova V, Altaner C: Stem cell based cancer gene therapy. Molecular Pharmaceutics. 2011;8(5):1480-87. DOI: 10.1021/mp200151a.

[80] Osaka M, Honmou O, Murakami T, Nonaka T, Houkin K, Hamada H, et al.: Intravenous administration of mesenchymal stem cells derived from bone marrow after contusive spinal cord injury improves functional outcome. Brain Research. 2010;1343:226-35. DOI: 10.1016/j.brainres.2010.05.011.

[81] Akiyama Y, Radtke C, Honmou O, Kocsis JD: Remyelination of the spinal cord following intravenous delivery of bone marrow cells. Glia. 2002;39(3):229-36. DOI: 10.1002/glia.10102.

[82] Bexell D, Svensson A, Bengzon J: Stem cell-based therapy for malignant glioma. Cancer Treatment Reviews. 2013;39(4):358-65. DOI: 10.1016/j.ctrv.2012.06.006.

[83] Morizono K, De Ugarte DA, Zhu M, Zuk P, Elbarbary A, Ashjian P, et al.: Multilineage cells from adipose tissue as gene delivery vehicles. Human Gene Therapy. 2003;14(1):59-66. DOI: 10.1089/10430340360464714.

[84] Keung EZ, Nelson PJ, Conrad C: Concise review: genetically engineered stem cell therapy targeting angiogenesis and tumor stroma in gastrointestinal malignancy. Stem Cells. 2013;31(2):227-35. DOI: 10.1002/stem.1269.

[85] Kucerova L, Poturnajova M, Tyciakova S, Matuskova M: Increased proliferation and chemosensitivity of human mesenchymal stromal cells expressing fusion yeast cytosine deaminase. Stem Cell Research. 2012;8(2):247-58. DOI: 10.1016/j.scr.2011.11.006.

[86] Dwyer RM, Khan S, Barry FP, O'Brien T, Kerin MJ: Advances in mesenchymal stem cell-mediated gene therapy for cancer. Stem Cell Research & Therapy. 2010;1(3):25. DOI: 10.1186/scrt25.

[87] Zhang J, Temin HM: Rate and mechanism of nonhomologous recombination during a single cycle of retroviral replication. Science. 1993;259(5092):234-38.

[88] Kucerova L, Altanerova V, Matuskova M, Tyciakova S, Altaner C: Adipose tissue-derived human mesenchymal stem cells mediated prodrug cancer gene therapy. Cancer Research. 2007;67(13):6304-13. DOI: 10.1158/0008-5472.CAN-06-4024.

[89] Cavarretta IT, Altanerova V, Matuskova M, Kucerova L, Culig Z, Altaner C: Adipose tissue-derived mesenchymal stem cells expressing prodrug-converting enzyme in-

hibit human prostate tumor growth. Molecular Therapy. 2010;18(1):223-31. DOI: 10.1038/mt.2009.237.

[90] Abrate A, Buono R, Canu T, Esposito A, Del Maschio A, Luciano R, et al.: Mesenchymal stem cells expressing therapeutic genes induce autochthonous prostate tumour regression. European Journal of Cancer. 2014;50(14):2478-88. DOI: 10.1016/j.ejca.2014.06.014.

[91] Kucerova L, Matuskova M, Pastorakova A, Tyciakova S, Jakubikova J, Bohovic R, et al.: Cytosine deaminase expressing human mesenchymal stem cells mediated tumour regression in melanoma bearing mice. The Journal of Gene Medicine. 2008;10(10):1071-82. DOI: 10.1002/jgm.1239.

[92] Kucerova L, Skolekova S, Demkova L, Bohovic R, Matuskova M: Long-term efficiency of mesenchymal stromal cell-mediated CD-MSC/5FC therapy in human melanoma xenograft model. Gene Therapy. 2014;21(10):874-87. DOI: 10.1038/gt.2014.66.

[93] Kucerova L, Matuskova M, Hlubinova K, Bohovic R, Feketeova L, Janega P, et al.: Bystander cytotoxicity in human medullary thyroid carcinoma cells mediated by fusion yeast cytosine deaminase and 5-fluorocytosine. Cancer Letters. 2011;311(1): 101-12. DOI: 10.1016/j.canlet.2011.07.014.

[94] Kucerova L, Feketeova L, Matuskova M, Kozovska Z, Janega P, Babal P, et al.: Local bystander effect induces dormancy in human medullary thyroid carcinoma model in vivo. Cancer Letters. 2013;335(2):299-305. DOI: 10.1016/j.canlet.2013.02.040.

[95] Bohovic R, Demkova L, Cihova M, Skolekova S, Durinikova E, Toro L, et al.: 3D multicellular models reflect the efficiency of MSC-directed enzyme/prodrug treatment. Neoplasma. 2015;62(4):521-30. DOI: 10.4149/neo_2015_063.

[96] Aboody KS, Najbauer J, Danks MK: Stem and progenitor cell-mediated tumor selective gene therapy. Gene Therapy. 2008;15(10):739-52. DOI: 10.1038/gt.2008.41.

[97] Altanerova V, Cihova M, Babic M, Rychly B, Ondicova K, Mravec B, et al.: Human adipose tissue-derived mesenchymal stem cells expressing yeast cytosinedeaminase::uracil phosphoribosyltransferase inhibit intracerebral rat glioblastoma. International Journal of Cancer. 2012;130(10):2455-63. DOI: 10.1002/ijc.26278.

[98] Altaner C, Altanerova V, Cihova M, Ondicova K, Rychly B, Baciak L, et al.: Complete regression of glioblastoma by mesenchymal stem cells mediated prodrug gene therapy simulating clinical therapeutic scenario. International Journal of Cancer. 2014;134(6):1458-65. DOI: 10.1002/ijc.28455.

[99] Matuskova M, Hlubinova K, Pastorakova A, Hunakova L, Altanerova V, Altaner C, et al.: HSV-tk expressing mesenchymal stem cells exert bystander effect on human glioblastoma cells. Cancer Letters. 2010;290(1):58-67. DOI: 10.1016/j.canlet.2009.08.028.

[100] de Melo SM, Bittencourt S, Ferrazoli EG, da Silva CS, da Cunha FF, da Silva FH, et al.: The anti-tumor effects of adipose tissue mesenchymal stem cell transduced with

HSV-Tk gene on U-87-driven brain tumor. PloS One. 2015;10(6):e0128922. DOI: 10.1371/journal.pone.0128922.

[101] Matuskova M, Kozovska Z, Toro L, Durinikova E, Tyciakova S, Cierna Z, et al.: Combined enzyme/prodrug treatment by genetically engineered AT-MSC exerts synergy and inhibits growth of MDA-MB-231 induced lung metastases. Journal of Experimental & Clinical Cancer Research: CR. 2015;34:33. DOI: 10.1186/s13046-015-0149-2.

[102] Tyciakova S, Matuskova M, Bohovic R, Polakova K, Toro L, Skolekova S, et al.: Genetically engineered mesenchymal stromal cells producing TNFalpha have tumour suppressing effect on human melanoma xenograft. The Journal of Gene Medicine. 2015;17(1-2):54-67. DOI: 10.1002/jgm.2823.

[103] Yan C, Yang M, Li Z, Li S, Hu X, Fan D, et al.: Suppression of orthotopically implanted hepatocarcinoma in mice by umbilical cord-derived mesenchymal stem cells with sTRAIL gene expression driven by AFP promoter. Biomaterials. 2014;35(9):3035-43. DOI: 10.1016/j.biomaterials.2013.12.037.

[104] Deng Q, Zhang Z, Feng X, Li T, Liu N, Lai J, et al.: TRAIL-secreting mesenchymal stem cells promote apoptosis in heat-shock-treated liver cancer cells and inhibit tumor growth in nude mice. Gene Therapy. 2014;21(3):317-27. DOI: 10.1038/gt.2013.88.

[105] Hong SH, Lee HJ, An J, Lim I, Borlongan C, Aboody KS, et al.: Human neural stem cells expressing carboxyl esterase target and inhibit tumor growth of lung cancer brain metastases. Cancer Gene Therapy. 2013;20(12):678-82. DOI: 10.1038/cgt.2013.69.

[106] Sugrue T, Lowndes NF, Ceredig R: Mesenchymal stromal cells: radio-resistant members of the bone marrow. Immunology and Cell Biology. 2013;91(1):5-11. DOI: 10.1038/icb.2012.61.

[107] Altaner C, Altanerova V: Stem cell based glioblastoma gene therapy. Neoplasma. 2012;59(6):756-60. DOI: 10.4149/neo_2012_95.

[108] Matuskova M, Baranovicova L, Kozovska Z, Durinikova E, Pastorakova A, Hunakova L, et al.: Intrinsic properties of tumour cells have a key impact on the bystander effect mediated by genetically engineered mesenchymal stromal cells. The Journal of Gene Medicine. 2012;14(12):776-87. DOI: 10.1002/jgm.2684.

[109] Lee JY, Lee DH, Kim HA, Choi SA, Lee HJ, Park CK, et al.: Double suicide gene therapy using human neural stem cells against glioblastoma: double safety measures. Journal of Neuro-Oncology. 2014;116(1):49-57. DOI: 10.1007/s11060-013-1264-6.

[110] Niess H, von Einem JC, Thomas MN, Michl M, Angele MK, Huss R, et al.: Treatment of advanced gastrointestinal tumors with genetically modified autologous mesenchymal stromal cells (TREAT-ME1): study protocol of a phase I/II clinical trial. BMC Cancer. 2015;15(1):237. DOI: 10.1186/s12885-015-1241-x.

Section 5

Endogenous Retroviruses

6

HERVs in Multiple Sclerosis — From Insertion to Therapy

Belén de la Hera and Elena Urcelay

Abstract

Genome-wide association studies (GWAS) have not been able to completely elucidate the genetic background of complex diseases. Part of it could lie in repetitive sequences not studied in the GWAS, as those corresponding to Human Endogenous Retroviruses (HERVs). In the present work, we aim to review the potential role of HERVs in the etiology of autoimmune diseases, especially in multiple sclerosis (MS); their potential pathogenic role and their putative consideration as a good target for new treatments. For this purpose, we carried out an in-depth literature review on HERVs, and we integrated our previous findings about HERV-W, HERV-K18, and HERV-Fc1 and MS susceptibility. The study was carried out by a systematic search from electronic databases using the keywords "HERV," "Multiple sclerosis," "HERV-W," "MSRV," "HERV-K," "HERV-Fc1," and "GNbAC1."

Keywords: Multiple sclerosis, HERV, MSRV, GNbAC1

1. Introduction

1.1. HERVs

The endogenous retroviruses (ERVs) could be defined as "genetic parasites" of vertebrates [1], given that their origin is very different from the one displayed by the rest of the genome. Their existence in the genome of mammals is only known since 1970 [2], although, they resulted from ancestral infections by exogenous retroviruses millions of years ago. During an infection, the exogenous retroviruses are able to integrate one copy of their genome (provirus) into the genome of the host. Thus, they can stay permanently associated with the host and be transmitted horizontally by the creation of new virions (the typical spread of an infectious virus). Only when they infect a germ line cell, the integrated DNA can become part of the gene pool and be transmitted in a Mendelian fashion like ERVs [1, 3-5], as shown in Figure 1. Those who are present in the human genome are named human endogenous retroviruses (HERVs).

The endogenization process profoundly impacts on the survival and evolution of the virus and the host. It results from the balance achieved between the immune surveillance and the virus virulence [6]. In this way, the HERVs must surpass the host's antiviral defense mechanisms and infect the germ cells without causing a cytotoxicity that would prevent persistence in the progeny of the host [6]. Furthermore, from this moment on, all host cells are carriers of an integrated provirus [6].

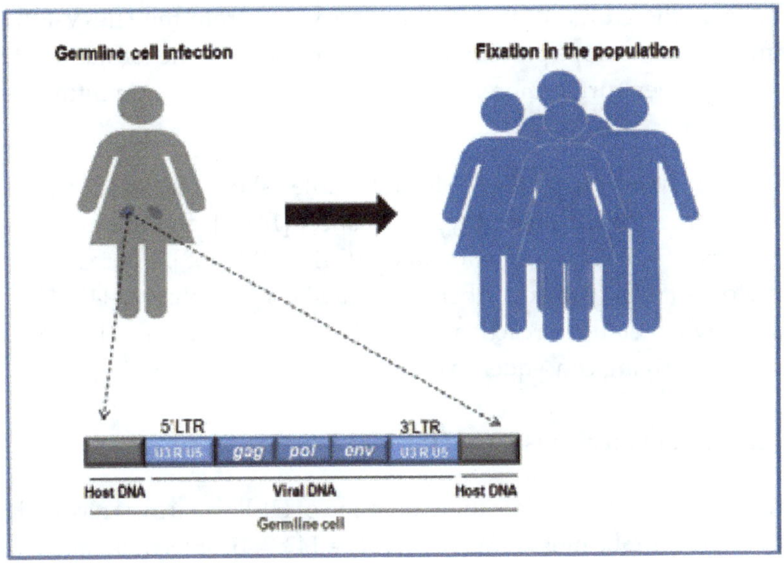

Figure 1. The endogenization process: once the retrovirus infects a germ line cell, vertical transmission in a Mendelian fashion occurs.

The retroviral insertion is aleatory, in the sense that no specific sites for retroviral integration exist in the host genome. Nonetheless, due to the epigenetic chromatin packaging, integrated HERVs elements are more commonly found within the transcriptionally active genome [6]. Currently, HERVs comprise nearly 8% of the human genome [7], distributed in approximately 31 independently acquired multigene families [8]. Even though no standard nomenclature has been defined for HERVs, they have been classified based on their homology with different groups of exogenous retroviruses. They are grouped as class I, class II, or class III retroviruses considering their homology with Gamma and Epsilon retroviruses, Betaretrovirus or Spumavirus, respectively [9, 10]. The family name is usually given by "HERV" followed by a one-letter amino acid code that corresponds to the tRNA specific of the site used to initiate reverse transcription [10]; consequently, the HERV-W family would use a tryptophan.

As mentioned, HERVs have a similar structure to proviruses of infectious retroviruses, with three principal genes, *gag*, *pol*, and *env*, flanked by two long terminal repeats (LTRs) [6]. The *gag* gene codes for the viral assembly proteins, including the nucleocapsid, matrix, and capsid proteins. The *pol* gene codes for the viral replication proteins, yielding the reverse transcriptase, protease, ribonuclease, and integrase proteins. Finally, the *env* gene codes for a viral glycoprotein, with both a surface and a transmembrane subunit. However, important changes are observed in the HERVs expression compared to that of exogenous retroviruses. Most HERVs encode incomplete proteins and accumulate mutations and recombinations. Furthermore, most HERVs with functional LTRs remain in a latent state under homeostatic conditions, owing to the epigenetic silencing of the provirus in heterochromatin [11]. Exceptionally, specific HERVs have been selected during evolution, provided that their biological functions could be beneficial for the host. In these cases, HERVs suffer a "domestication," meaning that a foreign gene can be used for cellular functions of the host [12]. In this group, we find proteins like Syncytin-1 from the HERV-W family, and Syncytin-2 from the HERV-FRD family [13]. These highly fusogenic envelope proteins are necessary to allow the formation of the placental syncytiotrophoblast layer; furthermore, they could be involved in the immune tolerance to the fetus [13].

In addition to these "domestic HERVs," several studies show reactivation of HERVs under pathologic conditions, such as different types of cancer [14-20]; autoimmune diseases including multiple sclerosis (MS) [21-37], rheumatoid arthritis (RA) [38], psoriasis [39], or systemic lupus erythematosus (SLE) [40]; and other diseases like schizophrenia [41, 42]. Nonetheless, we do not know whether their reactivation or increased expression is a causal effect, or conversely, is an underlying consequence of the disease.

1.2. Potential expression mechanisms of HERVs

Many factors can interfere or modulate the expression of HERVs, such as recombination events between two or more replication-defective HERVs [43, 44], infectious agents like Human herpesvirus 6 (HHV6) [34, 45] and Epstein–Barr virus (EBV) [46, 47], several transcription factors [31, 48], and the epigenomic context of the HERVs [6, 49, 50].

- Recombination events

Two or more replication-defective HERVs can restore their own defects through recombination events, resulting in a replication-competent retrovirus [5]. Even though this is an infrequent event, a study in mice points to a significantly increased frequency in specific immune deficiencies [44]. Furthermore, it has been demonstrated that recombination between three HERV-K defective proviruses is possible, leading to an infectious retrovirus [5, 43].

- Infectious agents

A putative explanation about the preferential expression of HERVs found in human brain samples could be the tropism of specific viruses and bacteria to the central nervous system (CNS). Neurotropic agents like herpesvirus [29, 51], Toxoplasma gondii [52], or certain strands of influenza virus [53] are able to cross the hematoencephalic barrier into the CNS. Usually, they are intercepted by cerebral macrophages leading to an abortive infection, but their transient presence in the CNS could activate the HERVs expression as a consequence of their immediate-early (IE) genes expression [4]. The expression of the IE genes of herpes simplex virus type 1 (HSV-1) and its interaction with the transcription factor binding sites situated in the U3 region of the LTR, such as AP-1 [54] and Oct-1 [55], lead to an activation of transcription in HERV-K and HERV-W families.

The herpesviruses are one of the best candidates: they may be neurotropic, remain latent, and can be reactivated. Furthermore, the expression of the Env epitopes in the surface of B cells and monocytes could be a consequence of the interaction between HERVs and herpesviruses [25]. Thus, the herpesviruses could play a dual role in neurodegenerative diseases, acting as pathological entities *per se* and as inducers of HERVs [6].

- Transcription factors

An important component of the antiviral innate immunity is the regulation of the expression and replication of HERVs by different transcription factors [48]. In HERV-W and HERV-K elements, both families previously related to multiple sclerosis (MS), different binding sites for transcription factors such as NF-Kβ [31, 48] are located in their promoter regions and could drive an increased expression of HERVs during inflammation.

- Epigenomic context

The chromatin state as well as the methylation state of GpC islands within the HERV promoter and regulatory regions seem to be crucial factors in the control of HERVs expression [49, 50]. Both play an important role as a part of the defense system against the potential effects of inserted sequences. Previously published studies describe how proviruses and solitary LTRs are densely methylated under physiological conditions, but hypomethylated in placenta [49, 50]. Thus, DNA hypomethylation, as observed in certain types of cancer, could allow reactivation of retroelements. In MS, HERVs have been described as susceptible elements to undergo epigenetic modifications, mainly due to modifications in the methylation state, resulting in activation of their expression and, consequently, inappropriate activation of the immune system.

1.3. Pathogenic mechanisms of HERVs

Even though most inserted copies in the human genome are defective copies, some HERVs could maintain the potential to cause or contribute to disease by different mechanisms [5]. As mentioned, HERVs may alter cellular functions by two ways, either acting as a genetic element or as a viral pathogen [6].

- Gene disruption

HERVs, like transposons, are able to experience transposition, recombination, and integration cycles. Some HERVs families include a high number of copies in the genome. It is believed that these families have been spread around the genome through the reintegration of a provirus. However, each new integration process increases the risk of a harmful insertion. They can disrupt genes present in their integration sites, for example, HERV-K integrations have been identified into tumor suppressor genes like *BRCA2* and into the repair *XRCC1* gene [6, 56].

- Modulation of gene expression

Some HERVs conserve regulatory sequences that can operate as functional promoters, enhancers, or polyadenylation signals, so they could change the expression of adjacent or distal genes [4]. They can also form part of regulatory RNAs: microRNAs (miRNA), small interfering RNAs (siRNA), and long intergenic noncoding RNAs (lincRNAs), contributing to the complex regulatory network of gene expression [5]. Furthermore, HERVs integrated into introns can provide alternative transcription start and termination sites [5].

- Pattern recognition receptors (PRRs)

The HERVs expression products, both nucleic acids and proteins, can modulate immune responses. They have the potential to interact with components involved in the immune innate response and to activate proinflammatory signaling pathways [57, 58]. Therefore, certain HERVs proteins could directly interact with specific toll-like receptors (TLRs), for example with TLR4, resulting in the production of TNFα and proinflammatory cytokines [58-60]. The nucleic acids derived from HERVs may also activate cytosolic PRRs; in this way, both an increased expression of RNA and the presence of cDNA in a nonfamilial compartment like the cytosol could activate PRRs [60]. Nonetheless, the human being has coevolved with endogenous retroelements and this could have shaped the sensibility of DNA sensors of the innate immune system, leading to an increased cDNA detection threshold to avoid an immune response against them. The cDNA levels are restrained by the action of gene products like Trex1 or SAMHD1 [60] and a loss-of-function mutation in these enzymes could result in the cDNA accumulation and the consequent sensors activation. This process would lead to a chronic immune response with release of pathogenic type I IFN and inflammatory mediators, similar to those observed in autoimmune diseases [60].

- Viral proteins: molecular mimicry, superantigen activity, or immunosuppressive proteins

HERVs proteins hold epitopes to B and T cells and molecular mimicry between viral proteins and certain autoantigens may exist, resulting in an autoimmune response. Moreover, some HERVs sequences are able to encode for superantigens. Superantigens combine with MHC class-II molecules to form ligands that stimulate T cells [61], and this may end in an abnormal activation of autoreactive T lymphocytes [62].

Alternatively, evidences exist of the immunosuppressive activity of certain HERVs Env proteins [63, 64]. This activity is reminiscent of their exogenous antecessors, which in this way increased the viability of the virions in the host. This capacity has suffered an adaptation process, and nowadays it might be implicated in the materno-fetal tolerance and could also prevent the immune response to exogen pathogens and tumors [60].

- Retroviral help for B cells

HERVs can also help B cells to quickly produce antibodies directed against pathogenic antigens [65]. The bacterial polysaccharide antigens and the carbohydrates linked to viral glycoproteins have the ability to stimulate B cells in the absence of T-cell help. These antigens are called thymus-independent antigens (TI), and they can be classified into two types: TI-1 or TI-2 antigens. TI-2 antigens cause extensive cross-linking of the BCR, leading to a quick differentiation of B cells into plasma cells. Finally, these plasma cells secrete protective antibodies, IgM and IgG [66]. However, the mechanism by which the TI-2 antigens activate B cells in the marginal zone without the help of T cells still remains poorly understood. It has been recently described that the cross-linking of B cells activates a signaling cascade, including the Bruton Tyrosine Kinase and the nuclear transcription factor NF-Kβ, allowing transcription of endogenous retroviral DNA [66]. The retroviral RNA may activate B cells by two complementary but different pathways: first, it could activate the retinoic-acid-inducible gene 1 receptor (RIG-1), resulting in a mitochondrial antiviral-signaling (MAVS); second, the RNA can be converted into DNA and can activate the cyclic GMP-AMP synthase (cGAS, cGAMP synthase). Finally, both signaling pathways would finish in the antigen-specific B-cell activation [65, 66].

2. HERVs and autoimmune diseases

HERVs represent the immunological limit between the self and the foreign. Their peculiar origin is very different from that of other genome elements, as they can share properties with infectious agents. Indeed, in case they would produce particles, these would not be so different from those originated from exogenous retroviruses. Therefore, they could activate the immune system and would induce autoimmunity [67]. As it has been previously discussed, HERVs have been associated, among other infectious or neurologic diseases, with different autoimmune diseases like MS [21-37], RA [38], psoriasis [39], T1D [68], or SLE [40], as shown in Table 1. Genome-wide association studies (GWAS) showed the existence of a genetic basis shared between different autoimmune diseases, discovering new immunogenic mechanisms implicated, and HERVs could be part of these shared genetic elements.

Class	Family	PBS	Related diseases	Expression mechanisms	Pathogenic mechanisms
I	HERV-W	Trp	MS Schizophrenia HIV Osteoarthitis	Herpesviruses Transcription factors Toxoplasma gondii Influenza A virus	Pro-inflammatory Env protein Superantigen activity OPCs differentiation interference Altered glial function
I	HERV-F	Phe	MS	Demethylating agents	Superantigen activity
I	HERV-H	His	3q13.31 microdeletion syndrome	N/A	Genetic deletion by recombination
I	HERV-E	Glu	SLE	Hypomethylation	Immunosupressor potential of Env
I	HERV-P	Pro	Cancer	Unkown	Unknown
II	HERV-K	Lys	MS ALS HIV	Herpersviruses HTLV-1 Type 1 IFN	Superantigen activity Neoepitopes Genes disruption

Class	Family	PBS	Related diseases	Expression mechanisms	Pathogenic mechanisms
			Schizophrenia T1D RA Juvenil arthitis Cancer	Transcription factors Hypomethylation	

MS, multiple sclerosis; HIV, human immunodeficiency virus; SLE, Systemic lupus erythematosus; ALS, amyotrophic lateral sclerosis; T1D, type 1 diabetes; RA, rheumatoid arthritis, HTLV-1, human T-lymphotropic virus-1; IFN, interferon; Env, envelope; OPCs, olygodendrocyte precursor cells; N/A, not applicable. Based on Douville and Nath, 2014 [6].

Table 1. HERVs families associated with different diseases

3. HERVs and MS

MS is one of the conditions more frequently related with HERVs. It is a chronic progressive disease characterized by neuroinflammation in the CNS accompanied by demyelination, axonal damage, and progressive neurologic dysfunction [69]. It is a complex disease, originated from the interaction of genetic, environmental, and epigenetic factors [70]. Recently, its incidence seems to be increased; at present MS affects 2.3 million people in the world [71]. However, many aspects of its pathogenesis are still poorly understood. GWAS have not completely explained the MS genetic background [72-74], albeit including the ImmunoChip Project [75] a total of 110 single nucleotide polymorphisms (SNPs) have been associated with MS susceptibility. Even considering the strongest risk factor, the HLA-DRB1*15:01 allele, each SNP has a modest effect and all together are able to explain only 20–28% of MS heritability [75]. Part of the missing heritability could reside on HERVs, as repetitive regions were not analyzed in the GWAS. Those repetitive regions were previously considered as "junk DNA" because it was thought that they had little or no physiological role. However, nowadays we know that these sequences could play an important role in the development of autoimmune diseases, including MS.

In 1989, Perron et al. [76] described the presence of extracellular virions associated with reverse transcriptase activity in a culture of leptomeningeal cells (LM7) obtained from the cerebrospinal fluid (CSF) of an MS patient. In the beginning, it was thought that those virions could correspond to the human T-lymphotropic virus (HTLV-1) due to the similarities between the tropical spastic paraparesis (a demyelinating progressive disease) caused by HTLV-1, and MS. However, a new retroviral element called MSRV (Multiple-sclerosis-associated retrovirus) was identified, the founder of the HERV-W family [77]. This multicopy family, consisting of approximately 650 loci around the human genome [35], comprises a total of 311 inserts (more or less complete proviruses or pseudogenes) and 343 additional HERV-W LTRs [78].

Only the *env* gene mapping on chromosome 7, encoding Syncytin, presents a complete open reading frame (ORF) and has been selectively conserved [35]. The MSRV *env* sequence can be differentiated from the one corresponding to Syncytin-1 by a 12-nucleotide insertion in the transmembrane moiety. Both genes are expressed in the brain of MS patients, but the MSRV-type *env* DNA copies were found sixfold more frequently in MS patients than in healthy controls, while comparable copy numbers of Syncytin-1 were observed [79]. Furthermore, Syncytin-1 is originated from the retroviral copy inserted in chromosome 7, and the pathogenic protein MSRV-type Env could be originated from several integrations in the human genome, or it could result from recombination events between insertions in different chromosomes [80, 81]. The genomic origin of HERV-W Env remains unknown although recent works consider the copy mapping to chromosome X one plausible candidate [4, 80, 81]. This copy, located on Xq22.3, would encode for an almost complete MSRV-type protein, truncated on its N-terminal end due to the presence of a stop codon mutation at position 39 [81]. *Ex vivo*, this copy still conserves coding capacity, as it is able to produce a truncated N-terminally Env protein [80]. Furthermore, the reversion of this stop codon would lead to a complete protein with signal peptide, expressed in the cellular surface in the same way that Syncytin [80].

Recently, a genetic screening was performed by specific PCR amplification followed by High Resolution Melting (HRM) analyses of the two MSRV-like env copies which show the ORF with the highest length similarity and homology to Syncytin (1614 bp), inserted in chromosome X (1428 bp) and in chromosome 20 (1419 bp). Both chromosomal origins show similar lengths of their respective ORFs, 10% shorter than the one measured for Syncytin, and could putatively originate functional proteins. The results pointed to the insertion in chromosome X, and not the one in chromosome 20, as an origin of MSRV. One polymorphism identified in chromosome X, rs6622139*T, was associated in women with MS susceptibility and severity [82], and it was also associated with higher MSRV-like *env* levels of expression (Mann–Whitney U test: $p=0.003$), while the two polymorphisms found in chromosome 20 did not show evidence of association [83].

Since it was described, several studies have associated the HERV-W family with MS: the presence of MSRV-type Env protein has been found in demyelinated acute lesions in MS patients [31], as well as an increased number of DNA copies [84] or a higher prevalence of MSRV-type RNA in serum and CSF of MS patients compared with patients suffering from other neurological diseases or healthy controls from all ethnic groups [24, 27, 28, 31, 84-86]. The MSRV presence in serum and CSF is correlated with the clinical progression, severity, and prognosis of MS [28, 46], while the absence of MSRV relates with a more stable course of the disease [28, 36]. The MSRV production is stimulated by cytokines like TNFα, IL6, and IFNγ[87], and current MS therapies like IFN-β and Natalizumab, which are able to reduce MS symptomatology, promote a diminution of MSRV virus load levels in blood [87-89].

HERV-W Env proteins, MSRV-type Env, and Syncytin have proinflammatory and superantigenic properties. They can cause neuroinflammation, neurodegeneration, immune system dysregulation, and endoplasmic reticulum stress [4, 21, 22, 58, 90, 91]. Their pathogenicity has been studied *in vitro* using different types of cell cultures and *in vivo* using a humanized Severe Combined Immunodeficiency Disease (SCID) animal model, showing neurotoxic effects in both settings [22, 92] and a reduced capacity of olygodendrocyte progenitor cells (OPC) differentiation, interfering in the remyelination process [57]. A recent study clarifies the possible pathogenic mechanisms of MSRV. In a human model of BBB, the endothelial cell line

HCMEC/D3, they show that MRSV-type Env interacts with TLR4 and induces a dose-dependent overexpression of ICAM1, as well as an induced IL6 and IL8 production; while the Env protein derived from Syncytin-1 did not show these effects [59]. Furthermore, they also described that the MSRV-type Env presence significantly stimulates the adhesion and migration of activated immune cells through the layer of endothelial cells. These results support the hypothesis that MSRV can be involved in MS pathogenesis, as well as in other chronic inflammatory diseases, at least in the maintenance of the underlying inflammatory condition [59]. Table 2 reflects the possible pathogenic mechanisms described for MSRV.

Cell type	Receptor	Pathogenic mechanisms
T lymphocytes	TCR	Superantigen activity, T lymphocytes proliferation and CK liberation
APC	TLR4	↑ pro-inflammatory CK ↑ costimulatory molecules
HCMEC/D3 endothelial cell line	TLR4	↑ ICAM1 expression IL6, IL8 expression
OPC	TLR4	OPCs differentiation interference (↑ iNOS, oxidative stress)

APC, antigen presenting cell; OPC, olygodendrocyte precursor cell; TCR, T-cell receptor; TLR4, toll-like receptor 4

Table 2. Known pathogenic mechanisms of MSRV

Even though the HERV-W family is one of the HERV families more related to MS, other families like HERV-K18 [37, 93] or HERV-Fc1 [29, 94, 95] have also been associated with MS susceptibility.

HERV-K is a multicopy family including approximately 332 copies dispersed through the human genome. It is the only known retroviral element that codes for all the structural and enzymatic proteins (Gag, Prt, Pol), as well as for the Env protein and for the accessory Rec protein [96]. This family has been related with different autoimmune diseases as MS [37], type-1diabetes (T1D) [68], or juvenile rheumatoid arthritis [38]; and different cancer types [14-17]. One specific member of this family, HERV-K18, has been associated with MS susceptibility and its expression is induced by herpesvirus [97, 98] and by EBV [99-102], both viruses previously proposed as potential environmental factors involved in MS development [45, 51, 97, 103-108]. Three different variants of the HERV-K18 copy mapping to chromosome 1 [37] have been described. They conform haplotypes within the first intron of CD48 that can be defined by two SNPs (18.1 SNP1*A/SNP2*A, 18.2 SNP1*G/SNP2*G, 18.3 SNP1*A/SNP2*G), all of them coding for an Env protein with superantigenic properties. However, only one of these variants (18.3) has been associated with a higher risk to MS [37] and with an overall higher susceptibility to autoimmune diseases, as described by a meta-analysis including a total of 2656 patients and 2016 controls [93].

Considering the HERV-Fc family, a total of 6 HERV-Fc elements and 11 LTRs have been identified across the human genome. Among them, only two elements correspond to a complete HERV-Fc provirus (Fc1env and Fc2master) [109]. Related to MS, it has been observed that the HERV-Fc1 RNA levels were significantly increased in the plasma of patients suffering from active MS, compared to nonactive MS or controls [30]. The HERV-Fc1 is an unusual provirus, because it includes a single copy in the genome, located on Xq21.33. Furthermore, it is a recent acquisition for the genome, only present in humans, chimpanzees, and gorillas [109]. Nexo et al. [94] were the first to describe that rs391745, located in the promoter region of HERV-Fc1, was associated with MS susceptibility in Danish cohorts and, then, a replication study was performed with a Norwegian cohort. The latter study also detailed that the association was only observed in the nonprimary-progressive MS forms [29], results validated in further studies [95]. Regarding the HERV-Fc1 expression mechanisms, it has been observed that the transcriptional expression levels of HERV-Fc1 RNA sequences are negatively correlated with the methylation levels of CpG islands on the 5′ LTR region and, therefore, a higher HERV-Fc1 expression involves DNA demethylation [11, 110].

4. HERVs as future treatment options

An increased expression of HERVs in several autoimmune diseases [21-40, 68] and different types of cancer [14-20], along with the decreased expression levels observed in successfully treated patients with immunomodulatory therapies [88, 89] or chemotherapy [111] point to the potential pathogenic role of HERVs and their putative consideration as a good target for new treatments.

A humanized monoclonal antibody anti-Env-SU MSRV/HERV-W, GNbAC1, has been studied as a putative MS treatment due to its potential neuroprotector effects [112-115]. The results of a phase IIa clinical trial [114] show that the GNbAC1 treatment blocks the transcription of proinflammatory genes mediated by Env, prevents the formation of nitrosantine, and restores OPC differentiation. Furthermore, GNbAC1 has advantages compared to other MS treatments, because the patients retain all their immune capacity. This treatment has also been studied in other diseases like diabetes and schizophrenia.

The proteins encoded by HERV-K *env* have been proposed as therapeutic targets for different types of cancer, due to the fact that a general hypomethylation of HERVs sequences has been observed, as well as an increased expression of Np9 and Rec proteins originating from HERV-K in different cancer cells [116]. Both proteins bind to the PLZF protein, a transcriptional repressor of the C-MYC proto-oncogen. The inflammation and the deregulation of proto-oncogen signaling caused by the HERV-K protein results in a protumorogenic microenvironment, which favors cell proliferation and metastasis [116]. The use of monoclonal antibodies against HERV-K Env protein inhibits tumoral growth and induces apoptosis in breast cancer cells *in vitro* [116]; therefore, it could be considered a good candidate as a therapy used together with other cancer treatments.

In addition to autoimmune diseases and cancer, the human immunodeficiency virus (HIV) has been also related to HERVs, particularly with the HERV-K family, raising the issue of potential beneficial effects of a therapy directed against HERVs in AIDS. Some studies report an increased expression of HERV-K provirus in HIV patients compared to controls [117, 118] and show that the immune responses against HERV-K decrease the HIV-1 viral load. *In vitro,* the use of an antibody directed against the HERV-K transmembrane protein (HML-2), HA-137, was able to eliminate the cells that displayed the antigen in their surface. This was carried out by an antibody-dependent cell-mediated cytotoxicity (ADCC) mechanism by natural killer (NK) cells. It has been described that the HIV-infected cells display this membrane antigen in their surface [119]; therefore, they would be potential targets of the antibody. The possibility of finding a target epitope different from those of the HIV virus could open up opportunities to the development of vaccines against this disease; a field that has been very limited due to the high rate of mutation of the HIV [119].

5. Conclusion

This work aimed to provide a systematic revision of HERVs, with particular emphasis on their potential pathogenic role in MS. Although many aspects of the etiology of this disease remain to be solved, different works support the relevance that HERVs may have in the etiopathogenesis of autoimmune diseases, and specifically in MS. HERVs may contribute to both, disease onset and maintenance, through an exacerbated activation of the immune system. Recently, the results of a phase IIa clinical trial that studies the effectiveness of a human monoclonal antibody (GNbAC1) as a therapeutic target in MS have been published with promising outcome. Thus, evidences support the role of HERVs as potential therapeutic armory in different autoimmune diseases, cancer, and HIV.

Author details

Belén de la Hera and Elena Urcelay*

*Address all correspondence to: elena.urcelay@salud.madrid.org

Instituto de Investigación Sanitaria del Hospital Clínico San Carlos, IdISSC, Madrid, Spain

References

[1] Ribet D, Heidman T. Formation et évolution des rétrovirus endogènes. *Virologie.* 2010;14:141-50. DOI:10.1684/vir.2010.0294.

[2] Coffin JM. Structure, replication, and recombination of retrovirus genomes: some unifying hypotheses. *J Gen Virol*. 1979;42:1-26.

[3] Chuong EB. Retroviruses facilitate the rapid evolution of the mammalian placenta. *Bioessays*. 2013;35:853-61. DOI: 10.1002/bies.201300059.

[4] Perron H, Bernard C, Bertrand JB, Lang AB, Popa I, Sanhadji K, et al. Endogenous retroviral genes, Herpesviruses and gender in Multiple Sclerosis. *J Neurol Sci*. 2009;286:65-72.

[5] Young GR, Stoye JP, Kassiotis G. Are human endogenous retroviruses pathogenic? An approach to testing the hypothesis. *Bioessays*. 2013;35:794-803. DOI: 10.1002/bies. 201300049.

[6] Douville RN, Nath A. Human endogenous retroviruses and the nervous system. *Handb Clin Neurol*. 2014;123:465-85. DOI:10.1016/B978-0-444-53488-0.00022-5.

[7] Lander ES, Linton LM, Birren B, Nusbaum C, Zody MC, Baldwin J, et al. Initial sequencing and analysis of the human genome. *Nature*. 2001;409:860-921. DOI: 10.1038/35057062.

[8] Belshaw R, Katzourakis A, Paces J, Burt A, Tristem M. High copy number in human endogenous retrovirus families is associated with copying mechanisms in addition to reinfection. *Mol Biol Evol*. 2005;22:814-7. DOI:10.1093/molbev/msi088.

[9] Voisset C, Weiss RA, Griffiths DJ. Human RNA "rumor" viruses: the search for novel human retroviruses in chronic disease. *Microbiol Mol Biol Rev*. 2008;72:157-96. DOI: 10.1128/MMBR.00033-07.

[10] Gifford R, Tristem M. The evolution, distribution and diversity of endogenous retroviruses. *Virus Genes*. 2003;26:291-315. DOI:10.1023/A:1024455415443.

[11] Laska MJ, Nissen KK, Nexo BA. (Some) cellular mechanisms influencing the transcription of human endogenous retrovirus, HERV-Fc1. *PLoS One*. 2013;8:e53895. DOI: 10.1371/journal.pone.0053895.

[12] Patel MR, Emerman M, Malik HS. Paleovirology - ghosts and gifts of viruses past. *Curr Opin Virol*. 2011;1:304-9. DOI:10.1016/j.coviro.2011.06.007.

[13] Dupressoir A, Heidmann T. [Syncytins – retroviral envelope genes captured for the benefit of placental development]. *Med Sci* (Paris). 2011;27:163-9. DOI:10.1051/medsci/2011272163.

[14] Buscher K, Trefzer U, Hofmann M, Sterry W, Kurth R, Denner J. Expression of human endogenous retrovirus K in melanomas and melanoma cell lines. *Cancer Res*. 2005;65:4172-80. DOI: 10.1158/0008-5472.CAN-04-2983

[15] Herbst H, Kuhler-Obbarius C, Lauke H, Sauter M, Mueller-Lantzsch N, Harms D, et al. Human endogenous retrovirus (HERV)-K transcripts in gonadoblastomas and go-

nadoblastoma-derived germ cell tumours. *Virchows Arch.* 1999;434:11-5. DOI:10.1007/s004280050298

[16] Lower R, Lower J, Frank H, Harzmann R, Kurth R. Human teratocarcinomas cultured in vitro produce unique retrovirus-like viruses. *J Gen Virol.* 1984;65 (Pt 5): 887-98.

[17] Wang-Johanning F, Liu J, Rycaj K, Huang M, Tsai K, Rosen DG, et al. Expression of multiple human endogenous retrovirus surface envelope proteins in ovarian cancer. *Int J Cancer.* 2007;120:81-90. DOI:10.1002/ijc.22256.

[18] Florl AR, Lower R, Schmitz-Drager BJ, Schulz WA. DNA methylation and expression of LINE-1 and HERV-K provirus sequences in urothelial and renal cell carcinomas. *Br J Cancer.* 1999;80:1312-21. DOI: 10.1038/sj.bjc.6690524.

[19] Menendez L, Benigno BB, McDonald JF. L1 and HERV-W retrotransposons are hypomethylated in human ovarian carcinomas. *Mol Cancer.* 2004;3:12. DOI: 10.1186/1476-4598-3-12.

[20] Wentzensen N, Coy JF, Knaebel HP, Linnebacher M, Wilz B, Gebert J, et al. Expression of an endogenous retroviral sequence from the HERV-H group in gastrointestinal cancers. *Int J Cancer.* 2007;121:1417-23. DOI:10.1002/ijc.22826.

[21] Antony JM, Ellestad KK, Hammond R, Imaizumi K, Mallet F, Warren KG, et al. The human endogenous retrovirus envelope glycoprotein, syncytin-1, regulates neuroinflammation and its receptor expression in multiple sclerosis: a role for endoplasmic reticulum chaperones in astrocytes. *J Immunol.* 2007;179:1210-24. DOI: 10.4049/jimmunol.179.2.1210

[22] Antony JM, van Marle G, Opii W, Butterfield DA, Mallet F, Yong VW, et al. Human endogenous retrovirus glycoprotein-mediated induction of redox reactants causes oligodendrocyte death and demyelination. *Nat Neurosci.* 2004;7:1088-95. DOI:10.1038/nn1319.

[23] Antony JM, Zhu Y, Izad M, Warren KG, Vodjgani M, Mallet F, et al. Comparative expression of human endogenous retrovirus-W genes in multiple sclerosis. *AIDS Res Hum Retroviruses.* 2007;23:1251-6. DOI:10.1089/aid.2006.0274..

[24] Arru G, Mameli G, Astone V, Serra C, Huang YM, Link H, et al. Multiple Sclerosis and HERV-W/MSRV: A Multicentric Study. *Int J Biomed Sci.* 2007;3:292-7.

[25] Brudek T, Christensen T, Aagaard L, Petersen T, Hansen HJ, Moller-Larsen A. B cells and monocytes from patients with active multiple sclerosis exhibit increased surface expression of both HERV-H Env and HERV-W Env, accompanied by increased seroreactivity. *Retrovirology.* 2009;6:104. DOI:10.1186/1742-4690-6-104.

[26] Dolei A. MSRV/HERV-W/syncytin and its linkage to multiple sclerosis: the usability and the hazard of a human endogenous retrovirus. *J Neurovirol.* 2005;11:232-5. DOI: 10.1080/13550280590952899.

[27] Dolei A, Perron H. The multiple sclerosis-associated retrovirus and its HERV-W endogenous family: a biological interface between virology, genetics, and immunology in human physiology and disease. *J Neurovirol.* 2009;15:4-13. DOI: 10.1080/13550280802448451.

[28] Dolei A, Serra C, Mameli G, Pugliatti M, Sechi G, Cirotto MC, et al. Multiple sclerosis-associated retrovirus (MSRV) in Sardinian MS patients. *Neurology*. 2002;58:471-3. DOI:10.1212/WNL.58.3.471.

[29] Hansen B, Oturai AB, Harbo HF, Celius EG, Nissen KK, Laska MJ, et al. Genetic association of multiple sclerosis with the marker rs391745 near the endogenous retroviral locus HERV-Fc1: analysis of disease subtypes. *PLoS One*. 2011;6:e26438. DOI:10.1371/journal.pone.0026438.

[30] Laska MJ, Brudek T, Nissen KK, Christensen T, Moller-Larsen A, Petersen T, et al. Expression of HERV-Fc1, a human endogenous retrovirus, is increased in patients with active multiple sclerosis. *J Virol*. 2012;86:3713-22. DOI:10.1128/JVI.06723-11.

[31] Mameli G, Astone V, Arru G, Marconi S, Lovato L, Serra C, et al. Brains and peripheral blood mononuclear cells of multiple sclerosis (MS) patients hyperexpress MS-associated retrovirus/HERV-W endogenous retrovirus, but not Human herpesvirus 6. *J Gen Virol*. 2007;88:264-74. DOI: 10.1099/vir.0.81890-0

[32] Mameli G, Astone V, Khalili K, Serra C, Sawaya BE, Dolei A. Regulation of the syncytin-1 promoter in human astrocytes by multiple sclerosis-related cytokines. *Virology*. 2007;362:120-30. DOI:10.1016/j.virol.2006.12.019.

[33] Rolland A, Jouvin-Marche E, Saresella M, Ferrante P, Cavaretta R, Creange A, et al. Correlation between disease severity and in vitro cytokine production mediated by MSRV (multiple sclerosis associated retroviral element) envelope protein in patients with multiple sclerosis. *J Neuroimmunol.* 2005;160:195-203. DOI:10.1016/j.jneuroim.2004.10.019.

[34] Ruprecht K, Obojes K, Wengel V, Gronen F, Kim KS, Perron H, et al. Regulation of human endogenous retrovirus W protein expression by herpes simplex virus type 1: implications for multiple sclerosis. *J Neurovirol.* 2006;12:65-71. DOI: 10.1080/13550280600614973.

[35] Schmitt K, Richter C, Backes C, Meese E, Ruprecht K, Mayer J. Comprehensive analysis of human endogenous retrovirus group HERV-W locus transcription in multiple sclerosis brain lesions by high-throughput amplicon sequencing. *J Virol.* 2013;87:13837-52. DOI:10.1128/JVI.02388-13..

[36] Sotgiu S, Arru G, Mameli G, Serra C, Pugliatti M, Rosati G, et al. Multiple sclerosis-associated retrovirus in early multiple sclerosis: a six-year follow-up of a Sardinian cohort. *Mult Scler*. 2006;12:698-703. DOI:10.1177/1352458506070773

[37] Tai AK, O'Reilly EJ, Alroy KA, Simon KC, Munger KL, Huber BT, et al. Human endogenous retrovirus-K18 Env as a risk factor in multiple sclerosis. *Mult Scler*. 2008;14:1175-80. DOI:10.1177/1352458508094641.

[38] Sicat J, Sutkowski N, Huber BT. Expression of human endogenous retrovirus HERV-K18 superantigen is elevated in juvenile rheumatoid arthritis. *J Rheumatol*. 2005;32:1821-31.

[39] Moles JP, Tesniere A, Guilhou JJ. A new endogenous retroviral sequence is expressed in skin of patients with psoriasis. *Br J Dermatol*. 2005;153:83-9. DOI:10.1111/j.1365-2133.2005.06555.x.

[40] Wu Z, Mei X, Zhao D, Sun Y, Song J, Pan W, et al. DNA methylation modulates HERV-E expression in CD4+ T cells from systemic lupus erythematosus patients. *J Dermatol Sci*. 2015;77:110-6. DOI:10.1016/j.jdermsci.2014.12.004.

[41] Perron H, Mekaoui L, Bernard C, Veas F, Stefas I, Leboyer M. Endogenous retrovirus type W GAG and envelope protein antigenemia in serum of schizophrenic patients. *Biol Psychiatry*. 2008;64:1019-23. DOI:10.1016/j.biopsych.2008.06.028.

[42] Yao Y, Schroder J, Nellaker C, Bottmer C, Bachmann S, Yolken RH, et al. Elevated levels of human endogenous retrovirus-W transcripts in blood cells from patients with first episode schizophrenia. *Genes Brain Behav*. 2008;7:103-12. DOI:10.1111/j.1601-183X.2007.00334.x.

[43] Lee YN, Bieniasz PD. Reconstitution of an infectious human endogenous retrovirus. *PLoS Pathog*. 2007;3:e10. DOI: 10.1371/journal.ppat.0030010.

[44] Young GR, Eksmond U, Salcedo R, Alexopoulou L, Stoye JP, Kassiotis G. Resurrection of endogenous retroviruses in antibody-deficient mice. *Nature*. 2012;491:774-8. DOI:10.1038/nature11599.

[45] Alvarez-Lafuente R, Garcia-Montojo M, De Las Heras V, Dominguez-Mozo MI, Bartolome M, Benito-Martin MS, et al. Herpesviruses and human endogenous retroviral sequences in the cerebrospinal fluid of multiple sclerosis patients. *Mult Scler*. 2008;14:595-601. DOI:10.1177/1352458507086425.

[46] Mameli G, Madeddu G, Mei A, Uleri E, Poddighe L, Delogu LG, et al. Activation of MSRV-type endogenous retroviruses during infectious mononucleosis and Epstein-Barr virus latency: the missing link with multiple sclerosis? *PLoS One*. 2013;8:e78474. DOI: 10.1371/journal.pone.0078474..

[47] Mameli G, Poddighe L, Mei A, Uleri E, Sotgiu S, Serra C, et al. Expression and activation by Epstein Barr virus of human endogenous retroviruses-W in blood cells and astrocytes: inference for multiple sclerosis. *PLoS One*. 2012;7:e44991. DOI:10.1371/journal.pone.0044991.

[48] Manghera M, Douville RN. Endogenous retrovirus-K promoter: a landing strip for inflammatory transcription factors? *Retrovirology*. 2013;10:16. DOI: 10.1186/1742-4690-10-16.

[49] Matouskova M, Blazkova J, Pajer P, Pavlicek A, Hejnar J. CpG methylation suppresses transcriptional activity of human syncytin-1 in non-placental tissues. *Exp Cell Res*. 2006;312:1011-20. DOI:10.1016/j.yexcr.2005.12.010.

[50] Reiss D, Zhang Y, Mager DL. Widely variable endogenous retroviral methylation levels in human placenta. *Nucleic Acids Res*. 2007;35:4743-54. DOI: 10.1093/nar/gkm455.

[51] Hawkes CH, Giovannoni G, Keir G, Cunnington M, Thompson EJ. Seroprevalence of herpes simplex virus type 2 in multiple sclerosis. *Acta Neurol Scand*. 2006;114:363-7. DOI:10.1111/j.1600-0404.2006.00677.x.

[52] Frank O, Jones-Brando L, Leib-Mosch C, Yolken R, Seifarth W. Altered transcriptional activity of human endogenous retroviruses in neuroepithelial cells after infection with Toxoplasma gondii. *J Infect Dis*. 2006;194:1447-9. DOI:10.1086/508496.

[53] Li F, Nellaker C, Sabunciyan S, Yolken RH, Jones-Brando L, Johansson AS, et al. Transcriptional derepression of the ERVWE1 locus following influenza A virus infection. *J Virol*. 2014;88:4328-37. DOI:10.1128/JVI.03628-13..

[54] Kwun HJ, Han HJ, Lee WJ, Kim HS, Jang KL. Transactivation of the human endogenous retrovirus K long terminal repeat by herpes simplex virus type 1 immediate early protein 0. *Virus Res*. 2002;86:93-100.

[55] Lee WJ, Kwun HJ, Kim HS, Jang KL. Activation of the human endogenous retrovirus W long terminal repeat by herpes simplex virus type 1 immediate early protein 1. *Mol Cells*. 2003;15:75-80.

[56] Misra A, Chosdol K, Sarkar C, Mahapatra AK, Sinha S. Alteration of a sequence with homology to human endogenous retrovirus (HERV-K) in primary human glioma: implications for viral repeat mediated rearrangement. *Mutat Res*. 2001;484:53-9. DOI: 10.1016/S0027-5107(01)00240-8.

[57] Kremer D, Schichel T, Forster M, Tzekova N, Bernard C, van der Valk P, et al. Human endogenous retrovirus type W envelope protein inhibits oligodendroglial precursor cell differentiation. *Ann Neurol*. 2013;74:721-32. DOI: 10.1002/ana.23970..

[58] Rolland A, Jouvin-Marche E, Viret C, Faure M, Perron H, Marche PN. The envelope protein of a human endogenous retrovirus-W family activates innate immunity through CD14/TLR4 and promotes Th1-like responses. *J Immunol*. 2006;176:7636-44. DOI: 10.4049/jimmunol.176.12.7636

[59] Duperray A, Barbe D, Raguenez G, Weksler BB, Romero IA, Couraud PO, et al. Inflammatory response of endothelial cells to a human endogenous retrovirus associat-

ed with multiple sclerosis is mediated by TLR4. *Int Immunol.* 2015;27. DOI:10.1093/intimm/dxv025.

[60] Hurst T, Magiorkinis G. Activation of the innate immune response by endogenous retroviruses. *J Gen Virol.* 2014. DOI:10.1099/jgv.0.000017.

[61] Herman A, Kappler JW, Marrack P, Pullen AM. Superantigens: mechanism of T-cell stimulation and role in immune responses. *Annu Rev Immunol.* 1991;9:745-72. DOI: 10.1146/annurev.iy.09.040191.003525.

[62] Zhang J, Vandevyver C, Stinissen P, Mertens N, van den Berg-Loonen E, Raus J. Activation and clonal expansion of human myelin basic protein-reactive T cells by bacterial superantigens. *J Autoimmun.* 1995;8:615-32. DOI:10.1016/0896-8411(95)90012-8.

[63] Mangeney M, Renard M, Schlecht-Louf G, Bouallaga I, Heidmann O, Letzelter C, et al. Placental syncytins: Genetic disjunction between the fusogenic and immunosuppressive activity of retroviral envelope proteins. *Proc Natl Acad Sci U S A.* 2007;104:20534-9. DOI:10.1073/pnas.0707873105.

[64] Morozov VA, Dao Thi VL, Denner J. The transmembrane protein of the human endogenous retrovirus--K (HERV-K) modulates cytokine release and gene expression. *PLoS One.* 2013;8:e70399. DOI:10.1371/journal.pone.0070399.

[65] Grasset EK, Cerutti A. Immunology. Retroviral help for B cells. *Science.* 2014;346:1454-5. DOI:10.1371/journal.pone.0070399.

[66] Zeng M, Hu Z, Shi X, Li X, Zhan X, Li XD, et al. MAVS, cGAS, and endogenous retroviruses in T-independent B cell responses. *Science.* 2013;346:1486-92. DOI: 10.1126/science.346.6216.1486.

[67] Nexo BA, Villesen P, Nissen KK, Lindegaard HM, Rossing P, Petersen T, et al. Are human endogenous retroviruses triggers of autoimmune diseases? Unveiling associations of three diseases and viral loci. *Immunol Res.* 2015. DOI:10.1007/s12026-015-8671-z.

[68] Marguerat S, Wang WY, Todd JA, Conrad B. Association of human endogenous retrovirus K-18 polymorphisms with type 1 diabetes. *Diabetes.* 2004;53:852-4. DOI: 10.2337/diabetes.53.3.852

[69] Noseworthy JH, Lucchinetti C, Rodriguez M, Weinshenker BG. Multiple sclerosis. *N Engl J Med.* 2000;343:938-52.

[70] Oksenberg JR, Baranzini SE. Multiple sclerosis genetics--is the glass half full, or half empty? *Nat Rev Neurol.* 2010;6:429-37. DOI:10.1038/nrneurol.2010.91.

[71] Browne P, Chandraratna D, Angood C, Tremlett H, Baker C, Taylor BV, et al. Atlas of Multiple Sclerosis 2013: A growing global problem with widespread inequity. *Neurology.* 2013;83:1022-4. DOI:10.1212/WNL.0000000000000768.

[72] IMSGC, Hafler DA, Compston A, Sawcer S, Lander ES, Daly MJ, et al. Risk alleles for multiple sclerosis identified by a genomewide study. *N Engl J Med*. 2007;357:851-62.

[73] IMSGC, Sawcer S, Hellenthal G, Pirinen M, Spencer CC, Patsopoulos NA, et al. Genetic risk and a primary role for cell-mediated immune mechanisms in multiple sclerosis. *Nature*. 2011;476:214-9. DOI:10.1038/nature10251.

[74] Gourraud PA, Harbo HF, Hauser SL, Baranzini SE. The genetics of multiple sclerosis: an up-to-date review. *Immunol Rev*. 2012;248:87-103. DOI: 10.1111/j.1600-065X.2012.01134.x.

[75] IMSGC, Beecham AH, Patsopoulos NA, Xifara DK, Davis MF, Kemppinen A, et al. Analysis of immune-related loci identifies 48 new susceptibility variants for multiple sclerosis. *Nat Genet*. 2013;45:1353-60. DOI:10.1038/ng.2770..

[76] Perron H, Geny C, Laurent A, Mouriquand C, Pellat J, Perret J, et al. Leptomeningeal cell line from multiple sclerosis with reverse transcriptase activity and viral particles. *Res Virol*. 1989;140:551-61.

[77] Perron H, Garson JA, Bedin F, Beseme F, Paranhos-Baccala G, Komurian-Pradel F, et al. Molecular identification of a novel retrovirus repeatedly isolated from patients with multiple sclerosis. The Collaborative Research Group on Multiple Sclerosis. *Proc Natl Acad Sci U S A*. 1997;94:7583-8.

[78] Pavlicek A, Paces J, Elleder D, Hejnar J. Processed pseudogenes of human endogenous retroviruses generated by LINEs: their integration, stability, and distribution. *Genome Res*. 2002;12:391-9. DOI:10.1101/gr.216902

[79] Mameli G, Poddighe L, Astone V, Delogu G, Arru G, Sotgiu S, et al. Novel reliable real-time PCR for differential detection of MSRVenv and syncytin-1 in RNA and DNA from patients with multiple sclerosis. *J Virol Methods*. 2009;161:98-106. DOI: 10.1016/j.jviromet.2009.05.024..

[80] Roebke C, Wahl S, Laufer G, Stadelmann C, Sauter M, Mueller-Lantzsch N, et al. An N-terminally truncated envelope protein encoded by a human endogenous retrovirus W locus on chromosome Xq22.3. *Retrovirology*. 2010;7:69. DOI: 10.1186/1742-4690-7-69.

[81] Laufer G, Mayer J, Mueller BF, Mueller-Lantzsch N, Ruprecht K. Analysis of transcribed human endogenous retrovirus W env loci clarifies the origin of multiple sclerosis-associated retrovirus env sequences. *Retrovirology*. 2009;6:37. DOI: 10.1186/1742-4690-6-37.

[82] Garcia-Montojo M, de la Hera B, Varade J, de la Encarnacion A, Camacho I, Dominguez-Mozo M, et al. HERV-W polymorphism in chromosome X is associated with multiple sclerosis risk and with differential expression of MSRV. *Retrovirology*. 2014;11:2.

[83] Varadé J, García-Montojo M, de la Hera B, Camacho I, García-Martíneza MA, Arroyo R, et al. Multiple sclerosis retrovirus-like envelope gene: Role of the chromosome 20 insertion. *BBA Clin*. 2015;3:162-7. DOI:10.1016/j.bbacli.2015.02.002.

[84] Garcia-Montojo M, Dominguez-Mozo M, Arias-Leal A, Garcia-Martinez A, De las Heras V, Casanova I, et al. The DNA copy number of human endogenous retrovirus-W (MSRV-type) is increased in multiple sclerosis patients and is influenced by gender and disease severity. *PLoS One*. 2013;8:e53623. DOI:10.1371/journal.pone.0053623.

[85] Garson JA, Tuke PW, Giraud P, Paranhos-Baccala G, Perron H. Detection of virion-associated MSRV-RNA in serum of patients with multiple sclerosis. *Lancet*. 1998;351:33. DOI:10.1016/S0140-6736(98)24001-3.

[86] Perron H, Germi R, Bernard C, Garcia-Montojo M, Deluen C, Farinelli L, et al. Human endogenous retrovirus type W envelope expression in blood and brain cells provides new insights into multiple sclerosis disease. *Mult Scler*. 2012;18:1721-36. DOI: 10.1177/1352458512441381..

[87] Serra C, Mameli G, Arru G, Sotgiu S, Rosati G, Dolei A. In vitro modulation of the multiple sclerosis (MS)-associated retrovirus by cytokines: implications for MS pathogenesis. *J Neurovirol*. 2003;9:637-43. DOI:10.1080/13550280390246462.

[88] Mameli G, Serra C, Astone V, Castellazzi M, Poddighe L, Fainardi E, et al. Inhibition of multiple-sclerosis-associated retrovirus as biomarker of interferon therapy. *J Neurovirol*. 2008;14:73-7. DOI:10.1080/13550280701801107.

[89] Arru G, Leoni S, Pugliatti M, Mei A, Serra C, Delogu LG, et al. Natalizumab inhibits the expression of human endogenous retroviruses of the W family in multiple sclerosis patients: a longitudinal cohort study. *Mult Scler*. 2014;20:174-82. DOI: 10.1177/1352458513494957.

[90] Antony JM, Deslauriers AM, Bhat RK, Ellestad KK, Power C. Human endogenous retroviruses and multiple sclerosis: innocent bystanders or disease determinants? *Biochim Biophys Acta*. 2011;1812:162-76. DOI: 10.1016/j.bbadis.2010.07.016..

[91] Perron H, Jouvin-Marche E, Michel M, Ounanian-Paraz A, Camelo S, Dumon A, et al. Multiple sclerosis retrovirus particles and recombinant envelope trigger an abnormal immune response in vitro, by inducing polyclonal Vbeta16 T-lymphocyte activation. *Virology*. 2001;287:321-32. DOI:10.1006/viro.2001.1045.

[92] Firouzi R, Rolland A, Michel M, Jouvin-Marche E, Hauw JJ, Malcus-Vocanson C, et al. Multiple sclerosis-associated retrovirus particles cause T lymphocyte-dependent death with brain hemorrhage in humanized SCID mice model. *J Neurovirol*. 2003;9:79-93. DOI:10.1080/13550280390173328.

[93] de la Hera B, Varade J, Garcia-Montojo M, Lamas JR, de la Encarnacion A, Arroyo R, et al. Role of the human endogenous retrovirus HERV-K18 in autoimmune disease

susceptibility: study in the Spanish population and meta-analysis. *PLoS One*. 2013;8:e62090. DOI:10.1371/journal.pone.0062090..

[94] Nexo BA, Christensen T, Frederiksen J, Moller-Larsen A, Oturai AB, Villesen P, et al. The etiology of multiple sclerosis: genetic evidence for the involvement of the human endogenous retrovirus HERV-Fc1. *PLoS One*. 2011;6:e16652. DOI: 10.1371/journal.pone.0016652.

[95] de la Hera B, Varade J, Garcia-Montojo M, Alcina A, Fedetz M, Alloza I, et al. Human endogenous retrovirus HERV-Fc1 association with multiple sclerosis susceptibility: a meta-analysis. *PLoS One*. 2014;9:e90182. DOI:10.1371/journal.pone.0090182.

[96] Kraus B, Fischer K, Sliva K, Schnierle BS. Vaccination directed against the human endogenous retrovirus-K (HERV-K) gag protein slows HERV-K gag expressing cell growth in a murine model system. *Virol J*. 2014;11:58. DOI:10.1186/1743-422X-11-58.

[97] Turcanova VL, Bundgaard B, Hollsberg P. Human herpesvirus-6B induces expression of the human endogenous retrovirus K18-encoded superantigen. *J Clin Virol*. 2009;46:15-9. DOI:10.1016/j.jcv.2009.05.015.

[98] Tai AK, Luka J, Ablashi D, Huber BT. HHV-6A infection induces expression of HERV-K18-encoded superantigen. *J Clin Virol*. 2009;46:47-8. DOI:10.1016/j.jcv.2009.05.019.

[99] Hsiao FC, Lin M, Tai A, Chen G, Huber BT. Cutting edge: Epstein-Barr virus transactivates the HERV-K18 superantigen by docking to the human complement receptor 2 (CD21) on primary B cells. *J Immunol*. 2006;177:2056-60. DOI: 10.4049/ jimmunol.177.4.2056

[100] Sutkowski N, Conrad B, Thorley-Lawson DA, Huber BT. Epstein-Barr virus transactivates the human endogenous retrovirus HERV-K18 that encodes a superantigen. *Immunity*. 2001;15:579-89. DOI:10.1016/S1074-7613(01)00210-2.

[101] Hsiao FC, Tai AK, Deglon A, Sutkowski N, Longnecker R, Huber BT. EBV LMP-2A employs a novel mechanism to transactivate the HERV-K18 superantigen through its ITAM. *Virology*. 2009;385:261-6. DOI:10.1016/j.virol.2008.11.025.

[102] Sutkowski N, Chen G, Calderon G, Huber BT. Epstein-Barr virus latent membrane protein LMP-2A is sufficient for transactivation of the human endogenous retrovirus HERV-K18 superantigen. *J Virol*. 2004;78:7852-60. DOI:10.1128/JVI.78.14.7852-7860.2004

[103] Buljevac D, van Doornum GJ, Flach HZ, Groen J, Osterhaus AD, Hop W, et al. Epstein-Barr virus and disease activity in multiple sclerosis. *J Neurol Neurosurg Psychiatry*. 2005;76:1377-81. DOI:10.1136/jnnp.2004.048504

[104] Christensen T. The role of EBV in MS pathogenesis. *Int MS J*. 2006;13:52-7.

[105] Nielsen TR, Pedersen M, Rostgaard K, Frisch M, Hjalgrim H. Correlations between Epstein-Barr virus antibody levels and risk factors for multiple sclerosis in healthy individuals. *Mult Scler*. 2007;13:420-3. DOI:10.1177/1352458506071470

[106] DeLorenze GN, Munger KL, Lennette ET, Orentreich N, Vogelman JH, Ascherio A. Epstein-Barr virus and multiple sclerosis: evidence of association from a prospective study with long-term follow-up. *Arch Neurol*. 2006;63:839-44. DOI:10.1001/archneur.63.6.noc50328.

[107] Hollsberg P, Kusk M, Bech E, Hansen HJ, Jakobsen J, Haahr S. Presence of Epstein-Barr virus and human herpesvirus 6B DNA in multiple sclerosis patients: associations with disease activity. *Acta Neurol Scand*. 2005;112:395-402. DOI:10.1111/j.1600-0404.2005.00516.x.

[108] Martinez A, Alvarez-Lafuente R, Mas A, Bartolome M, Garcia-Montojo M, de Las Heras V, et al. Environment-gene interaction in multiple sclerosis: human herpesvirus 6 and MHC2TA. *Hum Immunol*. 2007;68:685-9. DOI:10.1016/j.humimm.2007.05.005.

[109] Benit L, Calteau A, Heidmann T. Characterization of the low-copy HERV-Fc family: evidence for recent integrations in primates of elements with coding envelope genes. *Virology*. 2003;312:159-68. DOI:10.1016/S0042-6822(03)00163-6.

[110] Strissel PL, Ruebner M, Thiel F, Wachter D, Ekici AB, Wolf F, et al. Reactivation of codogenic endogenous retroviral (ERV) envelope genes in human endometrial carcinoma and prestages: Emergence of new molecular targets. *Oncotarget*. 2012;3:1204-19.

[111] Rhyu DW, Kang YJ, Ock MS, Eo JW, Choi YH, Kim WJ, et al. Expression of human endogenous retrovirus env genes in the blood of breast cancer patients. *Int J Mol Sci*. 2014;15:9173-83. DOI: 10.3390/ijms15069173.

[112] Kremer D, Forster M, Schichel T, Gottle P, Hartung HP, Perron H, et al. The neutralizing antibody GNbAC1 abrogates HERV-W envelope protein-mediated oligodendroglial maturation blockade. *Mult Scler*. 2014;21:1200-3. DOI: 10.1177/1352458514560926.

[113] Curtin F, Perron H, Kromminga A, Porchet H, Lang AB. Preclinical and early clinical development of GNbAC1, a humanized IgG4 monoclonal antibody targeting endogenous retroviral MSRV-Env protein. *MAbs*. 2015;7:265-75. DOI: 10.4161/19420862.2014.985021.

[114] Derfuss T, Curtin F, Guebelin C, Bridel C, Rasenack M, Matthey A, et al. A phase IIa randomised clinical study of GNbAC1, a humanised monoclonal antibody against the envelope protein of multiple sclerosis-associated endogenous retrovirus in multiple sclerosis patients. *Mult Scler*. 2014;21:885-93. DOI: 10.1177/1352458514554052.

[115] Curtin F, Lang AB, Perron H, Laumonier M, Vidal V, Porchet HC, et al. GNbAC1, a humanized monoclonal antibody against the envelope protein of multiple sclerosis-

associated endogenous retrovirus: a first-in-humans randomized clinical study. *Clin Ther*. 2012;34:2268-78. DOI:10.1016/j.clinthera.2012.11.006.

[116] Downey RF, Sullivan FJ, Wang-Johanning F, Ambs S, Giles FJ, Glynn SA. Human endogenous retrovirus K and cancer: Innocent bystander or tumorigenic accomplice? *Int J Cancer*. 2014;137:1249-57. DOI: 10.1002/ijc.29003.

[117] van der Kuyl AC. HIV infection and HERV expression: a review. *Retrovirology*. 2012;9:6. DOI:10.1186/1742-4690-9-6.

[118] Bhardwaj N, Maldarelli F, Mellors J, Coffin JM. HIV-1 infection leads to increased transcription of human endogenous retrovirus HERV-K (HML-2) proviruses in vivo but not to increased virion production. *J Virol*. 2014;88:11108-20. DOI:10.1128/JVI.01623-14.

[119] Michaud HA, SenGupta D, de Mulder M, Deeks SG, Martin JN, Kobie JJ, et al. Cutting edge: An antibody recognizing ancestral endogenous virus glycoproteins mediates antibody-dependent cellular cytotoxicity on HIV-1-infected cells. *J Immunol*. 2014;193:1544-8. DOI:10.4049/jimmunol.1302108.

7

Vaccines and Antiviral Agents

Hongxuan He

1. Introduction

There has been a long history of the battle between viral diseases and the mankind. The arms at our disposal against the virus invasion are continuously expanding its inventory. Most of them fall into the category of vaccines and antiviral and each of the two kinds of viral diseases intervention agents has its own advantages and limitations.

2. Vaccine

A vaccine is a biological preparation that improves immunity to a particular disease. A vaccine typically contains an agent that resembles a disease-causing microorganism, and is often made from weakened or killed forms of the microbe, its toxins or one of its surface proteins. The agent stimulates the body's immune system to recognize the agent as foreign, destroy it, and "remember" it, so that the immune system can more easily recognize and destroy any of these microorganisms that it later encounters.

Vaccines are dead or inactivated organisms or purified products derived from them.

There are several types of vaccines in use. These represent different strategies used to try to reduce risk of illness, while retaining the ability to induce a beneficial immune response.

2.1. Inactivated

Some vaccines contain killed, but previously virulent, micro-organisms that have been destroyed with chemicals, heat, radioactivity or antibiotics. Examples are the influenza vaccine, cholera vaccine,bubonic plague vaccine, polio vaccine, hepatitis A vaccine, and rabies vaccine.

2.2. Attenuated

Some vaccines contain live, attenuated microorganisms. Many of these are live viruses that have been cultivated under conditions that disable their virulent properties, or which use closely related but less dangerous organisms to produce a broad immune response [1]. They typically provoke more durable immunological responses and are the preferred type for healthy adults. Examples include the viral diseases yellow fever, measles, rubella, and mumps. Attenuated vaccines have some advantages and disadvantages. They have the capacity of transient growth so they give prolonged protection, and no booster dose is required. But they may get reverted to the virulent form and cause the disease.

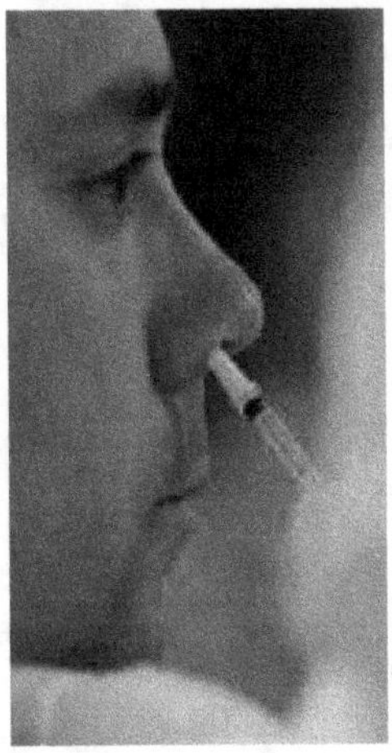

Figure 1. H1N1 flu nasal spray as an example of attenuated vaccine

2.3. Subunit

Protein subunit- rather than introducing an inactivated or attenuated micro-organism to an immune system (which would constitute a "whole-agent" vaccine), a fragment of it can create an immune response. Examples include the subunit vaccine against Hepatitis B virus that is composed of only the surface proteins of the virus (previously extracted from the blood stream of chronically infected patients, but now produced by recombination of the viral genes into yeast), the virus like particle (VLP) vaccine against human papillomavirus (HPV) that is composed of the viral major capsidprotein, and the hemagglutinin and neuraminidase subunits of the influenza virus. One method of production involves isolation of a specific protein from a virus and administering this by itself. A weakness of this technique is that isolated proteins can be denatured and will then bind to different antibodies than the proteins in the virus. A second method of subunit vaccine is the recombinant vaccine, which involves putting a protein gene from the targeted virus into another virus. The second virus will express the protein, but will not present a risk to the injector. This is the type of vaccine currently in use for hepatitis, and it is experimentally popular, being used to try to develop new vaccines for difficult to vaccinate viruses such as Ebola and HIV.

2.4. DNA Vaccine

In the past decade and a half, the DNA vaccine concept has been tested and applied against various pathogens and tumor antigens [2]. The optimized gene sequence of interest is delivered to the skin, subcutaneum or muscle by one of several delivery methods [3]. The expression of plasmid-encoded genes will produce foreign antigens and elicits immunological response. Until now, four DNA vaccine products have been approved, all in the area of veterinary medicine [4].

Vaccines are very effective on stable viruses, but are of limited use in treating a patient who has already been infected. It is also difficult to successfully deploy them against rapidly mutating viruses, such as influenza (the vaccine for which is updated every year) and HIV [5]. Antiviral drugs are particularly useful in these cases.

Vaccine Target	Product Name	Company involved	Date licensed and country	Target organism	Benefits
West Nile virus	West Nile Innovator	Centers for Disease Control and Prevention and Fort Dodge Laboratories	2005 USA	Horses	Protects against West Nile virus infection
Infectious Haematopoietic necrosis virus	Apex-IHN	Novartis	2005 Canada	Salmon	Improves animal welfare, increase food quality and quantity
Growth hormone releasing hormone	LifeTide-SWS	VGX Animal Health	2007 Austrilia	Swine and food Animals	Increases the number of piglets weaned in breeding sows
Melanoma	Canine Melanoma Vaccine	Merial, Memorial Sloan-Kettering Cancer Center and The Animal Medical Center of New York	2007 USA, conditional license	Dogs	Treats aggressive forms of cancer of the mouth, nail bed, foot pad or other areas as an alternative to radiation and surgery

Table 1. Current licensed DNA therapies (Adapted from Kutzler MA & Weiner et. al)

3. Antiviral agent

Antiviral drugs are a class of medication used specifically for treating viral infections. Like antibiotics for bacteria, specific antivirals are used for specific viruses. Unlike most antibiotics, antiviral drugs do not destroy their target pathogen; instead they inhibit their development [6].

Most of the antiviral drugs now available are designed to help deal with HIV, herpes viruses (best known for causing cold sores and genital herpes, but actually the cause of a wide range of other diseases, such as chicken pox), the hepatitis B and C viruses, which can cause liver cancer, and influenza A and B viruses. Researchers are working to extend the range of antivirals to other families of pathogens.

Designing safe and effective antiviral drugs is difficult, because viruses use the host's cells to replicate. This makes it difficult to find targets for the drug that would interfere with the virus without harming the host organism's cells. Moreover, the major difficulty in developing vaccines and anti-viral drugs is due to viral variation.

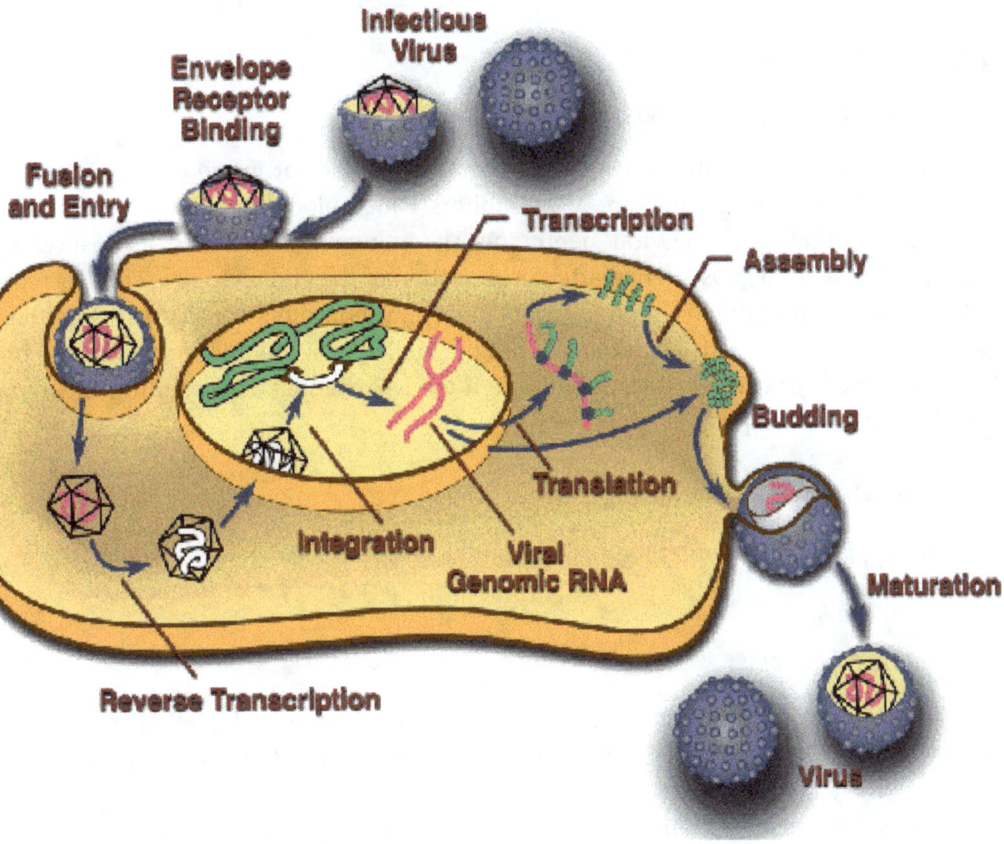

Figure 2. Virus life cycle and targets of antivirals

3.1. Before cell entry

One anti-viral strategy is to interfere with the ability of a virus to infiltrate a target cell. The virus must go through a sequence of steps to do this, beginning with binding to a specific "receptor" molecule on the surface of the host cell and ending with the virus "uncoating" inside the cell and releasing its contents. Viruses that have a lipid envelope must also fuse their envelope with the target cell, or with a vesicle that transports them into the cell, before they can uncoat [7].

3.1.1. Entry inhibitor

A very early stage of viral infection is viral entry, when the virus attaches to and enters the host cell [8]. A number of "entry-inhibiting" or "entry-blocking" drugs are being developed to fight HIV. HIV most heavily targets the immune system's white blood cells known as "helper T cells", and identifies these target cells through T-cell surface receptors designated "CD4" and "CCR5". Attempts to interfere with the binding of HIV with the CD4 receptor have failed to stop HIV from infecting helper T cells, but research continues on trying to interfere with the binding of HIV to the CCR5 receptor in hopes that it will be more effective.

3.1.2. Uncoating inhibitor

Inhibitors of uncoating have also been investigated.

Amantadine and rimantadine, have been introduced to combat influenza. These agents act on penetration/uncoating. They are M2 inhibitors which block the ion channel formed by the M2 protein that spans the viral membrane. The influenza virus enters its host cell by receptor-mediated endocytosis. Thereafter, acidification of the endocytotic vesicles is required for the dissociation of the M1 protein from the ribonucleoprotein complexes. Only then are the ribonucleoprotein particles imported into the nucleus via the nuclear pores. The hydrogen ions needed for acidification pass through the M2 channel. Amantadine and rimantadine block the channel [9].

3.2. During viral synthesis

A second approach is to target the processes that synthesize virus components after a virus invades a cell.

3.2.1. Reverse transcription

One way of doing this is to develop nucleotide or nucleoside analogues that look like the building blocks of RNA or DNA, but deactivate the enzymes that synthesize the RNA or DNA once the analogue is incorporated. This approach is more commonly associated with the inhibition of reverse transcriptase (RNA to DNA) than with "normal" transcriptase (DNA to RNA).

An improved knowledge of the action of reverse transcriptase has led to better nucleoside analogues to treat HIV infections. One of these drugs, lamivudine, has been approved to treat hepatitis B, which uses reverse transcriptase as part of its replication process. Researchers have gone further and developed inhibitors that do not look like nucleosides, but can still block reverse transcriptase.

Another target being considered for HIV antivirals include RNase H-which is a component of reverse transcriptase that splits the synthesized DNA from the original viral RNA.

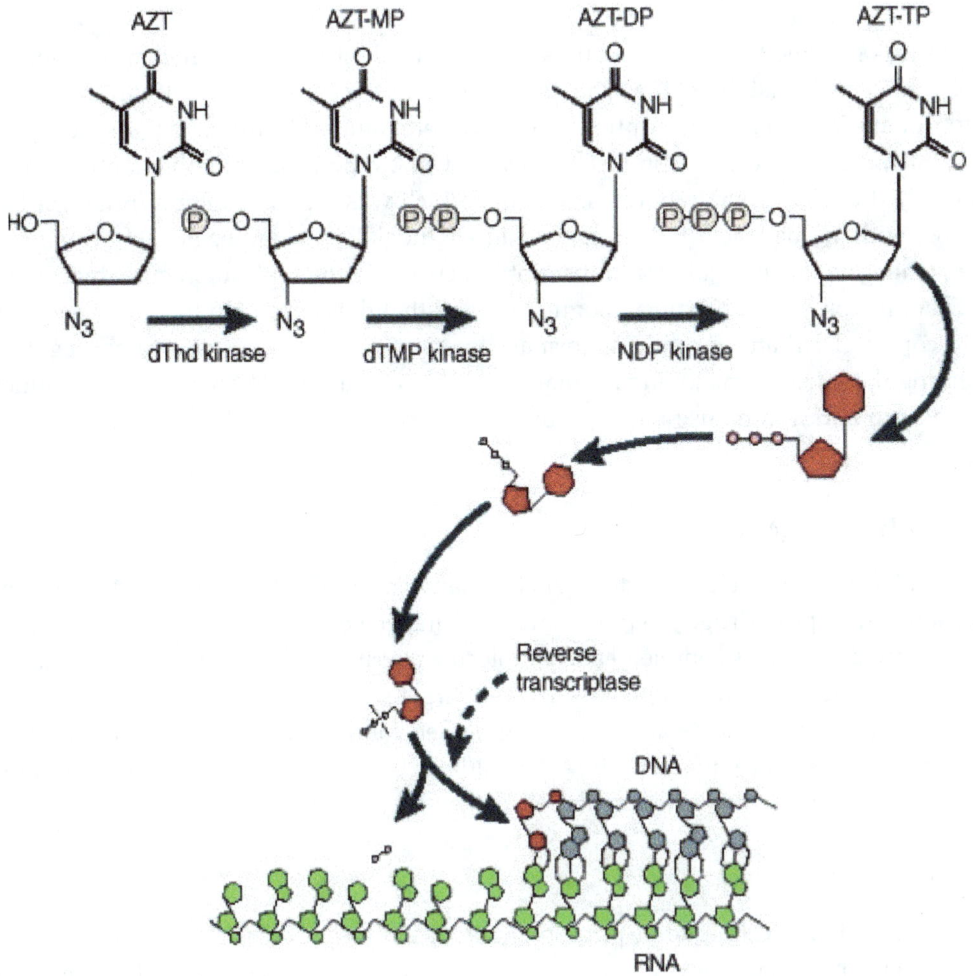

Figure 3. Example of the mechanisms of antivirals: Mechanism of action of azidothymidine (AZT). AZT needs to be phosphorylated, in three steps, to the triphospate form before it can interfere with the reverse transcriptase reaction

3.2.2. Integrase

Another target is integrase, which splices the synthesized DNA into the host cell genome. There appears to be no functional equivalent of the enzyme in human cells. The biochemical mechanism of integration of HIV DNA into the host cell genome involves a carefully defined sequence of DNA tailoring (3'-processing) and coupling (joining or integration) reactions [10]. In spite of some effort in this area targeted at the discovery of therapeutically useful inhibitors of this viral enzyme, there are no drugs for HIV/AIDS in clinical use where the mechanism of action is inhibition of HIV integrase. However there are several promising candidates in several classes of compounds, including nucleotides, dinucleotides, oligonucleotides and miscellaneous small molecules such as heterocyclic systems, natural products, diketo acids and sulfones, that have been discovered as inhibitors of HIV integrase.

3.2.3. Transcription

Once a virus genome becomes operational in a host cell, it then generates messenger RNA (mRNA) molecules that direct the synthesis of viral proteins. Production of mRNA is initiated by proteins known as transcription factors. Several antivirals are now being designed to block attachment of transcription factors to viral DNA. Kao et al. recently identified a compound called nucleozin via random screening, which was found to inhibit influenza by interacting with influenza NP. Nucleozin causes the NPs to aggregate abnormally, and consequently inhibits normal viral transcription, crippling the replication cycle by extension [11]. Examination of a nucleozin analogue revealed that the compound functions by binding to two copies of NP and forming abnormal dimers, causing the proteins to aggregate and preventing them from functioning normally. Nucleozin was also shown to inhibit influenza virus in vitro and in a mouse model, making it a promising candidate for a new antiviral drug.

3.2.4. Translation/antisense

Genomics has not only helped find targets for many antivirals, it has provided the basis for an entirely new type of drug, based on "antisense" molecules. These are segments of DNA or RNA that are designed as complementary molecule to critical sections of viral genomes, and the binding of these antisense segments to these target sections blocks the operation of those genomes. A phosphorothioate antisense drug named fomivirsen has been introduced, used to treat opportunistic eye infections in AIDS patients caused by cytomegalovirus, and other antisense antivirals are in development.

3.2.5. Translation/ribozymes

Yet another antiviral technique inspired by genomics is a set of drugs based on ribozymes, which are enzymes that will cut apart viral RNA or DNA at selected sites. In their natural course, ribozymes are used as part of the viral manufacturing sequence, but these synthetic ribozymes are designed to cut RNA and DNA at sites that will disable them.

A ribozyme antiviral to deal with hepatitis C has been suggested, and ribozyme antivirals are being developed to deal with HIV. An interesting variation of this idea is the use of genetically modified cells that can produce custom-tailored ribozymes. This is part of a broader effort to create genetically modified cells that can be injected into a host to attack pathogens by generating specialized proteins that block viral replication at various phases of the viral life cycle [12].

3.2.6. Protease inhibitors

Some viruses include an enzyme known as a protease that cuts viral protein chains apart so they can be assembled into their final configuration, such as Saquinavir (Figure 4). HIV includes a protease, and so considerable research has been performed to find "protease inhibitors" to attack HIV at that phase of its life cycle. Protease inhibitors became available in the 1990s and have proven effective, though they can have unusual side effects, for example causing fat to build up in unusual places. Improved protease inhibitors are now in development [13].

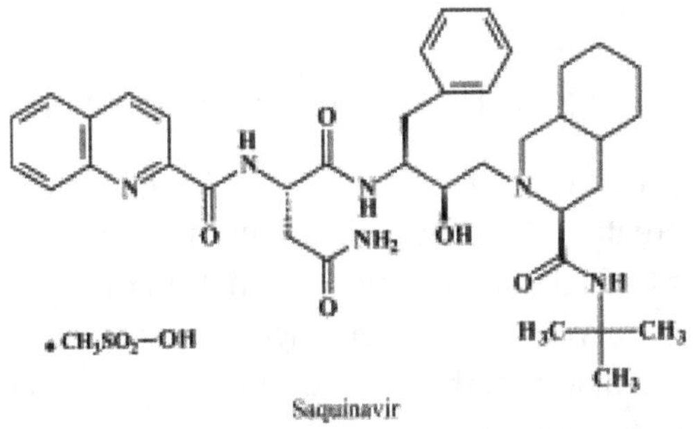

Figure 4. Protease inhibitor antiviral Saquinavir.

Structure: cis-N-tert-butyl-decahydro-2-[2(R)-hydroxy-4-phenyl-3(S)-[[N-2-quinolylcarbonyl-l-asparaginyl]-amino]butyl]-(4aS-8aS)-isoquinoline-3(S)-carboxamide methane sulfonate, hard gel capsules, Invirase®, also available as soft gelatin capsules (Fortovase®).

Activity spectrum: HIV (types 1 and 2).

Mechanism of action: transition-state, hydroxyethylene-based, peptidomimetic inhibitor of HIV protease.

3.3. Release phase

The final stage in the life cycle of a virus is the release of completed viruses from the host cell, and this step has also been targeted by antiviral drug developers. Two drugs named zanamivir (Relenza) and oseltamivir (Tamiflu) that have been recently introduced to treat influenza prevent the release of viral particles by blocking a molecule named neuraminidase that is found on the surface of flu viruses, and also seems to be constant across a wide range of flu strains [14].

3.4. Considerations in the clinical development of antiviral agents

A total of 37 antiviral compounds (not including interferons or immunoglobulins) have momentarily been licensed for the treatment of HIV, HBV, herpesvirus, influenza virus and/or HCV infections [15]. In the preceding sections these compounds have been discussed from the following viewpoints: chemical structure, activity spectrum, mechanism of action, principal clinical indication(s). Other points that need to be considered before the full clinical potential of any given drug could be appreciated, are: (i) duration of treatment, (ii) single- versus multiple-drug therapy, (iii) pharmacokinetics, (iv) drug interactions, (v) toxic side effects and (vi) development of resistance. A particular issue that may be important in the clinical setting is whether the listed anti-HIV agents would be equally suited for the treatment of HIV-2 and HIV-1 infections.

As to the duration of treatment, this may vary from a few days (HSV, VZV, influenza virus infections) to several months or years (HIV, HBV and HCV infections), depending on whether we are dealing with an acute (primary (i.e. influenza) or recurrent (i.e. HSV, VZV) infection or chronic, persistent (i.e. HIV, HBV, HCV) infection. For HIV infections it is still being evaluated whether long-term treatment can be interrupted, without loss of benefit (or increased benefit) to the patient (structured treatment interruption, STI) [16].

While the short-term treatment (5–7 days) of HSV, VZV and influenza virus infections, and even the more prolonged treatment of CMV infections, can be based on single-drug therapy, for the long-term treatment of HIV infections combination of several drugs in a triple-drug cocktail (also referred to as HAART for 'highly active anti-retroviral therapy') has become the standard procedure, and likewise, the long-term treatment of HBV infections may in the future also evolve from single- to dual- or triple-drug therapy [17].

Pharmacokinetic parameters to be addressed, when evaluating the therapeutic potential, include bioavailability (upon either topical, oral or parenteral administration), plasma protein binding affinity, distribution through the organism (penetration into the CNS, when this is needed), metabolism through the liver (i.e. cytochrome P-450 drug-metabolizing enzymes) and elimination through the kidney. Particularly when concocting the multiple-drug combinations for the treatment of HIV infection, possible drug–drug interactions should be taken into account: i.e. some compounds act as P-450 inhibitors and others as P-450 inducers, and this may greatly influence the plasma drug levels achieved, especially in the case of NNRTIs and PIs [18].

Toxic side effects, both short and long-term, must be considered when the drugs have to be administered for a prolonged period, as in the treatment of HIV infections. These side effects may seriously compromise compliance (adherence to drug intake), and could, at least in part, be circumvented by a reduction of the pill burden to, ideally, once-daily dosing.

Finally, resistance development may be an important issue, again for those compounds that have to be taken for a prolonged period, as is generally the case for most of the NRTIs, NNRTIs and PIs currently used in the treatment of HIV infections. Yet, the nucleoside phosphonate analogues (NtRTIs) tenofovir and adefovir do not readily or rapidly lead to resistance development, even after more than 1 year of therapy (for HIV and HBV, respectively). Resistance has been noted with HBV against lamivudine after long-term therapy (>6 months), but, if resistant to lamivudine, HBV infections remain amenable to treatment with adefovir dipivoxil. As has been occasionally observed in immunosuppressed patients, HSV may develop resistance to acyclovir, and CMV to ganciclovir, but, if based on ACV TK or CMV PK deficiency, these resistant viruses remain amenable to treatment with foscarnet and/or cidofovir [19]. In immunocompetent patients, treated for an acute or episodic HSV, VZV or influenza virus infection, short-term therapy is unlikely to engender any resistance problems

The evolution of viral vaccines from the time of Jennerian prophylaxis to today's recombinant technology has been a continuing story of success. From the relatively crude or "first generation" vaccines for smallpox, rabies, and yellow fever followed a second and third generation of improved or new viral vaccines. The application of techniques for attenuating, inactivating, and partially purifying candidate viruses yielded safe, effective vaccines against influenza, poliomyelitis, measles, mumps, and rubella. With the advent of effective national immunization programs in the United States and other areas of the world to promote wide scale use of these vaccines, we have seen a dramatic decrease in incidence of the viral infections of children. The new biotechnology serves as the cornerstone for a fourth generation of vaccines and has already provided a licensed recombinant yeast human hepatitis B vaccine. The prospects for a wide spectrum of new or improved vaccines are highly encouraging, not only because of the recent technical advances but also because vaccine development has been recognized as a priority area of research. Under the National Institute of Allergy and Infectious Diseases' Program for Accelerated Development of New Vaccines, support is being provided for developmental vaccine studies with hepatitis A and B, influenza A and B, rabies, rotavirus, varicella, and respiratory syncytial virus. The outlook for antivirals is equally optimistic. The same technologies that have provided greater insight into the genetics and molecular biology of viruses and hence the means to fashion subunit or even synthetic vaccines have yielded data that can be applied to successful development of targeted antiviral compounds.

Author details

Hongxuan He*

Address all correspondence to: hehx@ioz.ac.cn

National Research Center for Wildlife Borne Diseases, Institute of Zoology, Chinese Academy of Sciences, Beijing, PR China

References

[1] Steel J, et al. (2009) Live attenuated influenza viruses containing NS1 truncations as vaccine candidates against H5N1 highly pathogenic avian influenza. Journal of virology 83[4]:1742-1753.

[2] Kutzler MA & Weiner DB (2008) DNA vaccines: ready for prime time? Nature reviews. Genetics 9[10]:776-788.

[3] Ramakrishna L, Anand KK, Mohankumar KM, & Ranga U (2004) Codon optimization of the tat antigen of human immunodeficiency virus type 1 generates strong immune responses in mice following genetic immunization. Journal of virology 78[17]: 9174-9189.

[4] Tacket CO, et al. (1999) Phase 1 safety and immune response studies of a DNA vaccine encoding hepatitis B surface antigen delivered by a gene delivery device. Vaccine 17[22]:2826-2829.

[5] Roose K, Fiers W, & Saelens X (2009) Pandemic preparedness: toward a universal influenza vaccine. Drug news & perspectives 22[2]:80-92.

[6] De Clercq E (2004) Antiviral drugs in current clinical use. Journal of clinical virology : the official publication of the Pan American Society for Clinical Virology 30[2]: 115-133.

[7] Teissier E, Penin F, & Pecheur EI (2011) Targeting cell entry of enveloped viruses as an antiviral strategy. Molecules 16[1]:221-250.

[8] Doms RW & Moore JP (2000) HIV-1 membrane fusion: targets of opportunity. The Journal of cell biology 151[2]:F9-14.

[9] Hay AJ, Wolstenholme AJ, Skehel JJ, & Smith MH (1985) The molecular basis of the specific anti-influenza action of amantadine. The EMBO journal 4[11]:3021-3024.

[10] Mathe C & Nair V (1999) Potential inhibitors of HIV integrase. Nucleosides & nucleotides 18(4-5):681-682.

[11] Kao RY, et al. (2010) Identification of influenza A nucleoprotein as an antiviral target. Nature biotechnology 28[6]:600-605.

[12] Olmstead AD, Knecht W, Lazarov I, Dixit SB, & Jean F (2012) Human subtilase SKI-1/S1P is a master regulator of the HCV Lifecycle and a potential host cell target for developing indirect-acting antiviral agents. PLoS pathogens 8[1]:e1002468.

[13] Lamarre D, et al. (2003) An NS3 protease inhibitor with antiviral effects in humans infected with hepatitis C virus. Nature 426(6963):186-189.

[14] Moscona A (2005) Neuraminidase inhibitors for influenza. The New England journal of medicine 353[13]:1363-1373.

[15] B. Kronenberger SZ (2012) Antiviral tagets in HCV. Chronic Hepatitis C Virus, ed Shiffman ML (Springer, New York), pp 203-225.

[16] Mancini E, Castiglione F, Bernaschi M, de Luca A, & Sloot PM (2012) HIV reservoirs and immune surveillance evasion cause the failure of structured treatment interruptions: a computational study. PloS one 7[4]:e36108.

[17] Regidor DL, et al. (2011) Effect of highly active antiretroviral therapy on biomarkers of B-lymphocyte activation and inflammation. AIDS 25[3]:303-314.

[18] Flentge CA, et al. (2009) Synthesis and evaluation of inhibitors of cytochrome P450 3A (CYP3A) for pharmacokinetic enhancement of drugs. Bioorganic & medicinal chemistry letters 19[18]:5444-5448.

8

Viral Counter Defense X Antiviral Immunity in Plants: Mechanisms for Survival

Alessandra Tenório Costa, Juliana Pereira Bravo,
Rodrigo Kazuo Makiyama,
Alessandra Vasconcellos Nunes and Ivan G. Maia

1. Introduction

The history of plant virology dates to the late 19th century, when Iwanowski and Beijerinck, who were investigating the cause of a mysterious disease of tobacco, independently described an unusual agent that caused tobacco mosaic disease. This agent was later named *Tobacco mosaic virus* (TMV) [1]. During this period, viruses including *Potato virus* X (PVX), *Potato virus* Y (PVY) and *Lettuce mosaic virus* (LMV) were described. These viruses could be distinguished based on their transmission and method of disease induction. In addition, numerous techniques for the study of viruses were developed.

Viruses are among the most agriculturally important groups of plant pathogens, causing serious economic losses in many major crops by reducing yield and quality. A virus can be defined as a set of one or more nucleic acid template molecules, often encased in a protective coat of protein or lipoprotein, which is able to organise its own replication only within suitable host cells [1]. Because the genetic information encoded by viral genomes is limited, viruses depend entirely on host cells to replicate their genome and produce infectious progeny. Both plant and animal viruses can be classified according to the type of nucleic acid that makes up their genome. In plants, the vast majority of viruses have positive-sense (+) RNA genomes (i.e., the RNA genome has the same polarity as cellular mRNA), although negative-sense (−) RNA and double-stranded RNA viral genomes also exist. Other plant viruses have DNA genomes; the DNA can be double-stranded (caulimoviruses) or single-stranded (geminiviruses) [1, 2]. Cell-to-cell movement of plant viruses occurs through cytoplasmic "bridges" between cells called plasmodesmata, and viruses are able to move systemically throughout plants via the phloem [1].

Viruses depend on other organisms (vectors) to transmit them from diseased to healthy plants. These vectors are often sap-sucking insects such as aphids, thrips, whiteflies, leaf-feeding beetles, plant-feeding mites, soil-inhabiting nematodes or fungal pathogens. Some viruses can be mechanically transmitted, on pruning knives or gardeners hands; or by grafting material, and a relatively small number of species can pass through infected seed [3]. Viruses use a variety of strategies that frequently induce disease in the plants they infect. Different viruses induce distinct diseases, and this can be true even for different strains of the same virus. Virus infection can profoundly alter the physiology of the host due to the interaction with cellular components. In plants, the severity of viral diseases varies considerably depending on the host genotype, the stage of infection and the environmental conditions. Diseases caused by viruses can vary broadly in intensity, from very mild symptoms observed in tolerant plants up to very severe symptoms and plant death [1, 4, 5].

Each plant virus encodes an average of 4-10 proteins necessary to coordinate the complex biochemical and intermolecular interactions required for viral infection cycles. The cycle of infection includes viral genome replication, cell-to-cell and systemic movement and transmission [6]. For efficient viral infection to occur, viral proteins must be able to interact with factors in the host cell, thereby manipulating metabolic pathways and coordinating biochemical interactions that promote infection. Thus, during co-evolution between viruses and their hosts, a variety of complex interactions have developed that involve several distinct mechanisms of plant defence and virus attack.

During evolution, plants have developed diverse defence mechanisms that are activated during viral infection. One of these is the hypersensitive response (HR), which activates initial defence responses that prevent the infection from spreading further and then kills the cells within the infected zone. The onset of a second mechanism, systemic acquired resistance (SAR), then protects the plant together with HR against new attacks by the same pathogen. SAR is induced by a variety of agents after initial infection and can provide resistance to a wide range of pathogens for days [7-10]. The HR and SAR responses are accompanied by changes in gene expression that include the production of pathogenesis-related (PR) proteins and of several proteins involved in cell signalling [11].

Plants also possess other antiviral defence mechanisms such as RNA silencing, a remarkable type of gene regulation based on sequence-specific targeting and degradation of RNA [12], and the more recently described ubiquitin/26S proteasome system (UPS), which plays a central role in the degradation of proteins. The latter system is involved in almost all phases of the defence mechanisms of plants, regardless of the type of pathogen [13]. In addition to its proteolytic activity against ubiquitinated pathogen proteins, which directs their degradation by the 20S proteasome, the degradation of viral RNA can also occur via the ribonuclease activity of the 20S proteasome [14, 15]. While the proteasome as a structure, and RNA silencing as a mechanism, are two conserved features among eukaryotes, several lines of evidence suggest that the proteasome-linked RNase activity is most likely not directly related to RNA silencing. The selective degradation of viral RNAs by the 20S complex can represent an alternative pathway of host defence that occurs in parallel to RNA silencing and reinforces cellular antiviral defence in plants [14].

Over time, the strategies used by the virus to overcome these elaborate host defences can lead to a number of fundamental changes in the plant's physiology. Such changes, including structural modifications in the host cell, may give rise to intranuclear inclusions of various types and may affect the nucleolus or the size and shape of the nucleus. Within the mitochondria, abnormal membrane systems may develop [16, 17]. In plants infected with *Turnip yellow mosaic virus* (TYMV - Tymovirus), abnormalities such as clumping of chloroplasts and abnormal size and number of starch grains in leaf cells may occur, and small vesicles near the periphery and chloroplasts may become greatly enlarged and filled with starch grains [18]. These abnormalities were also observed in squash infected by the *Zucchini yellow mosaic potyvirus* (ZYMV - Potyvirus) [17]. In plant cell walls, three types of abnormality have been observed: abnormal thickening due to the deposition of callose near the edges of virus-induced lesions; cell wall protrusions involving the plasmodesmata (these protrusions may have one or more canals and may be quite short or of considerable length); paramural bodies, which are depositions of electron-dense material between the cell wall and the plasma membrane, may appear and extend over substantial areas of the cell wall, or be limited in extent occurring in association with plasmodesmata. Moreover, in the cytoplasm of an infected cell, virus particles may accumulate in sufficient numbers to form three-dimensional crystalline arrays. The ability to form crystals within the host cell cytoplasm depends on properties of the virus itself, and is not related to the overall concentration reached in the tissue or to the ability of the purified virus to form crystals [1].

Intriguingly, in carrot plants (*Daucus carota* L.) infected by *Cucumber mosaic virus* (CMV) some cytological and physiological changes were observed due to alterations in various host metabolites. With respect to cytological changes, scattered metaphase was observed in the diseased plant cells. The mitotic index of the diseased cells was decreased, while the nucleus / cytoplasm ratio was increased. Chromatin bridges were also observed at anaphase I and II. Physiological changes resulting in decreased carbon, nitrogen and protein content and increased phosphorous content of the virally infected plants have been observed [19].

Other viral counterattack mechanisms involve changes in the plant cell cycle. In plants, as in all eukaryotes, the four phases of the mitotic cell cycle (G1, S, G2 and M) are conserved. During development, plant cells leave the cell division cycle, and in mature plants, DNA replication and the corresponding enzymes are confined to meristematic tissues [20]. Geminiviruses are good models for the study of the relationship between the cell cycle and viral DNA replication because they replicate in differentiated cells, such as mature cells of the leaves, stems and roots, in which most of the cellular factors required for viral DNA replication are normally absent. These cells have left the cell division cycle and no longer contain detectable levels of plant DNA replication enzymes necessary for geminivirus replication [21]. Due to the requirement for cellular factors, geminiviral DNA replication must be coupled to a special state of the infected cell, suggesting that the virus may have evolved mechanisms that affect the expression of cellular genes involved in S-phase progression and G1/S transition [22]. One such a mechanism involved in regulating changes of the host cell cycle appears to be the inactivation of the retinoblastoma protein (pRb), which negatively regulates the G1/S transition in cells. The Rep protein, which is encoded by all geminiviruses and is the only viral protein necessary for viral DNA replication, has been found to induce expression and also interacts with the host "proliferating cell nuclear antigen" (PCNA), an auxiliary protein of DNA polymerases required during replication and repair in non-dividing plant cells. This observation suggests that Rep protein can provide the necessary stimulus to induce the dedifferentiation process. Mechanisms other than sequestering plant pRb most likely contribute to the multiple effects of geminivirus proteins on cellular gene expression, cell growth control and cellular DNA replication [21, 22].

Ultimately, RNA silencing suppressor (RSS) proteins are able to block or attenuate plant host defence mechanisms, particularly post-transcriptional gene silencing and efficiently inhibit host antiviral responses by interacting with the key components of cellular silencing machinery, often mimicking their normal cellular functions [23]. Viral suppressors of RNA silencing have been identified from almost all plant virus genera; these VSRs are surprisingly diverse, exhibiting no obvious sequence similarities. Most identified VSRs are multifunctional; in addition to being RNA silencing suppressors, they often perform essential roles, functioning as coat proteins, replicases, movement proteins, helper components for viral transmission, proteases or transcriptional regulators. The first viral RNA silencing suppressor identified was the helper component-proteinase (HC-Pro) of potyviruses [24]; currently, many different suppressors are known.

The aim of this chapter review is to discuss the current status of knowledge regarding various components of host silencing machinery and viral suppression. We also pretend to describe how the defence response in plants is directed against the virus and, in particular, how the virus can sidetrack the host's defence response. Relevant topics on the molecular bases of the induction and suppression of the RNA silencing mechanism, as well as the applications and perspectives of the use of silencing suppression in plant biotechnology, will be emphasised.

2. The importance of silencing pathways

A major breakthrough in the history of biology was the discovery of an RNA-induced silencing response in the nematode *Caenorhabditis elegans* [25]. Prior to that discovery, this RNA-induced silencing had been interpreted as a defence mechanism against viruses and other invading nucleic acids; the discovery of endogenous small regulatory RNAs in many species led to the realisation that gene silencing is a fundamental genetic regulatory mechanism in eukaryotic organisms [26].

Several studies in genetics, biochemistry and the development of novel techniques in molecular biology have helped identify different components of the RNA silencing machinery and have confirmed that RNA interference (RNAi), co-suppression and virus-induced gene silencing share mechanistic similarities [27]. RNA silencing is a conserved eukaryotic pathway involved in the suppression of gene expression via sequence-specific interactions that are mediated by 21–23 nucleotide (nt) RNA molecules [28]. Organisms utilise RNA silencing for three purposes: 1) creating and maintaining heterochromatin at repetitive DNA and transposons; 2) regulating development, stress responses and other endogenous regulatory functions; 3) defending against viral and bacterial infections [29]. Silencing is utilised in developmental pathways and in cellular differentiation to repress genes whose products are not required at specific stages of development or in specific cell types; in plants silencing is also used to respond to internal and external stresses by changing the expression of specific genes involved in the response. In some situations, tissue- or cell-specific silencing is desirable.

Initially discovered in transgenic plants, especially in those created to acquire virus resistance, RNA silencing is now also believed to be responsible for various epigenetic effects and their maintenance; the silencing of transgenic loci in plants most likely results from the activation of defence mechanisms. A number of silencing studies with different plant systems have explored transgenes as indicators of silencing pathways; these works have received important attention in part because silencing reduces the reliability of transgenic approaches in biotechnology of several agriculturally important cultures.

RNA silencing can be classified into two major categories: transcriptional gene silencing (TGS) and post-transcriptional gene silencing (PTGS). TGS is defined as inhibition of transcription, whereas PTGS involves the post-transcriptional degradation of RNA species but does not affect the transcription rate [30]. TGS occurs when double-stranded RNA molecules (dsRNA) containing promoter sequences are present, and PTGS occurs when dsRNA comprise open reading frames (ORF). Together, TGS and PTGS depend on small interfering RNAs (siRNA) or microRNAs (miRNA) that are produced from dsRNA precursors [31]. Because RNA silencing, mainly PTGS, also contributes to antiviral immunity in plants, fungi and invertebrates, it is an important part of innate immunity [32]. The silencing may persist over many cell divisions or plant generations [33].

The basic steps in common to all RNA silencing pathways (Figure 1) include: (i) formation of a dsRNA; (ii) processing of the dsRNA by an RNase III–like enzyme named Dicer (DCL) to shorter (20-30 nucleotide) dsRNA duplexes, the so called siRNA (iii) binding of the small RNA duplexes to a protein from the Argonaute (AGO) family; and (iv) targeting of the RNA-induced complex to mRNA (or DNA) guided by strand complementary to the small dsRNA, which is called the guide [34]. At present, there is good evidence for the existence of at least four different types of RNA silencing pathways in plants. These pathways involve different types of small RNA molecules, specially siRNA and miRNA. Heterochromatin-associated siRNA (hc-siRNA), trans-acting siRNA (tasiRNA) and viral siRNA (primary and secondary) are also important in silencing [35]. A better understanding of silencing pathways is very important because of the potential usefulness of silencing as a powerful tool for gene function studies and crop improvement.

2.1. Post-transcriptional gene silencing

As already mentioned, PTGS is essential to antiviral immunity in plants, thus our focus will be concentrated in this phenomenon. PTGS was first observed in 1990, and initially referred to as 'co- suppression', it was first discovered in transgenic petunia plants in which the introduction of the gene for chalcone synthase created a block in anthocyanin biosynthesis, resulting in variegated pigmentation [36].

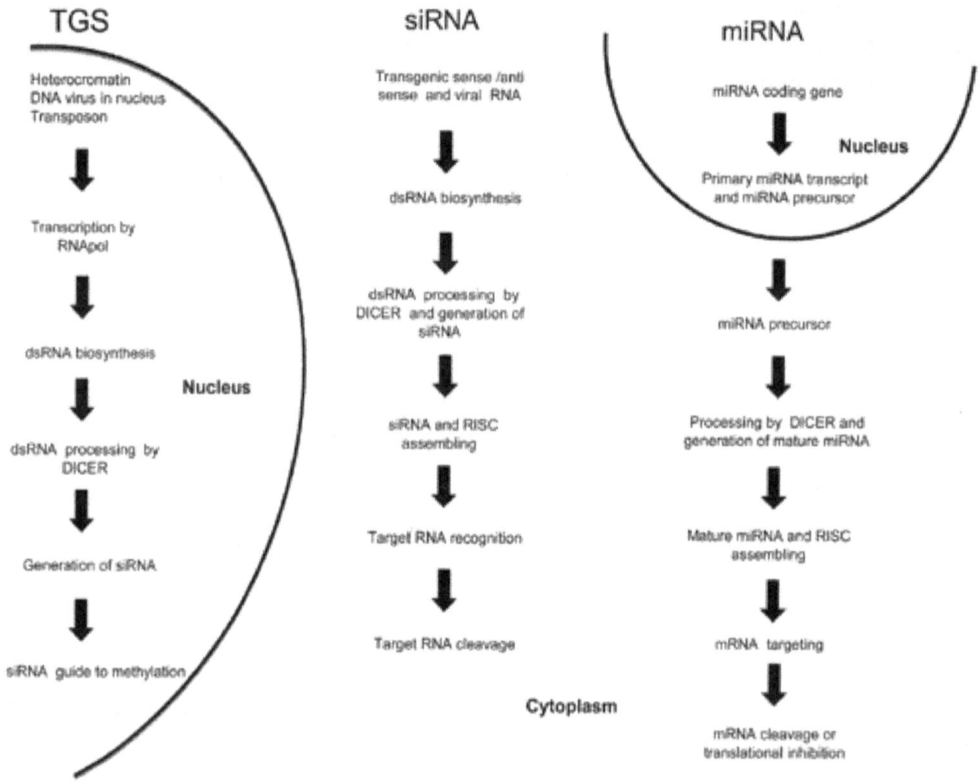

Figure 1. Overview of the transcriptional gene silencing (TGS), siRNA and miRNA pathways. The important steps of each pathway are depicted.

PTGS was also detected in transgenic plants engineered for virus resistance, and associated with the phenomenon of recovery of a host plant from viral infection. This recovery process was soon understood to be associated with the plant's inherent RNA silencing mechanism, which is an evolutionarily conserved antiviral system. The first report related to virus-induced gene silencing was published as long ago as 1929 by McKinney, who reported that tobacco plants infected with the "green" strain of TMV were protected against infection by a closely related second virus (TMV "yellow" strain). This phenomenon was later described as "cross-protection" [37]. However, a remarkable explanation of cross-protection was provided when it was shown that virus infection prevents infection by a second virus if the two viruses possess homologous sequences. Importantly, this viral-RNA-mediated cross-protection was functionally equivalent to post-transcriptional gene silencing [38].

Viruses are not only targets of transgene induced RNA silencing but also elicit silencing themselves. Transgenic plants expressing a truncated version of the coat protein of *Tobacco etch virus* (TEV) were initially susceptible to infection and showed symptoms. However, a few weeks after the transgenic plants recovered from the TEV infection, newly developed leaves were symptomless and virus-free. Remarkably, the recovered leaves were resistant against a second TEV infection but were susceptible to infection by the heterologous PVY [39].

PTGS is a mechanism that has been preserved among eukaryote kingdoms. It is a genetic regulatory mechanism that is involved in several processes including defence of the genome against mobile DNA elements, establishment of heterochromatin, control of plant and animal development, and downregulation of gene expression by specific cleavage and translational repression of target mRNAs that contain complementary sites to miRNAs or siRNAs [40].

The involvement of short RNAs in PTGS was discovered when ~25 nt RNA molecules with sequence homology to a transgene were detected only in plants in which the corresponding transgene was silenced [41]. These molecules are also generated against replicating viruses. They are loaded in an active silencing complex called RISC (RNA-induced silencing complex). RISC shows slight variability in composition from one RNA silencing pathway to the next and from species to species; however, all RISCs contain a guide RNA strand bound to an AGO protein. AGO imparts endonuclease activity to the RISC (the so called slicer activity) being responsible for target RNA cleavage.

During PTGS, RNA is degraded predominantly in the cytoplasm. The strongest evidence in support of this comes from study of PTGS-based degradation of RNA viruses that are expressed exclusively in the cytoplasm [42]. However, recent work has demonstrated that the enzymes involved in PTGS are localised in both the cytoplasm and the nucleus. Overall, the available data suggest that PTGS activity in plant cells occurs in both the cytoplasmic and nuclear compartments. Nuclear PTGS would allow regulation of a potentially larger set of endogenous targets that cannot be regulated through cytoplasmic PTGS [31].

Some aspects of genomic silencing remain unknown. For a more complete understanding of genomic silencing, supplementary approaches are needed. This is especially important because gene silencing has the potential for use as a potent sequence-specific gene inactivation tool in functional genomics and plant biotechnology.

3. RNA silencing suppression

RNA silencing is known to serve as a mechanism for plant defence against pathogens. To counteract this mechanism, viruses have evolved the ability to avoid or suppress RNA silencing. Using this strategy, viruses protect their genomes from degradation through the production of proteins that act as suppressors of RNA silencing. These viral proteins act through a variety of molecular mechanisms including, particularly, blockage of the intercellular and systemic spread of mobile small silencing RNA molecules. The ability of viruses to infect cells can have a profound impact on host endogenous RNA silencing regulatory pathways and can result in alterations in endogenous short RNA expression profile and gene expression [43]. A general overview of the RNA silencing pathway discriminating the different steps targeted by different VSR is provided in Figure 2.

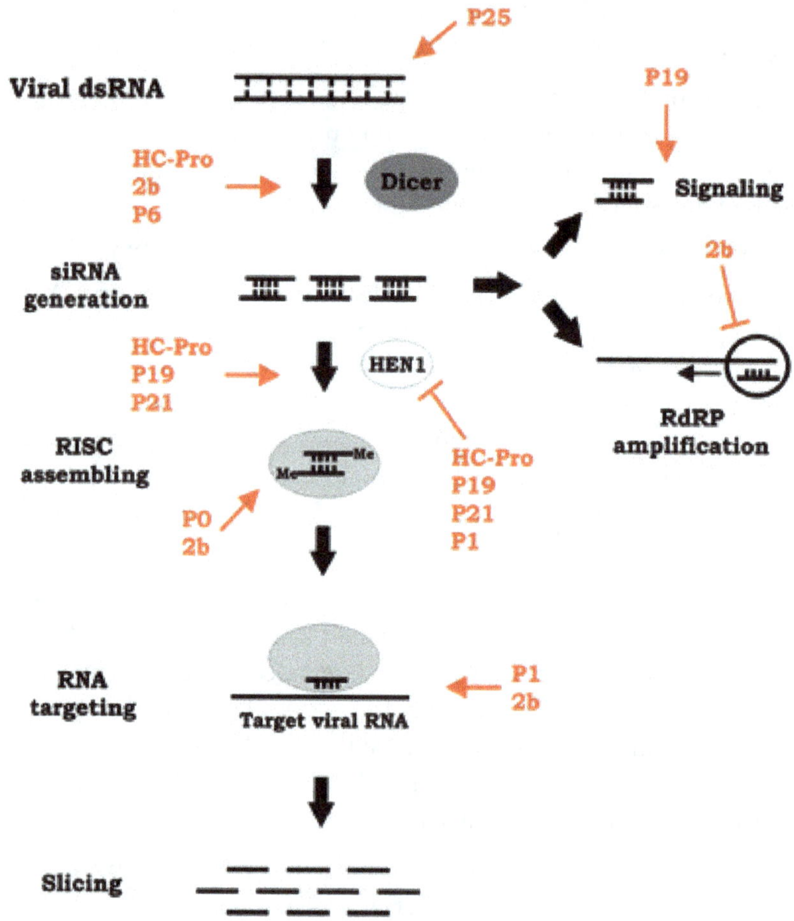

Figure 2. Schematic representation of the RNA silencing pathway triggered by double stranded RNA molecules (dsRNA) of virus origin. The steps of the pathway that are targeted by different viral suppressor proteins are depicted. Amplification of the silencing signal requires an RNA-dependent RNA polymerase (RdRP). Both strands of the small RNA duplexes (3′ termini) are methylated by HEN1 before RISC loading.

VSRs were first evident from the analysis of potyviral synergistic interactions with other viruses. It was shown that this synergism is the result of suppression of a host defence mechanism by the potyviral HC-Pro [44]. Subsequent studies established that HC-Pro is a suppressor of PTGS and provided a link between PTGS and antiviral defence [24, 45]. At the same time, analysis of a second viral protein, the 2b protein of CMV, identified this protein as a suppressor of PTGS in *Nicotiana benthamiana*. Intringuily, HC-Pro and 2b do not target the silencing mechanism in the same way; HC-Pro suppresses silencing in tissues in which it is already established, whereas the 2b protein only affects silencing initiation [45]. In 1999, a seminal study [46] showed definitively that suppression of RNA silencing is an anti-defence strategy commonly used by plant viruses with DNA or RNA genomes, such as Geminivirus (through protein AC2), Sobemovirus (through protein P1), Tombusvirus (by means of the "19K" protein) and others (Comovirus, Tobamovirus and Tobravirus).

Silencing suppression has also been documented in virus capable of infecting other organisms such as insect and fungus, including flock-house virus (FHV), cricket paralysis virus *Drosophila* C virus, and *Cryphonectria parasitica* hypovirus [47]. In insect cells, functional similarities between the CMV 2b protein and the suppressor B2 protein from FHV were described. Deletion of the B2 ORF from FHV results in a drastic loss of virus accumulation in *Drosophila melanogaster* S2 cells, and this loss can be rescued by decreasing the cellular content of AGO2. B2 therefore seems to suppress the effect of the AGO2-dependent silencing response that normally restricts FHV accumulation [48].

Viral suppressors are considered to be of recent evolutionary origin, and they are often encoded by out-of-frame ORFs within more ancient genes. They are surprisingly diverse within and across kingdoms, with no obvious sequence homology [47]. VSRs are variously positioned on the viral genome and can be expressed using different strategies such as subgenomic RNAs, transcriptional read-through, ribosomal leaky-scanning or proteolytic maturation of polyproteins. Due to their molecular evolution, many of the viral suppressors identified to date are multifunctional, *i.e.*, besides being RNA silencing suppressors; they also perform essential roles in other steps of the viral life cycle [23].

Viruses have developed three efficient silencing suppression strategies to counter host antiviral immunity. The first is related to the inhibition of key components of RNA silencing pathways; the mechanisms involved in this strategy have already been well characterised for some viral proteins and will be described forward. A second silencing-suppression strategy, which will also be described later, involves the recruitment of endogenous negative regulators of RNA silencing. For example, yeast two-hybrid system results showed that HC-Pro is able to interact with the tobacco calmodulin-related rgs-CaM, a cellular suppressor of PTGS [49]. The third strategy relies on modification of the host transcriptome and is supported by studies of the geminivirus transcriptional-activator proteins (TrAPs), which have been identified as silencing suppressors (50).

VSRs are essential for viruses to replicate in host cells and to achieve systemic infection [47]. Although they do not share any obvious sequence or structural similarity across viral families and groups, they have having been initially identified as pathogenicity determinants causing developmental defects in host plants, or as host range determinants [51, 52]. Viral silencing suppressors could cause developmental defects in plants because they act in miRNA and siRNA pathways that are mechanistically similar to developmental pathways. In a study involving transgenic expression of the HC-Pro of *Turnip mosaic virus* (TuMV) in *Arabidopsis thaliana,* this protein was shown to alter the accumulation of miRNAs and to prevent the endonucleolytic cleavage of a number of their cellular targets. This effect coincided with the occurrence of morphological defects resembling those of Dicer-like partial mutants called dcl-1. Remarkably, similar defects were observed upon TuMV infection, providing support for the idea that some of the symptoms caused by this virus are actually the result of alteration of miRNA-guided functions by HC-Pro [53]. Analyses of *N. benthamiana* and *Nicotiana tabacum* plants expressing well characterized silencing suppressors derived from 6 different virus genera: P1 of *Rice yellow mottle virus* (RYMV) and *Cocksfoot mottle virus* (CfMV), P19 of TBSV, P25 of PVX, HC-Pro of PVY, 2b of CMV and AC2 of *African cassava mosaic virus* (ACMV) were performed. Interestingly, some of the silencing suppressors promoted specific phenotypic effects. HC-Pro caused a severely distorted growth habit in both *Nicotiana* spp., while the P25 protein of PVX caused a specific flower malformation and an early senescence phenotype in the *N. benthamiana* plants, although not affecting *N. tabacum*. Moreover, P19 expressing *N. benthamiana* plants had blistered leaf epidermis, hairy and serrated leaves in one of the lines and occasional bending of the flower stalks while in *N. tabacum* caused occasional malformation of flowers [54].

From almost all virus genus that infect plants, over 50 individual VSRs have been identified, strongly suggesting that successful virus infection requires their expression [47, 55]. The data available in the literature suggest that almost all viruses encode at least one suppressor, but in many cases, viruses encode more than one [47]. Virus-encoded suppressors seem to have primordial RNA-binding properties and often show preference for a specific RNA molecule [56, 57].

Studies comparing the activities of three distinct RNA silencing suppressors (P19, P21 and HC-Pro) *in vitro* and *in vivo* showed that all three silencing suppressors are dsRNA-binding proteins that interact physically with siRNA duplexes [57]. These three suppressors inhibit siRNA-directed target RNA cleavage in a *D. melanogaster in vitro* RNA silencing system. Moreover, P19, HC-Pro and P21 uniformly inhibit the siRNA-initiated RISC assembly pathway by preventing RNA silencing initiator complex formation through siRNA sequestration. None of these silencing suppressors inhibit pre-assembled RISC activity *in vitro* or *in vivo*.

Suppression can vary in degree and spatial detail ranging, for example, from suppression in all tissues of all infected leaves to suppression only in the veins of newly emerged leaves. This suggests that different suppressors might be targeted to different parts of the gene silencing mechanism [58] such as viral RNA recognition, dicing, RISC assembly, RNA targeting and amplification [20]. For example, DCL function was indirectly blocked by the *Cauliflower mosaic virus* encoded P6 protein (viral translational transactivator protein). Transgenic P6 expression in *A. thaliana* reduced levels of DCL4-dependent 21-nt siRNAs (DCL4 converts non-coding RNA precursors into 21-nt tasiRNAs controlling developmental timing and organ polarity), similar to the effect of inactivating *A. thaliana* DRB4 (dsRNA-binding protein that physically and specifically interact with Dicers). Moreover, immunoprecipitation assays demonstrated that P6 physically interacts with DRB4 and that the ability of P6 to move within cellular compartments (nucleus and cytoplasm) was important for its silencing suppression activity [59].

Direct interaction between the V2 protein of *Tomato yellow leaf curl virus* (TYLCV) with SlSGS3, the tomato functional homolog of the *A. thaliana* SGS3 protein (AtSGS3), which is a coiled-coil protein involved in siRNA signal amplification, interferes with RNA silencing. Furthermore, the fact that a V2 mutant is unable to bind SGS3 loses its ability to suppress silencing indicates that the V2–SGS3 interaction may represent one of the key events in V2-induced RNA-silencing suppression in TYLCV-infected plant cells [60]. The HC-Pro protein can also act in a different manner as a viral suppressor of RNA silencing and might additionally be involved in sequestration of RNA duplexes. It was demonstrated that the FRNK amino acid motif in the central domain of HC-Pro is a probable point of contact involved in siRNA and miRNA duplex sequestration [61]. Mutations of FRNK (severe strains) to FINK (attenuated strains) caused attenuation of symptoms in squash leaves upon infection by *Zucchini yellow mosaic virus* (ZYMV). A decrease in miRNA accumulation was also observed. This raises the hypothesis that interactions of the FRNK box with different plant miRNAs directly influences their accumulation and endogenous regulatory functions, thereby contributing to symptom development.

Viral suppressors can interfere with the activity of methyltransferase HEN1. Studies performed with transgenic *A. thaliana* expressing P21 of BYV, P19 of tombusviruses, or P1/HC-Pro of TuMV demonstrated that, in addition to affect miRNA duplexes, these proteins interfere with short RNA stabilisation by blocking HEN1 methylation. Because miRNA precursors are supposedly cleaved in the nucleus, the fact that they are blocked for methylation by cytoplasmic viral suppressors could be explained in three ways: first, they may compete with HEN1 for substrate miRNA/miRNA* duplexes (sequestration by the suppressors could prevent HEN1 from interacting with duplexes or prevent HEN1 access to the 2′ OH of the 3′ terminal nucleotide); second, the viral silencing suppressors may bind directly to HEN1 and inhibit its activity, or interact with other factors required for HEN1 function; and third, viral suppressors may affect the subcellular localisation of HEN1 [62].

Diverse VSRs have been shown to bind AGO proteins. The first protein identified that binds AGO1 and AGO4 *in vivo* was 2b protein encoded by CMV. AGO1 is the major effector in both miRNA-directed and virus-induced RNA silencing. The 2b protein co-localises with AGO1 both in the cytoplasm and in the nucleolus [63]. The direct interaction of 2b protein with the PAZ and PIWI domains of AGO1, leading to the inhibition of its slicer activity, was verified by bimolecular fluorescence complementation (BiFC) and co-immunoprecipitation (Co-IP) assays [64]. The 2b protein also interacts directly with AGO4 in the nucleolus [65]. AGO4 binds to 24-nt long repeat-associated siRNA (ra-siRNA) to participate in RNA-directed DNA methylation (RdDM) [66]. In this case, 2b competes with AGO4 for binding to 24-nt ra-siRNA, suppressing the DNA methylation mediated by AGO4 [65]. However, the effects of inhibition by 2b of the RNA-dependent DNA methylation phenomena on virus replication and spread remain to be investigated.

RISC activity also undergoes interference by viral suppressors. In agro-infiltrated leaves of *N. benthamiana* containing a GFP transgene, the P0 protein encoded by *Beet western yellows virus* (BWYV) was identified as having strong silencing suppressor activity [67]. Further studies on two *A. thaliana* infecting poleroviruses revealed that P0 contains a conserved minimal F-box motif that interacts with homologues of S-phase kinase related protein 1 (SKP1), a core subunit of the multi-component SCF family of ubiquitin E3 ligases. Mutations in the F-box motif interrupt the interaction between P0 and a SKP1 homolog in *N. benthamiana*, causing a decrease in virus pathogenicity. In transgenic *A. thaliana* plants, expression of P0 caused severe developmental defects similar to those observed in mutants affected in miRNA pathways. Downregulation of a SKP1 homolog in *N. benthamiana* resulted in plant resistance to polerovirus infection. These results support a model in which P0 acts as an F-box protein that targets an essential component of the host post-transcriptional gene silencing machinery [68]. The results of subsequent investigation of the molecular mechanism by which P0 impairs PTGS showed that P0 expression does not affect the biogenesis of primary siRNAs, but it does affect their activity. Furthermore, in transformed *A. thaliana* plants P0 expression leads to various developmental abnormalities reminiscent of mutants affected in miRNA pathways. In this system, P0 expression is accompanied by enhanced levels of several miRNA-target transcripts, suggesting that P0 acts at the level of RISC. It was also revealed that P0 physically interacts with AGO1 to trigger AGO1 protein decay in plants [69].

There are also viral suppressors with unspecified function, such as the triple gene block protein 1 (TGBp1) of PVX, cysteine-rich proteins encoded by hordeiviruses, tobraviruses, furoviruses, pecluviruses and carlaviruses, and the b protein of *Barley stripe mosaic virus* (BSMV) [52].

A number of assays have been established to verify the silencing suppression activity of a given viral gene and/or to identify multiple VSRs encoded by a single virus. In the case of plants, two assays have been widely used. The first is based on the transient, mixed expression of two transgenes in leaves co-infiltrated with two *Agrobacterium tumefaciens* strains. One strain induces RNA silencing of a reporter gene such as the green fluorescent protein (GFP) in the infiltrated leaf (local silencing); the subsequent spread of silencing into upper non-infiltrated tissues in transgenic plants that carry a homologous, integrated transgene (systemic silencing) is measured. The potential silencing suppressor is identified by the ability of the transiently expressed viral gene to enhance and/or sustain visibly higher levels of expression of the reporter gene. However, this assay was not capable of identifying several viral suppressors, including the CMV 2b gene, P25 of PVX, and the coat protein (CP) of *Citrus tristeza virus* (CTV) because they display very low suppression activity in agro-infiltration assays. Thus, their suppressor activities were confirmed by alternative approaches [40, 51]. The second type of assay, which is based on the use of grafting experiments, enables the identification of VSRs that are active against systemic silencing but not local silencing. In these experiments, selected transgenic plants stably expressing a candidate VSR are genetically crossed with a transgenic plant line that carries an autonomously silencing reporter transgene, such as 35S-GUS (β-glucoronidase) in tobacco line such as 6b5. Spreading expression of the viral protein suppresses reporter transgene silencing in the resulting F1 progeny (6b5xVSR) and can be determined by grafting reporter scions onto rootstocks made from the F1 plants. The reporter scions are from another transgenic plant line that expresses the reporter GUS transgene at high levels. The reporter gene becomes silenced a few weeks after grafting onto 6b5 rootstocks owing to the importation of a sequence-specific silencing signal from the silencing rootstock. Silencing does not occur in the scions if the VSR can inhibit either the synthesis of the mobile silencing signal in the F1 rootstocks or its export from rootstock to scion. Analysis of expression of the reporter transgene in the F1 progeny can also reveal whether the VSR suppresses local silencing, DNA methylation of the reporter transgene, or both [40].

Viruses are able to spread through infected plant cells using two ways of movement: cell-to-cell movement and long-distance movement. To combat this distribution, plants emit a silencing signal that spreads between cells. Because the effect of spreading is nt-sequence specific, the nature of the signal is likely to incorporate a siRNA or other RNA species [70]. Long-distance spreading depends on an RNA-dependent RNA polymerase (RDR), whereas short-distance movement of the signal does not [70]. A large number of suppressors of RNA silencing, including some effectors of long-distance virus movement through the phloem, are involved in these movements. For example, the P19 of tombusvirus is a suppressor of silencing that is not required for virus replication in isolated cells but is required for extensive viral invasion of systemic leaves [46]. P19 blocks the intercellular movement of the silencing signal by binding DCL4-dependent 21-nt siRNA [40]. Likewise, the potyviral HC-Pro and cucumoviral 2b proteins are suppressors [24] required for systemic virus infection [71]. *Cucumovirus* 2b protein inhibits the systemic movement of RNA silencing by either binding dsRNA/siRNA or inhibiting the slicer activity of AGO1 [23]. A site-directed mutation strategy involving the HC-Pro protein of TEV showed a correlation between silencing suppression and the ability to mediate long-distance virus movement [72]. Reevaluation of the role of *Tomato bushy stunt virus* (TBSV) P19 in the systemic invasion of *N. benthamiana* by the virus revealed a silencing suppression role for the structural protein (CP) of *Turnip crinkle virus* (TCV). The authors showed that a TBSV P19 deletion mutant, while capable of systemic movement in the plants, accumulated progressively less viral RNA in the systemic leaves due to loss of silencing suppressor ability. When the TBSV structural protein was replaced with TCV CP to create a chimeric virus, it restored close to wild type levels of virus accumulation in systemic leaves. This result shows that both of these genes participate in efficient systemic TBSV infection and suggests that TCV CP not only provides structural protein but also complements the silencing suppressor function of TBSV P19. Moreover, it is also suggestive that assembled virions are likely important for the effective unloading of viruses from the vascular system into the leaf mesophyll. This work provides direct evidence that P19 primarily enhances systemic invasion by suppressing the host PTGS responsible for eliminating viral RNAs in the infected plants [51]. The P25 protein of three 'triple gene block' (TGB) proteins of potexviruses is another example of suppressor that is required for cell-to-cell movement of the virus; it is an RNA helicase that moves cell-to-cell and modifies plasmodesmata [73, 74].

As previously mentioned, in many cases viruses encode more than one VSR. A good example is the closterovirus CTV, which encodes three different silencing suppressors [P23, coat protein (CP) and P20] and exhibits distinctive features related to silencing suppression. CTV has a plus-strand RNA genome of approximately 20 kilobases (kb) in length. Its P20 and P23 proteins, but not CP, suppressed RNA silencing in an agro-infiltration assay and were able to reverse transgene silencing. In addition, P20 and CP, but not P23, prevented intercellular silencing spread. It was suggested that P23 appears similar to HC-Pro because, although both are potent suppressors of intracellular silencing, neither prevents intercellular silencing or DNA methylation of the target transgene. On the other hand, the suppressor activity of P20 shares features with silencing suppression mediated by CMV 2b, i.e., both are potent but incomplete suppressors of intercellular silencing, and suppression of intercellular silencing is not associated with reduced DNA methylation of a target GUS transgene. In the case of CP, suppression of intercellular silencing spread is not associated with suppression of intracellular silencing, unlike P20, CMV 2b and P25 of PVX, which are known to interfere with intercellular silencing [75].

Silencing suppressors may confer biased protection against viral RNA and subviral parasites. It was shown that the P1/HC-Pro proteins of TEV caused an increase in the accumulation of the negative strand viral RNA of PVX [44], suggesting that negative-strand viral RNAs are more susceptible to the RNA silencing based host defence while positive-strand viral RNAs are better protected.

Viral infection is greatly influenced by changes in environmental temperature. A general explanation for this phenomenon is that RNA silencing-mediated plant defence is temperature dependent. Generally, at low temperature (15°C), both virus- and transgene-induced RNA silencing is inhibited; the level of virus- or transgene-derived siRNAs is dramatically reduced, leading to enhanced host susceptibility to virus infection and loss of silencing-mediated transgenic phenotypes. In contrast, with increasing temperature (27°C), RNA silencing is activated, and the amount of siRNA gradually increases. However, accumulation of miRNAs, which play a critical role in developmental regulation is temperature independent [76]. Because the replication of viruses does not appear to be disproportionately inhibited by higher temperature, one can assume that the activity of viral silencing suppressors is relatively constant over the temperature range that permits viral systemic infection. Thus, the level of silencing suppression activity should be relatively constant over this temperature range and therefore more readily overcome at higher temperature due to enhancement of the RNA silencing pathway. Conversely, it can be predicted that at low temperatures, the weakened RNA silencing would be more readily overcome by viral silencing suppressors [51].

3.1. Mechanisms of suppression

3.1.1. Viral suppressors

HC-Pro

As described above, HC-Pro was one of the first viral proteins to be identified as a suppressor of transgene- and virus-induced RNA silencing [77, 78]. HC-Pro is produced by plant viruses of the Potyvirus genus, family Potyviridae, the most important group of plant pathogenic viruses. HC-Pro has attracted renewed attention in recent years due to its multifunctionality and involvement in different steps of the potyvirus life cycle [79, 80]. Potyviruses, like the majority of plant viruses, have a single-stranded, positive-sense RNA genome that consists of approximately 10.080 nucleotides and is polyadenylated on its 3 'end and surrounded by a capsid [81]. The genomic RNA has a single ORF located between two noncoding regions, which are called 5'NTR and 3'NTR (non-translated region). Translation of the single ORF produces a polyprotein with molecular weight between 340 and 370 kilodaltons (kDa). This polyprotein is cleaved into functional proteins of the virus through the proteolytic activity of three proteases of viral origin (P1, HC-Pro, and NIa), resulting in 8-10 final products. HC-Pro and P1 act in *cis*, each carrying out its own cleavage, and NIa catalyses its own cleavage and that of six other polypeptides [80, 82].

A typical potyviral HC-Pro consists of approximately 460 amino acids and has a molecular weight of approximately 52 kDa, it performs a surprisingly large number of functions; in fact, among proteins produced by potyviruses, it is the protein for which the greatest number of features has been described [80]. Apart its role as silencing suppressor, HC-Pro plays several other roles as a proteinase, participates in aphid transmission, acts as an auxiliary viral replication factor and participates in virus cell-to-cell and long distance movement [44, 80, 83-88].

HC-Pro can be divided into three functional regions, a N-terminal region that is essential for transmission, a C-terminal region that is responsible for its proteolytic activity and a central region involved in all other functions described. However, recent studies show that most functions overlap along its primary amino acid sequence [89].

Concerning its ability to suppress silencing, HC-Pro was shown to restore GFP expression in both old and new leaves of post-transcriptionally silenced transgenic plants (reviewed in [90]). HC-Pro suppresses PTGS via interaction with one or more cellular proteins that are either components of the silencing machinery or regulators of the silencing pathway. Studies have shown that HC-Pro interferes with the accumulation of the small RNAs associated with silencing. These small RNAs derive from the cleavage of dsRNA by Dicer and HC-Pro may target the process at this step [91, 92]. Dicer could be blocked by HC-Pro in several ways: HC-Pro can prevent the small RNAs from being produced by preventing the enzyme from binding to the dsRNA template, thus blocking the cleavage step; alternatively, it could block at a step downstream of cleavage, preventing incorporation of siRNAs and making the silencing unstable. A model in which HC-Pro suppression of PTGS occurs upstream of accumulation of small RNAs has been proposed [93]. Furthermore, HC-Pro has been shown to transactivate the replication, and enhance the pathogenicity, of a broad range of heterologous plant viruses [44].

Since HC-Pro prevents accumulation of siRNAs of silenced genes, it prevents silencing in a universal manner; however, in tobacco HC-Pro was shown to increase the *in vivo* accumulation of several miRNAs, namely, *miR167*, *miR164* and *miR156* [94]. In addition, HC-Pro is not able to inhibit the systemic silencing signal, suggesting that HC-Pro works downstream from production of the systemic signal. It was suggested that HC-Pro works at the point of RISC assembly and that it most likely unwinds miRNA duplexes [95]. The specificity of HC-Pro binding to small RNAs was tested by the use of synthetic 21-nt or 24-nt siRNA duplexes and 19-nt or 21-nt blunt-ended RNA duplexes. The results showed that HC-Pro binds with specificity to 21-nt siRNA duplexes. Moreover, it has higher binding affinity for duplexes with 2-nt overhangs than for small single-stranded RNAs or blunt-ended small RNA duplexes [57].

HC-Pro is often mentioned in conflicting reports in the literature that address the relationship between PTGS and DNA methylation. In some instances, a good correlation between HC-Pro suppression of PTGS and the decrease of DNA methylation is observed. When introduced in a GUS-silenced tobacco line, for example, HC-Pro affected the accumulation of small RNAs of the PTGS pathway and reduced methylation of the corresponding GUS locus [96], suggesting that silencing is directly related to DNA methylation. In contrast, another study showed that HC-Pro increased DNA methylation of the promoter sequence of a silenced DNA target gene when silencing was induced by dsRNA directed against the promoter region [97]. In the same study, it was shown that the amount of promoter-derived siRNA molecules increased five-fold in the presence of HC-Pro.

P25

P25, the product of the first gene of the "triple gene block" (also known as TGBp1) encoded by PVX, is an RNA helicase that induces plasmodesmal gating. P25 promotes cell-to-cell movement of the virus and is also associated with suppression of RNA silencing [98]. P25 was one of the first VSRs to be identified and shown to inhibit transgene sense- or dsRNA-induced RNA silencing. The mechanism of action of P25 contrasts with that of HC-Pro, which acts at a downstream cellular signalling step. For this reason, mixed infections of PVX with any other potyvirus (which encode HC-Pro) normally result in synergistic disease [86]. Such diseases are common and often occur in plants as a result of the interaction between viruses that suppress silencing at various points of the silencing pathway [99, 100]. P25 is the only suppressor so far described that affects gene silencing but fails to recover silenced GFP expression post-transcriptionally. Surprisingly, P25 does not interfere with silencing of viral-induced sites [101].

Antiviral silencing suppression by P25 is required for cell-to-cell movement of the virus but has no apparent effect on viral accumulation in protoplasts, unlike most known VSRs such as cucumoviral 2b, tombusviral P19, and carmoviral P38. The analysis of a variety of random mutants of P25 showed that all produced defects in the suppression of silencing and in cell-to-cell movement. Some P25 mutants, defective in suppression activity, could be supplemented by heterologous viral suppressors. However, other mutants showed silencing suppression activity but were not functional as movement proteins. These results demonstrate a crucial role for P25 in cell-to-cell movement of the virus and also suggest the importance of an additional function of P25 in these activities [98].

P25 exhibits strong activity against silencing produced by both sense and inverted repeat transgenes in leaves of *N. benthamiana* and in transgenic *A. thaliana*. These observations indicate that P25 targets a downstream step in the synthesis of dsRNA [98, 101, 102]. As reported above, P25 inhibits systemic silencing but does not inhibit gene silencing induced by viruses in locally infected leaves. Moreover, it reduces the accumulation of both primary and secondary siRNAs but has no effect on the accumulation of endogenous miRNA and siRNA. It has been speculated that P25 does not interfere with programmed RISC [101, 102].

Co-immunoprecipitation assays indicate that P25 interacts with various members of the AGO family, including AGO1, AGO2, AGO3 and AGO4, but not AGO5 or AGO7. Furthermore, P25 promotes the proteasome-dependent degradation of AGO1 [103], indicating that its suppressor activity is dependent on AGO1 degradation. It is not currently known whether P25 inhibits the local motion silencing signal targeting AGO1 to promote movement of the virus [104].

P19

The tombusviral 19 kDa protein, P19, is one of the best studied viral silencing suppressors. The hypothesis that P19 is a viral suppressor arose in 1995 when Scholthof and colleagues reported that the 19 K protein of TBSV is a pathogenicity determinant. TBSV is a virus with a broad host range that induces a variety of symptoms in different hosts [105]. This virus contains a single copy of a positive-sense single-stranded RNA genome of 4800 nucleotides [106]. Five major ORFs are encoded by the TBSV genome. Two small nested genes located near the 3' terminus of the genome are expressed via a second subgenomic mRNA that directs synthesis of a 22 kDa protein (P22) and a 19 kDa protein (P19) [107]. P19 can act both as an elicitor of the HR response in *N. tabacum* or as an inductor of systemic necrosis in *N. benthamiana* [108]. Due to its activity as a host-specific symptom determinant, the P19 was suggested to play a role in overcoming host defence systems [108, 109]. This hypothesis was confirmed by the inoculation of silenced GFP tobacco plants with a recombinant PVX carrying the 19K coding region [48]. In these assays, plants infected with PVX-19K showed severe symptoms two weeks after inoculation while those already infected with PVX-m19K (with a nontranslatable P19 RNA) showed mild mosaic symptoms. Suppression of silencing occurred in PVX-19K infected plants but was manifested only in new emerging tissues and was most pronounced in the veins. However, symptoms of PVX-19K were visible on all areas of the leaves. Interestingly, P19 restores GFP expression in PTGS inactivated transgenic plants only around the veins of new emerging leaves [46], even though TBSV accumulates to a high concentration in the whole leaf [108].

Several recent studies report a breakthrough in understanding the molecular mechanism of the suppressor activity of P19. This suppressor prevents incorporation of siRNAs into effectors such as the RISC complex by binding specifically to 21-nt siRNAs *in vitro* and *in vivo* [110, 111]. This model was confirmed by three-dimensional structural resolution of the P19-siRNA complex showing that P19 acts as a clamp for dsRNA binding to the ends of the siRNA duplex [112, 113]. However, it was also reported that after the RISC complex is formed, P19 is no longer effective, being unable to bind to siRNA and miRNA [57].

P19 inhibits the onset of transgene-induced local and systemic silencing [110]. It does not interfere with the location of virus-induced silencing, but it can prevent systemic silencing. It was suggested that P19 depletes PTGS generated 21-25-nt dsRNAs, thus inhibiting the development of transgene-induced silencing and preventing the production of active signal complex. Interestingly, transgenic plants expressing biologically active P19 showed an altered phenotype, suggesting that the P19-targeted PTGS pathway might also have a role in developmental regulation. Low level expression of P19 altered leaf morphology in transgenic plants. In addition to leaf curling, some severely affected plants also showed delayed appearance of developed secondary stems. Although it is possible that developmental abnormalities in transgenic plants are not related to the silencing suppressor activity of P19, these findings are suggestive that the P19-targeted PTGS pathway plays a role in plant development.

The silencing suppressor activity of P19 is also observed in other hosts. Since siRNA binding by P19 does not require host factors *in vitro*, and that these short RNAs are specificity determinants of silencing effector complexes, P19 could be used to inhibit RNA silencing in heterologous systems, including *D. melanogaster*, worms and mammals [110].

The P19 protein of *Cymbidium ringspot virus* (CymRSV), a relative of the TBSV P19 protein, specifically binds to siRNAs *in vitro*, and two reports show co-crystallisation of P19 homodimers with siRNA [112, 113]. P19 also binds RNA duplexes with a blunt end and with a 2-nt 3′ overhanging end. In animals, Dicer digests from the ends of long dsRNAs [114] and therefore might produce long dsRNAs with 2-nt 3′ overhangs. Although it is possible that P19 competes with Dicer-related proteins for the 2-nt 3′ overhanging ends of long dsRNAs, the high level of 21-25-nt RNAs in CymRSV infected cells suggests that P19 fails to suppress Dicer-like activity.

A study using mutants of CymRSV demonstrated that lack of P19 suppressor did not affect most basic viral functions, including genome replication, cell-to-cell movement and phloem long-distance transport [109]. In contrast, the systemic infection of plants inoculated with a silencing suppressor mutant of CymRSV was seriously compromised and led to the development of a recovery phenotype, suggesting that P19 suppressor targets a non-cell-autonomous step of RNA silencing [110].

In CymRSV infected plants, siRNAs are present in P19–siRNA complexes, while in plants infected with the P19-defective mutant in which P19 was inactivated (termed Cym19stop), siRNAs were found as free molecules. P19 apparently does not affect virus-induced cell-autonomous silencing because CymRSV and Cym19stop viral RNAs, as well as siRNAs derived from these viruses, accumulate to the same levels in transfected single cells [110]. In addition, the P19 protein was shown to repress the accumulation of all size classes of siRNA produced in agroinfiltration assays [110, 115]. While CymRSV infects *N. benthamiana* systemically and typically kills the host within two weeks, infection with the mutant virus results in a recovery-like phenotype showing mild symptoms and low virus levels in the upper leaves [116]. Moreover, P19-deficient and wild type CymRSV accumulate at similar levels in both protoplasts and inoculated leaves, indicating that this protein does not prevent RISC from degrading viral RNAs by sequestering viral derived siRNAs (vsiRNA) [37]. In systemic leaves, P19-deficient CymRSV accumulates only in the vascular bundles and exhibits defects in invading the surrounding tissues suggesting that blocking the local movement of RNA silencing by P19 is essential for systemic virus infection [117].

In this context, studies have shown that P19 specifically sequesters the DCL4-dependent 21-nt siRNAs derived from transgene RNAs; these siRNAs normally move into the neighbouring recipient cells and act as a silencing signal [118]. These results imply that P19 promotes systemic virus infection by sequestering vsiRNA, thus preventing the signal for RNA silencing from spreading out of vascular bundles into neighbouring cells [104]. Therefore, when P19 is absent, the systemic signal moves faster than the virus in the infected plant, thereby establishing antiviral silencing in cells ahead of the infection front. As a result, any virus entering these cells is immediately controlled by silencing-mediated RNA degradation. In conclusion, the presence of the silencing suppressor is essential for the development of systemic virus infection [37].

3.1.2. Endogenous suppressors

In addition to the numerous viral suppressors of RNA silencing, endogenous RNA silencing suppressors have also been reported in eukaryotes. The endogenous suppression of RNA silencing negatively controls the presence of siRNAs and miRNAs in different ways. The generation and control of such siRNAs and miRNAs are essential for normal development of plants and animals [119-121].

The first endogenous RNA silencing suppressor was identified in *N. tabacum* and was named rgs-CaM (regulator of gene silencing CaM). This protein was found in a screen for proteins interacting with the viral suppressor HC-Pro. Expression of rgs-CaM can be induced in leaves of *N. tabacum* when HC-Pro is expressed either from a transgene or from infection with a virus that encodes HC-Pro. When expressed at high levels in *N. benthamiana,* rgs-CaM suppresses both PVX–induced gene silencing and sense transgene–mediated PTGS (S-PTGS) [49]. A recent study, however, demonstrated that rgs-CaM is not an endogenous suppressor of silencing [122]. In fact, this protein acts as an endogenous pattern recognition receptor able to bind to several viral silencing suppressors through their RNA-binding domains. Thus, rgs-CaM activity confers a countermeasure against viral suppressors.

In addition, an inhibitor protein of *A. thaliana* RNase L activity, called RLI2, was also described as having a silencing suppressor activity when expressed at high levels in transgenic *N. benthamiana* [123]. Another known endogenous suppressor, the *A. thaliana* exoribonuclease XRN4, suppresses silencing by promoting the degradation of aberrant, uncapped RNAs that constitute possible templates for an RNA dependent RNA polimerase (RdRP) pathway involved in silencing. These aberrant molecules represent important activators of silencing, serving as templates for the production of new dsRNAs by the action of the RdRP. Indeed, mutations in the gene *xrn4* promote RdRP-dependent silencing [124] and lead to over-accumulation of miRNA-generated cleavage products [125]. Three other suppressor proteins, the exoribonucleases XRN2, XRN3 and FRY1, were identified in *A. thaliana,* thus complementing existing knowledge of the suppression of silencing involving XRN4 [126]. While XRN4 is cytoplasmic, XRN2 and XRN3 are nuclear exoribonucleases [127]. XRN2 and XRN3 contribute to the suppression of RNA silencing by degrading miRNA-derived loops excised during miRNA maturation in the nucleus. In contrast, XRN4 acts exclusively in the cytoplasm, promoting degradation of uncapped messages such as miRNA target cleavage products [124-126]. Fry1 acts as a fine-tuning modulator of the activities of XRN2, XRN3 and XRN4 [126]. Interestingly a family of exoribonucleases known as small RNA degrading nucleases (SDN) degrades mature miRNA molecules in *A. thaliana,* acting specifically on single-stranded miRNAs [121].

3.1.3. Modifications of the host transcriptome

Viruses can counterattack RNA silencing immunity not only by acting directly on gene products that are required for silencing *per se* but also by inducing stress and plant defence responses that interfere with antiviral silencing [128, 129]. An interestingly example include the RAV2/EDF2 protein, which belongs to the RAV/EDF family of transcription factors. This protein is required for suppression of silencing by potyvirus HC-Pro and carmovirus P38, two viruses that belong to unrelated families. RAV2 is required for suppression of silencing in a direct way that involves blocking the activity of primary siRNAs as well as indirectly by its effects on upregulation of some stress and defence response genes [128].

The induction of biotic or abiotic stress activates other defence responses that can divert the host from antiviral silencing [129]. Therefore, RAV2 is a critical control factor for carmovirus and potyvirus suppressors [128].

Other viruses make use of alternative mechanisms for suppression of silencing. The TrAP geminiviral protein AC2 upregulates a gene coding for the cold- and abscisic acid-inducible protein KIN1 as well as five additional known or putative cold-regulated genes [50]. As already mentioned, the efficiency of RNA silencing is dependent on temperature; at low temperatures, inhibition of silencing occurs and the plant becomes susceptible to viral infection [76]. The inhibition of silencing at low temperature is a pathway used by AC2 to accomplish the suppression of silencing [50]. Another strategy exploited by geminivirus is up-regulation of an endogenous RNA silencing suppressor, Werner exonuclease-like 1 (WEL1), which is mediated by AC2. Interestingly, the related proteins MUT-7 (mutate 7) and Werner syndrome-like exonuclease (WEX) have been identified as positive regulators of RNA silencing in *C. elegans* and *A. thaliana,* respectively [130, 131]. Thus, AC2 up-regulation of Wel-1 results in interference with, or competition for, factors that are required for normal WEX function. Transient expression of a WEL-1 transcription unit is sufficient to suppress RNA silencing in *N. benthamiana* [50].

4. Applications

4.1. Virus-Induced Gene Silencing (VIGS)

Virus-induced gene silencing (VIGS) is a technique derived from the knowledge of RNA silencing. It uses recombinant viruses to specifically reduce or knock-down endogenous gene activity; it is based on post-transcriptional gene silencing (PTGS) [132]. When used to infect plants, recombinant viral vectors carrying segments of host genes produce siRNAs that are specific to host mRNA. The RISC complex mediates the degradation of target host mRNAs, leading to downregulation of gene expression. Thus, the infected plant has a phenotype similar to a loss-of-function mutant of the gene of interest [133].

VIGS is used as a tool for turning down host gene expression, especially in plants. In principle, a plant gene of interest can be silenced by infecting the plant with a viral vector that has been modified to express a nucleic acid sequence homologous to the host gene. As a proof of concept, several endogenous genes have been silenced using VIGS. The plant gene *phytoene desaturase* (PDS), a regulator of carotenoid biosynthesis, was silenced in *N. benthamiana* plants by the use of a recombinant TMV vector. As a result, degradation of the host PDS mRNA and resultant alterations in the pigment synthesis pathway were observed [133].

There are four main reasons for the popularity of VIGS. First, the methodology is simple, often involving agroinfiltration or biolistic inoculation of plants. Second, the results are obtained rapidly, typically within two to three weeks of inoculation. Third, the technology bypasses transformation steps and hence is applicable to a number of plant species that are recalcitrant to transformation. Fourth, the method has the potential to silence multi-copy genes [134].

Efficient silencing depends mainly on the choice of VIGS vector. There are many factors to be considered when choosing the virus to be used for VIGS. Among the factors to be considered are (1) the virus must produce few or no symptoms during infection, thereby facilitating easy visualisation and interpretation of the mutant phenotype; (2) it must induce persistent silencing, thus viruses with strong silencing suppressors are to be avoided because they can interfere with the establishment of silencing; (3) it is advantageous to have infectious cDNA clones of the virus for cloning purposes; and (4) the virus must retain infectivity after insertion of foreign DNA. The virus should also show uniform spread, infect most cell types including the meristem, and preferably show a broad host range [133].

Several RNA and DNA viruses have been modified to create VIGS vectors. The gene to be silenced is cloned in an infectious derivative of a viral DNA (DNA virus-based vectors) or cDNA (RNA virus-based vectors) derived from viral RNA. Plant inoculation with viral vectors is most commonly achieved via *A. tumefaciens* infection, but can also be achieved by mechanical inoculation of *in vitro* synthesized transcripts, or for DNA-based vectors, by biolistic delivery methods. During the course of viral infection, either double-stranded RNA or RNA with a high degree of secondary structure is often produced; both of these are efficient initiators of RNA silencing directed against the infecting viral RNA. Other factors that play an important role in gene silencing in VIGS are the orientation of the insert (inverted repeats are more efficient than antisense orientation, which, in turn is more efficient than same sense orientation) and systemic spread of the silencing effect (the silencing signal is believed to spread independently of the VIGS vector to other parts of the plant) [134]. More than 30 VIGS vectors have been developed, and these vectors have been widely used to study the functions of genes involved in basic cellular functions, metabolic pathways, development, plant-microbe interaction, and abiotic mechanisms [132].

The first viral vector used for VIGS was TMV. Shortly thereafter, another vector was produced based on another RNA virus, PVX carrying a cDNA fragment derived from the *PDS* gene [135]. However, although these first vectors were effective, they have intrinsic disadvantages. First, the VIGS phenotype is superimposed and sometimes complicated by chlorosis, leaf distortion and necrotic symptoms of virus infection. A second disadvantage of these viral vectors is their inability to invade every cell, such that cells in which the target gene is not silenced may obscure VIGS phenotypes [136]. A novel VIGS vector based on TRV was then established. TRV was shown to induce more efficient silencing of transgenes and endogenous genes. It could spread more vigorously throughout the entire plant, including meristem tissue, and the symptoms induced by TRV are much milder than those produced by other viruses [136].

A primary limitation of VIGS technology is that a viral vector can be used only in plants that are hosts of the virus used. The first VIGS vectors (e.g., PVX) do not infect the model plant *A. thaliana*. Therefore, new vectors such as the TRV-based vector [136] were developed to overcome this difficulty. TRV is also one of the few viruses that have been modified into a highly efficient cloning and expression system for use in large-scale functional genomics screening. TRV vectors can induce VIGS in a number of solanaceous hosts like *N. benthamiana*, tomato, potato, pepper, petunia, poppy (Eudicot species), and the model system *A.*

thaliana (family Brassicaceae) [133, 137]. VIGS vectors have been applied not only in dicotyledonous plants but also in monocotyledonous plants. For this, a modified VIGS vector based on *Brome mosaic virus* (BMV) was developed and validated in barley, rice and maize [138].

The VIGS system can be helpful in assessing gene function, especially for genes that cause zygotic/embryonic lethality when mutated and in species that are recalcitrant to genetic transformation. As aforementioned, it can be designed to silence multiple members of a gene family, thereby circumventing the problem of functional redundancy of genes [133].

4.2. Use of viral suppressors

The discovery of RNA silencing, and its derived technology (RNA interference; RNAi), has increased our knowledge of gene regulation and function. RNAi opened up novel avenues in biology, making it possible to develop fascinating strategies for application in genetic analysis, plant protection, and many other areas related to crop improvement [139]. In this context, a large number of silencing suppressor proteins have been described, and the discovery of the molecular basis of silencing suppression has inspired new concepts about the molecular basis of symptoms caused by viruses in plants [37].

Many biotechnological applications involving plants require high levels of protein expression. Generally, stably transformed plants are the preferred platform for large-scale production. To try to increase expression levels, transgenic lines that encode a replicating RNA virus vector carrying a gene of interest, a technology coined 'amplicon', have been exploited. The rationale of this method involves increasing the accumulation of the product of interest through transcription of an amplicon transgene that initiates viral RNA replication and gene expression. However, the strategy failed because the transformants consistently exhibited RNA silencing of the amplicon transgene [140]. The viral dsRNA replication intermediates produced in every cell of the transgenic plants were recognised as potent triggers of the silencing-based defence mechanism that is normally elicited in the course of natural infections. Based on those findings, it was subsequently reasoned that co-expression of viral suppressors might prevent this adverse response and permit the high levels of gene expression initially envisioned with the use of amplicons [141].

To test this idea, in reference [94] crossed transgenic tobacco plants expressing TEV HC-Pro with amplicon lines designed to express a GUS reporter gene from the PVX genome. Pairing the suppressor and the amplicon locus resulted in a dramatic increase in virus accumulation and gene expression such that the leaves of mature plants accumulated the GUS protein up to 3% of total soluble protein. Remarkably, in spite of high virus accumulation, the plants did not suffer from viral disease and remained symptomless.

As opposed to stable, transgenic expression, transient expression is of interest for achieving expression of useful proteins. In plants, recombinant strains of *A. tumefaciens* can be used for transient gene expression. In principle, this system could allow high levels of gene expression; however, its utility has thus far been limited because ectopic protein expression usually ceases after 2– 3 days [141]. RNA silencing is, in fact, a major cause of this lack of efficiency. It was therefore anticipated that co-delivery of *A. tumefaciens* cultures with silencing suppressors would enhance expression of the genes of interest [101]. Studies with the P19 protein of TBSV were among those that provided the best results. Expression of a range of proteins was enhanced 50-fold or more in the presence of this suppressor, and experiments with GFP indicated that the co-infiltrated tissues accumulated the protein up to 7% of total soluble protein [142]. Due to its simplicity and rapidity, the P19-enhanced expression system is currently used in industrial production as well as used as a research tool for the isolation and biochemical characterisation of a broad range of proteins without the need for the time-consuming regeneration of stably transformed plants [141].

5. Perspectives

The molecular basis of the silencing suppression of VSR proteins is quite complex and is currently incompletely understood. By the way, the discovery of the mode of action of different viral suppressors has demonstrated the existence of a complex interaction between VSR and plant silencing-regulated networks. For example, in addition to sequestering siRNA duplexes, the P19 protein of tombusviruses specifically controls antiviral AGO1 expression through enhanced miR168 expression, which arrests AGO1 translation [23]. It is likely that many other VSRs interact in diverse ways with RNA-silencing pathways. Many of these interactions remain to be discovered, and there are several gaps in our knowledge regarding the effectors of plant silencing machinery. Until very recently, the mechanisms of plant si/miRNA RISC assembly or the components of the plant RISC, which may also be potential targets of VSRs, were little known. The recently developed system of plant in vitro RISC [143] will likely accelerate the exploration of plant RISC assembly and RNA-targeting mechanisms mediated by this effector. This system will enable exploration of the mechanisms by which VSRs interact with one or more of the RISC components and prevent its assembly.

More information about the replication, subcellular localisation and regulation of the expression of viral genes, including VSRs, is required so that we may better understand the molecular mechanisms of VSR-mediated silencing suppression for the many plant viruses for which they are still not known. Because many VSRs have multiple functions in the virus life cycle, separate analysis of their silencing suppressor activities can lead to misinterpretations; thus, it is essential that VSRs be studied in their natural virus backgrounds [23].

Although common mechanisms of silencing suppression exist, there is also great variation in suppression mechanisms, likely driven by evolution and fitness, this variation has yielded viral strains with different properties. It is likely that additional differences will be found when plant viruses and their suppressors are tested in several plant species. This will provide us with a greater understanding of the parameters associated with the natural host range of a virus and may possibly lead to new strategies for crop protection [52].

Some of the already well described VSRs can be used as powerful tools for better understanding silencing pathways because they target specific steps of silencing machinery. Indeed, the P19 protein was recently used to demonstrate that siRNA duplexes function as mobile silencing signals between plant cells, in addition that P1, P38 and P0 proteins may prove to be powerful tools for studying the still unknown components of RISCs [118].

6. Conclusions

The discovery of the mechanisms involved in RNA silencing and in silencing suppression by virus has helped researchers to investigate several aspects of the plant-virus co-evolution. Our understanding of the underlying mechanisms served as a basis for the development of different technologies aiming plant manipulation to generate novel traits or pathogen resistance. Elucidation of the mode of action of different plant viral suppressors provided fundamental contributions to the comprehension of the RNA silencing phenomena. In this context, strategies exploiting viral suppressors to prevent the occurrence of RNA silencing in genetically engineered plants have been developed [144]. High levels of transgene expression should be expected through the implementation of such strategies. This review has addressed some important topics in the areas of RNA silencing and silencing suppression. Ongoing studies aimed at further clarification of the main points of these processes are currently being conducted by many different groups.

Acknowledgements

We are grateful to the Brazilian funding agencies CNPq, FAPESP and CAPES, for their financial support. A.T. Costa, R. K. Makiyama and A. V. Nunes are recipients of doctoral fellowships from CNPq (140299/2010-6), FAPESP (2010/03001-0), CAPES, respectively, and J.P. Bravo is the recipient of a CNPq/PNPD postdoctoral fellowship (558413/2008).

Author details

Alessandra Tenório Costa, Juliana Pereira Bravo, Rodrigo Kazuo Makiyama,
Alessandra Vasconcellos Nunes and Ivan G. Maia

*Address all correspondence to: alebiologyt@hotmail.com

São Paulo State University "Júlio Mesquita Filho" Department of Genetics, Institute of Biosciences – UNESP, Botucatu Campus SP, Brazil

References

[1] Hull R. Matthew's Plant Virology (4ª ed.). Academic Press; 2002.

[2] Laliberté J-F, Sanfaçon H. Cellular Remodeling During Plant Virus Infection. Annual Review of Phytopathology 2010;48:69-91.

[3] Rodrigues SP, Lindsey GG, Fernandes PMB. Biotechnological Approaches for Plant Viruses Resistance: From General to the Modern RNA Silencing Pathway. Brazilian Archives of Biology and Technology 2009;52(4):795-808.

[4] Collmer CW, Marston MF, Taylor JC, Jahn M. The I Gene of Bean: A Dosage-Dependent Allele Conferring Extreme Resistance, Hypersensitive Resistance, or Spreading Vascular Necrosis in Response to the Potyvirus Bean common mosaic virus. Phytopathology 2000;13:1266-1270.

[5] Krause-Sakate R, Redondo E, Richard-Forget F, Jadao AS, Houvenaghel MC, German-Retana S, Pavan MA, Candresse T, Zerbini FM, Le Gall O. Molecular Mapping of the Viral Determinants of Systemic Wilting Induced by a Lettuce mosaic virus (LMV) Isolate in Some Lettuce Cultivars. Virus Research 2005;109:175-180.

[6] Whitham S, Wang Y. Roles for Host Factors in Plant Viral Pathogenicity. Current Opinion in Plant Biology 2004;7:365-371.

[7] Baker B, Zambryski P, Staskawicz B, Dineshkumar SP. Signaling in Plant-Microbe Interactions. Science 1997;276:726-733.

[8] Durrant WE, Dong X. Systemic Acquired Resistance. Annual Review of Phytopathology 2004;42:185-209.

[9] Seo YS, Gepts P, Gilbertson RL. Genetics of Resistance to the Geminivirus, Bean dwarf mosaic virus, and the Role of the Hypersensitive Response in Common Bean. Theoretical and Applied Genetics 2004;108:786-793.

[10] Palukaitis P, Carr JP. Plant Resistance Responses to Viruses. Journal of Plant Pathology 2008;90 (2):153-171.

[11] Cooper B. Collateral Gene Expression Changes Induced by Distinct Plant Viruses During the Hypersensitive Resistance Reaction in Chenopodium amaranticolor. The Plant Journal 2001;26(3):339-349.

[12] Vance V, Vaucheret H. RNA Silencing in Plants-Defense and Counterdefense. Science 2001;292:2277-2280.

[13] Dielen A-S, Badaoui S, Candresse T, Germa-Retana S. The Ubiquitin/26S Proteasome System in Plant–Pathogen Interactions: A Never-Ending Hide-and-Seek Game. Molecular Plant Pathology 2010;11:293-308.

[14] Ballut L, Drucker M, Pugniere M, Cambon F, Blanc S, Roquet F, Candresse T, Schmid HP, Nicolas P, Gall OL, Badaoui S. HcPro a Multifunctional Protein Encoded by a Plant RNA Virus, Targets the 20S Proteasome and Affects Its Enzymic Activities. Journal of General Virology 2005;86:2595-2603.

[15] Zeng lR, Vega-Sanchez ME, Zhu T, Wang, GL. Ubiquitination-Mediated Protein Degradation and Modification: An Emerging Theme in Plant-Microbe Interactions. Cell Research 2006;16:413-426.

[16] Francki,RIB. Responses of plant cells to virus infection with special references to the sites of RNA. In: Positive strand RNA viruses (M.A. Brinton and R.R. Rueckert, eds.) Plan R. Liss, New York, 1987:423-436.

[17] El-Hoseny ME El-Fallal A A, A.K.El-S;,A.D, A.S.S. Biology Cytopathology and Molecular Identification of an Egyptian Isolate of Zucchini yellow mosaic Potyvirus (ZYMV-EG). Pakistan Journal of Biotechnology 2010;7(1-2):75-80.

[18] Hatta T, Bullivant S,Matthews REF. Fine Structure o f Vesicles Induced in Chloroplasts of Chinese Cabbage Leaves by Infection with Turnip yellow mosaic virus. Journal of General Virology,1973; 20, 37-50.

[19] Afrren B, Gulfishan M, Baghel G, Fatma M, Khan AA, Naqvi QA. Molecular Detection of a Virus Infecting Carrot and its Effect on Some Cytological and Physiological Parameters. African Journal of Plant Science 2011;5(7):407-411.

[20] Oakenfull EA, Riou-Khamlichi C, Murray JA. Plant D-type Cyclins and the Control of G1 Progression. Philosophical Transactions of the Royal Society B 2002;357:749-760.

[21] Chapa-Oliver AM, Guevara-González RG, González-Chavira MM, Ocampo-Velázquez RV, Feregrino-Pérez AA, Mejía-Teniente L, Herrera-Ruíz G, Torres-Pacheco I. Analogies between Geminivirus and Oncovirus: Cell Cycle Regulation. African Journal of Biotechnology 2011;10(55):11327-11332.

[22] Gutierrez C. DNA Replication and Cell Cycle in Plants: Learning from Geminiviruses. The EMBO Journal 2000;19(5):792-799.

[23] Burgyán J, Havelda Z. Viral Suppressors of RNA Silencing. Trends in Plant Science 2011;16(5):265-272.

[24] Anandalakshmi R, Pruss GJ, Ge X, Marathe R, Mallory AC, Smith TH, Vance VB. A Viral Suppressor of Gene Silencing in Plants. Proceedings of the National Academy of Sciences of the United States of America 1998;95:13079-13084.

[25] Fire A, Xu S, Montgomery MK, Kostas AS, Driver SE, Mello CC. Potent and Specific Genetic Interference by Double-stranded RNA in Caenorhabditis elegans. Nature 1998;391:806-811.

[26] Mlotshwa S, Pruss GJ, Peragine A, Endres MW, Li J, Chen X, Scott PR, Bowman LH, Vance V. DICER-LIKE2 Plays a Primary Role in Transitive Silencing of Transgenes in Arabidopsis. PLos ONE 2008;3(3):e1755. doi:10.1371/journal. Pone.0001755.

[27] Hannon G J. RNA interference. Nature 2002;418(11):244-251.

[28] Song L, Gao S, Jiang W, Chen S, Liu Y, Zhou L, Huang W. Silencing Suppressors: Viral Weapons for Countering Host Cell Defenses. Protein Cell 2011;2(4):273-281.

[29] Hohn T, Vazquez F. RNA Silencing Pathways of Plants: Silencing and its Suppression by Plant DNA Viruses. Biochimica et Biophysica Acta 2011;1809:588-600.

[30] Fagard M, Vaucheret H. (Trans) gene Silencing in Plants: How Many Mechanisms? Annual Review of Plant Physiology and Plant Molecular 2000;51:167-194.

[31] Hoffer P, Ivashuta S, Pontes O, Vitins A, Pikaard C, Mroczka A, Wagner N, Voelker T. Posttranscriptional Gene Silencing in Nuclei. PNAS 2011;108(1):409-414.

[32] Lin J, Chun-Hong W, Yi L. Viral Suppression of RNA Silencing. Science China 2012;55(2):109-118.

[33] Paszkowski J, Whitham SA. Gene Silencing and DNA Methylation Processes Current Opinion in Plant Biology 2001;4:123-129.

[34] Avramova, Z. Epigenetic Regulatory Mechanisms in Plants Handbook of Epigenetics: The New Molecular and Medical Genetics. Chapter 16 Elsevier 2011.

[35] Melnyk CW, Molnar A, Baulcombe DC. Intercellular and Systemic Movement of RNA Silencing Signals. The EMBO Journal 2011;30:3553-3563.

[36] Napoli C, Lemieux C, Jorgensen R. Introduction of a Chimeric chalcone synthase Gene into Petunia Results in Reversible Co-suppression of Homologous Genes in trans. Plant Cell 1990;2:279-289.

[37] Burgyán J. Virus Induced RNA Silencing and Suppression: Defence and Counter-Defence. Journal of Plant Pathology 2006;88(3):233-244.

[38] Ratcliff FG, MacFarlane SG, Baulcombe DC. Gene Silencing without DNA: RNA-Mediated Cross-Protection between Viruses. The Plant Cell 1999;11:1207-1215.

[39] Lindbo JA, Silva-Rosales L, Proebsting WM, Dougherty WG. Induction of a Highly Specific Antiviral State in Transgenic Plants: Implications for Regulation of Gene Expression and Virus Resistance. Plant Cell 1993;5:1749-1759.

[40] Li F, Ding S. Virus Counter Defense: Diverse Strategies for Evading the RNA-Silencing Immunity. Annual Review Microbiology 2006;60:503-531.

[41] Hamilton AJ, Baulcombe DC. A Species of Small Antisense RNA in Posttranscriptional Gene Silencing in Plants. Science 1999;286: 950-952.

[42] Sijen T, Vijn I, Rebocho A, van Blokland R, Roelofs D, Joseph NM, Kooter JM. Transcriptional and Posttranscriptional Gene Silencing are Mechanistically Related. Current Biology 2001;11:436-440.

[43] Shimura H, Pantaleo V. Viral Induction and Suppression of RNA Silencing in Plants. Biochimica et Biophysica Acta 2011;1809:601-612.

[44] Pruss G, Ge X, Shi X M, Carrington JC, Vance VB. Plant Viral Synergism: The Potyviral Genome Encodes a Broad-Range Pathogenicity Enhancer that Transactivates Replication of Heterologous Viruses. The Plant Cell 1997;9:859-868.

[45] Brigneti G, Voinnet O, Li W-X, Ji L-H, Ding S-W, Baulcombe DC. Viral Pathogenicity Determinants are Suppressors of Transgene Silencing in Nicotiana benthamiana. The EMBO Journal 1998;17(22):6739-6746.

[46] Voinnet O, Pinto YM, Baulcombe DC. Suppression of Gene Silencing: A general Strategy Used by Diverse DNA and RNA Viruses of Plants. PNAS 1999;96(24): 14147-14152.

[47] Ding S-W, Voinnet O. Antiviral Immunity Directed by Small RNAs. Cell 2007;130:413-426.

[48] Voinnet O. Induction and Suppression of RNA Silencing: Insights from Viral Infections. Nature Reviews 2005;6:206-221.

[49] Anandalakshmi R, Marathe R, Ge X, Herr JM, Jr Mau C, Mallory A, Pruss G, Bowman L, Vance VB. A Calmodulin-Related Protein that Suppresses Post-Transcriptional Gene Silencing in Plants. Science 2000;290:142-144.

[50] Trinks D, Rajeswaran R, Shivaprasad PV, Akbergenov R, Oakeley EJ, Veluthambi K, Hohn T, Pooggin MM. Suppression of RNA Silencing by a Geminivirus Nuclear Protein, AC2, Correlates with Transactivation of Host Genes. Journal of Virology 2005;79(4):2517-252750

[51] Qu F, Morris T J. Suppressors of RNA Silencing Encoded by Plant Viruses and Their Role in Viral Infections. FEBS Letters 2005;579:5958-5964.

[52] Alvarado V, Scholthof HB. Plant Responses against Invasive Nucleic Acids: RNA Silencing and Its Suppression by Plant Viral Pathogens. Seminars in Cell & Developmental Biology 2009;20:1032-1040.

[53] Kasschau KD, Xie Z, Allen E, Llave, C, Chapman EJ, Krizan KA, Carrington JC. P1/HC-Pro, a Viral Suppressor of RNA Silencing, Interferes with Arabidopsis Development and miRNA Function. Developmental Cell 2003;4:205-217.

[54] Siddiqui SA Sarmiento C Truve E,Lehto H, Lehto K. Phenotypes and Functional Effects Caused by Various Viral RNA Silencing Suppressors in Transgenic N. benthamiana and N. tabacum. Molecular Plant-Microbe Interactions 2008; 21:178–187.

[55] Díaz-Pendón JA, Ding S-W. Direct and Indirect Roles of Viral Suppressors of RNA Silencing in Pathogenesis. Annual Review Phytopathology 2008;46:303-26.

[56] Mérai Z, Kerényi Z, Kertész S, Magna M, Lakatos L, Silhavy D. Double-Stranded RNA Binding May be a General Plant RNA Viral Strategy to Suppress RNA Silencing. Journal of Virology 2006;80(12):5747–5756.

[57] Lakatos L, Csorba T, Pantaleo V, Chapman, EJ, Carrington, JC, Liu, Y-P, Dolja, VV, Calvino, LF, López-Moya, JJ, Burgyán, J. Small RNA Binding is a Common Strategy

to Suppress RNA Silencing by Several Viral Suppressors. The EMBO Journal 2006;25(12):2768-2780.

[58] Kasschau KD, Carrington JC. A Counterdefensive Strategy of Plant Viruses:Suppression of Posttranscriptional Gene Silencing. Cell 1998;95:461–470.

[59] Haas G, Azevedo J, Moissiard G, Geldreich A, Himber C, Bureau M, Fukuhara T, Keller M, Voinnet O. Nuclear Import of CaMV P6 is Required for Infection and Suppression of the RNA Silencing Factor DRB4. The EMBO Journal 2008;27(15): 2102-2112.

[60] Glick E, Zrachya A, Levy Y, Mett A, Gidoni D, Belausov E, Citovsky V, Gafni Y. Interaction with Host SGS3 is Required for Suppression of RNA Silencing by Tomato yellow leaf curl virus V2 Protein. PNAS 2008;105(1):157-161.

[61] Shiboleth YM, Haronsky E, Leibman D, Arazi, T, Wassenegger M, Whitham SA, Gaba V, Gal-On A. The Conserved FRNK Box in HC-Pro, a Plant Viral Suppressor of Gene Silencing, Is Required for Small RNA Binding and Mediates Symptom Development. Journal of Virology 2007;81(23):13135-13148.

[62] Yu B, Chapman EJ, Yang Z, Carrington JC, Chen X. Transgenically Expressed Viral RNA Silencing Suppressors Interfere with microRNA Methylation in Arabidopsis. FEBS Letters 2006;580:3117-3120.

[63] González I, Martínez L, Rakitina DV, Lewsey MG, Atencio FA, Llave C, Kalinina NO, Carr J P, Palukaitis P, Canto T. Cucumber mosaic virus 2b Protein Subcellular Targets and Interactions: Their Significance to RNA Silencing Suppressor Activity. Molecular Plant-Microbe Interactions 2010;23(3):294-303.

[64] Zhang X, Yuan Y-R, Pei Y, Lin S-S, Tuschl T, Patel DJ, Chua N-H. Cucumber mosaic virus-encoded 2b Suppressor Inhibits Arabidopsis Argonaute1 Cleavage Activity to Counter Plant Defense. Genes & Development 2006;20:3255-3268.

[65] Hamera S, Song X, Su L, Chen X, Fang R. Cucumber mosaic virus Suppressor 2b Binds to AGO4-Related Small RNAs and Impairs AGO4 Activities. The Plant Journal 2012;69:104-115.

[66] Mi S, Cai T, Hu Y, Chen Y, Hodges E, Ni F, Wu L, Li S, Zhou H, Long C, Chen S, Hannon GJ, Qi Y. Sorting of Small RNAs into Arabidopsis Argonaute Complexes is Directed by the 5′Terminal Nucleotide. Cell 2008;133:116-127.

[67] Pfeffer S, Dunoyer P, Heim F, Richards KE, Jonard G, Ziegler-Graff V. P0 of Beet Western Yellows Virus is a Suppressor of Posttranscriptional Gene Silencing. Journal of Virology 2002;76(13):6815-6824.

[68] Pazhouhandeh M, Dieterle M, Marrocco K, Lechner E, Berry B, Brault V, Hemmer O, Kretsch T, Richards KE, Genschik P, Ziegler-Graff V. F-Box-Like Domain in the Polerovirus Protein P0 is Required for Silencing Suppressor Function. PNAS 2006;103(6): 1994-1999.

[69] Bortolamiol D, Pazhouhandeh M, Marrocco K, Genschik P, Ziegler-Graff V. The Polerovirus F Box Protein P0 Targets ARGONAUTE1 to Suppress RNA Silencing. Current Biology 2007;17:1615-1621.

[70] Himber C, Dunoyer P, Moissiard G, Ritzenthaler C, Voinnet O. Transitivity Dependent and Independent Cell-to-Cell Movement of RNA Silencing. The EMBO Journal 2003;22:4523-4533.

[71] Ding S-W, Li W-X, Symons RH. A Novel Naturally Occurring Hybrid Gene Encoded by a Plant RNA Virus Facilitates Long Distance Virus Movement. The EMBO Journal 1995;14(23):5762-5772.

[72] Kasschau KD, Carrington JC. Long-Distance Movement and Replication Maintenance Functions Correlate with Silencing Suppression Activity of Potyviral HC-Pro. Virology 2001;285:71-81.

[73] Angell SM, Davies C, Baulcombe DC. Cell-to-Cell Movement of Potato virus X is Associated with a Change in the Size Exclusion Limit of Plasmodesmata in Trichome Cells of Nicotiana clevelandii. Virology 1996;215:197-201.

[74] Kalinina NO, Rakitina DV, Solovyev AG, Schiemann J, Morozov SY. RNA Helicase Activity of the Plant Virus Movement Proteins Encoded by the First Gene of the Triple Gene Block. Virology 2002;296:321-329.

[75] Lu R, Folimonov A, Shintaku M, Li W-X, Falk BW, Dawson WO, ' S-W. Three Distinct Suppressors of RNA Silencing Encoded by a 20-kb Viral RNA Genome. PNAS 2004;101(44):15742-15747.

[76] Szittya G, Silhavy D, Molnár A, Havelda Z, Lovas Á, Lakatos L, Bánfalvi Z, Burgyán J. Low Temperature Inhibits RNA Silencing-Mediated Defence by the Control of siRNA Generation. The EMBO Journal 2003;22(3):633-640.

[77] Wezel RV, Liu H, Wu Z, Stanley J, Hong Y. Contribution of the Zinc Finger to Zinc and DNA Binding by a Suppressor of Posttranscriptional Gene Silencing. Journal of Virology 2003;77(1):696-700.

[78] Hartitz M D, Sunter G, Bisaro DM. The tomato golden mosaic virus transactivator (TrAP) is a Single-Stranded DNA and zinc-binding Phosphoprotein with an Acidic Activation Domain. Virology 1999; 263:1-14.

[79] Riechmann JL, Lain S, Garcia, JA. Highlights and Prospects of Potyvirus Molecular Biology. Journal of General Virology 1992;73:1-16.

[80] Maia IG, Haenni A, Bernardi F. Potyviral HC-Pro: A Multifunctional Protein. Journal Genetics Virology 1996;77(7):1335-1341.

[81] Fauquet CM, Mayo MA, Maniloff J, Desselberger V, Ball LA. Virus Taxonomy. Eigth Report of the International Commitee on Taxonomy of Viruses. Academic Press; 2005.p1259.

[82] Carrington JC, Freed DD. Cap-Independent Enhacement of Translation by a Plant Potyvirus 5′ Nontranslated Region. Journal of Virology 1990;64(4):1590-1597.

[83] Carrington JC, Freed DD, Sanders TC. Autocatalytic Processing of the Potyvirus Helper Component Proteinase in Escherichia coli and in vitro. Journal Virology 1989;63:4459-4463.

[84] Cronin S, Verchot J, Haldeman-Cahill R, Schaad MC, Carrington JC. Long-distance Movement Factor: a Transport Function of the potyvirus Helper Component Proteinase. Plant Cell 1995;7:549-559.

[85] Kasschau KD. Carrington JC. Requirement for HC-Pro Processing during Genome Amplification of Tobacco etch potyvirus. Virology 1995;209(1):268-273.

[86] Vance V B, Berger P H, Carrington J C, Hunt A G, Shi X M. 5' Proximal Potyviral Sequences Mediate Potato virus X Potyviral Synergistic Disease in Transgenic Tobacco. Virology 1995;206: 583-590.

[87] Rojas MR, Zerbini FM, Allison RF, Gilbertson RL, Lucas WJ. Capsid Protein and Helper Component-Proteinase Function as Potyvirus Cell-to-Cell Movement Proteins. Virology 1997;237:283-295.

[88] Urcuqui-Inchima S, Maia IG, Arruda P, Haenni AL, Bernardi F. Deletion Mapping of the Potyviral Helper Component-Proteinase Reveals two Regions Involved in RNA Binding. Virology 2000;268:104-111.

[89] Plisson C, Drucker M, Blanc S, German-Retana S, Le Gall O, Thomas D, Bron P. Structural Characterization of HC-Pro, a Plant Virus Multifunctional Protein. The Journal of Biological Chemistry 2003;278:23753-23761.

[90] Li WX, Ding SW. Viral Suppressors of RNA Silencing. Current Opinion in Biotechnology 2001;12:150-154.

[91] Bass B. Double-Stranded RNA as a Template for Gene Silencing. Cell 2000;101:235-238.

[92] Zamore PD, Tuschl T, Sharp PA, Bartel DP. RNAi: Double-stranded RNA Directs the ATP-dependent Cleavage of mRNA at 21 to 23 Nucleotide Intervals. Cell 2000;101:25–33.

[93] Mallory AC, Ely L, Smith TH, Marathe R, Anandalakshmi R, Fagard M, Vaucheret H, Pruss G, Bowman L, Vance VB. HC-Pro Supression of Transgene Silencing Eliminates the Small RNAs but not Transgene Methylation or the Mobile Signal. Plant Cell 2001;13(3):571-583.

[94] Mallory AC, Parks G, Endres MW, Baulcombe D, Bowman LH, Pruss GJ, Vance VB. The Amplicon-Plus System for High-Level Expression of Transgenes in Plants. Nature Biotechnology 2002;20(6):622-625.

[95] Chapman EJ, Prokhnevsky AI, Gopinath K, Dolja VV, Carrington JC. Viral RNA Silencing Suppressors Inhibit the microRNA Pathway at an Intermediate Step. Genes Development 2004;18:1179-1186.

[96] Llave C, Kasschau KD, Carrington JC. Virus Encoded Suppressor of Post-Transcriptional Gene Silencing Targets a Maintenance Step in Silencing Pathway. Proceedings of the National Academy of Science 2000;97:13401-13406.

[97] Mette MF, Matzke AJ, Matzke MA. Resistance of RNA-mediated TGS to HC-Pro, a Viral Suppressor of PTGS, Suggests Alternate Pathways for dsRNA Processing. Current Biology 2001;11:1119-1123.

[98] Bayne EH, Rakitina DV, Morozov SY, Baulcombe DC. Cell-to-Cell Movement of Potato potexvirus X is Dependent on Suppression of RNA Silencing. Plant Journal 2005;44:471-82.

[99] Marathe R, Anandalakshmi R, Smith TH, Pruss GJ, Vance VB. RNA Viruses as Inducers, Suppressors and Targets of Post-Transcriptional Gene Silencing. Plant Molecular Biology 2000;43(2-3):295-306.

[100] Matzke M, Matzke A, Pruss GJ, Vance VB. RNA-Based Silencing Strategies in Plants. Current Opinion in Genetics & Development 2001;11(2):221-227.

[101] Voinnet O, Lederer C, Baulcombe DC. A Viral Movement Protein Prevents Spread of the Gene Silencing Signal in Nicotiana benthamiana. Cell 2000;103:157-167.

[102] Moissiard G, Parizotto EA, Himber C, Voinnet O. Transitivity in Arabidopsis can be Primed, Requires the Redundant Action of the Antiviral Dicer-like 4 and Dicer-like 2, and is Compromised by Viral-Encoded Suppressor Proteins. RNA 2007;13:1268-1278.

[103] Chiu MH, Chen IH, Baulcombe DC, Tsai CH. The Silencing Suppressor P25 of Potato virus X Interacts with Argonaute1 and Mediates Its Degradation through the Proteasome Pathway. Molecular Plant Pathology 2010;11(5):641-649.

[104] Jiang L, Wei CH, Li Y. Viral Suppression of RNA Silencing. Science China Life Sciences 2012;55:109-118.

[105] Martelli GP, Gallitelli D, Russo M. Tombusviruses. In: The Plant Viruses. (ed.) Koenig R 1998.13-72.

[106] Hearne PQ, Knorr DA, Hillman BI, Morris TJ. The Complete Structure and Synthesis of Infectious RNA from Clones of Tomato bushy stunt virus. Virology 1990;177(1): 141-151.

[107] Hayes RJ, Petty ITD, Coutts RHA, Buck KW. Gene Amplification and Expression in Plants by a Replicating Geminivirus Vector. Nature 1988;334:179-182.

[108] Scholthof HB, Scholthof K-BG, Jackson AO. ldentification of Tomato bushy stunt virus Host-Specific Symptom Determinants by Expression of Individual Genes from a Potato virus X Vector. Plant Cell 1995;7(8):1157-1172.

[109] Dalmay T, Rubino L, Burgyán J, Kollar A, Russo M. Functional Analysis of Cymbidium ringspot virus Genome. Virology 1993;194(2):697-704.

[110] Silhavy D, Molnár A, Lucioli A, Szittya G, Hornyik C, Tavazza M, Burgyán J. A Viral Protein Suppresses RNA Silencing and Binds Silencing-Generated, 21- to 25-Nucleotide Double-Stranded RNAs. The EMBO Journal 2002;21, 3070-3080.

[111] Lakatos L, Szittya G, Silhavy D, Burgyán, J. Molecular Mechanism of RNA Silencing Suppression Mediated by p19 Protein of Tombusviruses. The EMBO Journal 2004;23:876-884.

[112] Vargason J, Szittya G, Burgyán J, Hall TM. Size Selective Recognition of siRNA by an RNA Silencing Suppressor. Cell 2003;115(7):799-811.

[113] Ye K, Malinina L, Patel DJ. Recognition of Small Interfering RNA by a Viral Suppressor of RNA Silencing. Nature 2003;426:874-878.

[114] Elbashir S, Lendeckel W, Tuschl T. RNA Interference is Mediated by 21- and 22-Nucleotide RNAs. Genes & Development 2001;15(2):188-200.

[115] Hamilton A, Voinnet O, Chappell L, Baulcombe DC. Two Classes of Short Interfering RNA in RNA Silencing. The EMBO Journal 2002;21(17):4671-4679.

[116] Szittya G, Molnar A, Silhavy D, Hornyik C, Burgyán J. Short Defective Interfering RNAs of Tombusviruses are not Targeted but Trigger Post-Transcriptional Gene Silencing Against Their Helper Virus. The Plant Cell 2002;14(2):359-372.

[117] Havelda Z, Hornyik C, Crescenzi A, Burgyán J. In situ Characterization of Cymbidium Ringspot Tombusvirus Infection-Induced Posttranscriptional Gene Silencing in Nicotiana benthamiana. Journal of Virology 2003;77:6082-6086.

[118] Dunoyer P, Schott G, Himber C, Meyer D, Takeda A, Carrington JC, Voinnet O. Small RNA Duplexes Function as Mobile Silencing Signals between Plant Cells. Science 2010;328:912-916.

[119] Carrington JC, Ambros V. Role of microRNAs in Plant and Animal Development. Science 2003;301(5631):336-338.

[120] Mallory AC, Vaucheret H. Functions of microRNAs and Related Small RNAs in Plants. Nature Genetics 2006; 38:S31-S36.

[121] Ramachandran V, Chen X. Degradation of microRNAs by a Family of Exoribonucleases in Arabidopsis. Science 2008;321:1490-1492.

[122] Nakahara KS, Masuta C, Yamada S, Shimura H, Kashihara Y, Wada TS, Meguro A, Goto K, Tadamura K, Sueda K, Sekiguchi T, Shao J, Itchoda N, Matsumura T, Igarashi M, Ito K, Carthew RW, Uyeda I. Tobacco Calmodulin-Like Protein Provides Secondary Defense by Binding to and Directing Degradation of Virus RNA Silencing Suppressors. PNAS 2012;doi:10.1073/pnas.1201628109.

[123] Sarmiento C, Nigul L, Kazantseva J, Buschmann M, Truve E. AtRLI2 is an Endogenous Suppressor of RNA Silencing. Plant Molecular Biology 2006;61:153-163.

[124] Gazzani S, Lawrenson T, Woodward C, Headon D, Sablowski, R. A Link between mRNA Turnover and RNA Interference in Arabidopsis. Science 2004;306(5698): 1046-1048.

[125] Souret FF, Kastenmayer JP, Green PJ. AtXRN4 Degrades mRNA in Arabidopsis and Its Substrates Include Selected miRNA Targets. Molecular Cell 2004;15:173-183.

[126] Gy I, Gasciolli V, Lauressergues D, Morel JB, Gombert J, Proux F, Proux C, Vaucheret H, Mallory AC. Arabidopsis FIERY1, XRN2, and XRN3 are Endogenous RNA Silencing Suppressors. The Plant Cell 2007;19:3451-3461.

[127] Kastenmayer JP, Green PJ. Novel Features of the XRN-Family in Arabidopsis: Evidence that AtXRN4, One of Several Orthologs of Nuclear Xrn2p/Rat1p, Functions in the Cytoplasm. Proceedings of the National Academy of Sciences USA 2000;97(25): 13985-13990.

[128] Endres MW, Gregory BD, Zhihuan G, Foreman AW, Sizolwenkosi M, Xin G, Pruss GJ, Ecker JR, Bowman LH, Vance VB. Two Plant Viral Suppressors of Silencing Require the Ethylene-Inducible Host Transcription Factor RAV2 to Block RNA Silencing. Plos Pathogens 2010;6(1):1-12.

[129] Taliansky M, Kim SH, Mayo MA, Kalinina NO, Fraser G, McGeachy KD, Barker H. Escape of a Plant Virus from Amplicon-Mediated RNA Silencing is Associated with Biotic or Abiotic Stress. Plant Journal 2004;39(2):194-205.

[130] Glazov E, Phillips K, Budziszewski GJ, Meins FJr, Levin JZ. A Gene Encoding an RNaseD Exonuclease-Like Protein is Required for Post-Transcriptional Silencing in Arabidopsis. Plant Journal 2003;35(3):342-349.

[131] Ketting RF, Haverkamp TH, Luenen HG, Plasterk RH. Mut-7 of C. elegans, Required for Transposon Silencing and RNA Interference, is a Homolog of Werner Syndrome Helicase and RNaseD. Cell 1999;99:133-141.

[132] Becker A, Lange M. VIGS – Genomics goes Functional. Trends Plant Science 2010;15(1):1-4.

[133] Padmanabhan MS, Dinesh-Kumar SP. Virus-Induced Gene Silencing (VIGS). Elsevier Ltd; 2008.375-380.

[134] Purkayastha A, Dasgupta I. Virus-Induced Gene Silencing: A Versatile Tool for Discovery of Gene Functions in Plants. Plant Physiology and Biochemistry 2009;47:967-976.

[135] Ruiz MT, Voinnet O, Baulcombe DC. Initiation and Maintenance of Virus-Induced Gene Silencing. The Plant Cell 1998;10:937-946.

[136] Ratcliff F, Martin-Hernandez AM, Baulcombe DC. Tobacco rattle virus as a Vector for Analysis of Gene Function by Silencing. The Plant Journal 2001;25(2):237-245.

[137] Huang C, Qian Y, Li Z, Zhou X. Virus-Induced Gene Silencing and Its Application in Plant Functional Genomics. Science China Life Sciences 2012;55(2):99-108.

[138] Ding XS, Schneider WL, Chaluvadi SR, Mian MAR, Nelson RS. Characterization of a Brome mosaic virus Strain and Its Use as a Vector for Gene Silencing in Monocotyledonous Hosts. Molecular Plant-Microbe Interactions 2006;19:1229-1239.

[139] Jagtap UB, Gurav RG, Bapat, VA. Role of RNA Interference in Plant Improvement. Naturwissenschaften 2011;98:473-492.

[140] Angell SM, Baulcombe DC. Consistent Gene Silencing in Transgenic Plants Expressing a Replicating Potato virus X RNA. The EMBO Journal 1997;16(12):3675-3684.

[141] Moissiard G, Voinnet O. Viral Suppression of RNA Silencing in Plants. Molecular Plant Pathology 2004;5(1):71-82.

[142] Voinnet O, Rivas S, Mestre P, Baulcombe DC. An Enhanced Transient Expression System in Plants Based on Suppression of Gene Silencing by the P19 Protein of Tomato bushy stunt virus. The Plant Journal 2003;33:949-956.

[143] Iki T, Yoshikawa M, Nishikiori M, Jaudal MC, Matsumoto-Yokoyama E, Mitsuhara I, Meshi T, Ishikawa M. In Vitro Assembly of Plant RNA-Induced Silencing Complexes Facilitated by Molecular Chaperone HSP90. Molecular Cell 2010;39:282-291.

[144] Scholthof HB. Heterologous Expression of Viral RNA Interference Suppressors: RISC Management. Plant Physiology 2007;145(4):1110-1117.

Permissions

All chapters in this book were first published in AMR, by InTech Open; hereby published with permission under the Creative Commons Attribution License or equivalent. Every chapter published in this book has been scrutinized by our experts. Their significance has been extensively debated. The topics covered herein carry significant findings which will fuel the growth of the discipline. They may even be implemented as practical applications or may be referred to as a beginning point for another development.

The contributors of this book come from diverse backgrounds, making this book a truly international effort. This book will bring forth new frontiers with its revolutionizing research information and detailed analysis of the nascent developments around the world.

We would like to thank all the contributing authors for lending their expertise to make the book truly unique. They have played a crucial role in the development of this book. Without their invaluable contributions this book wouldn't have been possible. They have made vital efforts to compile up to date information on the varied aspects of this subject to make this book a valuable addition to the collection of many professionals and students.

This book was conceptualized with the vision of imparting up-to-date information and advanced data in this field. To ensure the same, a matchless editorial board was set up. Every individual on the board went through rigorous rounds of assessment to prove their worth. After which they invested a large part of their time researching and compiling the most relevant data for our readers.

The editorial board has been involved in producing this book since its inception. They have spent rigorous hours researching and exploring the diverse topics which have resulted in the successful publishing of this book. They have passed on their knowledge of decades through this book. To expedite this challenging task, the publisher supported the team at every step. A small team of assistant editors was also appointed to further simplify the editing procedure and attain best results for the readers.

Apart from the editorial board, the designing team has also invested a significant amount of their time in understanding the subject and creating the most relevant covers. They scrutinized every image to scout for the most suitable representation of the subject and create an appropriate cover for the book.

The publishing team has been an ardent support to the editorial, designing and production team. Their endless efforts to recruit the best for this project, has resulted in the accomplishment of this book. They are a veteran in the field of academics and their pool of knowledge is as vast as their experience in printing. Their expertise and guidance has proved useful at every step. Their uncompromising quality standards have made this book an exceptional effort. Their encouragement from time to time has been an inspiration for everyone.

The publisher and the editorial board hope that this book will prove to be a valuable piece of knowledge for researchers, students, practitioners and scholars across the globe.

List of Contributors

Shailendra K. Saxena and Sai V. Chitti
CSIR-Centre for Cellular and Molecular Biology (CCMB), Hyderabad, India

Lokeswara S. Balakrishna and Anand K. Kondapi
Department of Biotechnology and Bioinformatics, School of Life Sciences, University of Hyderabad, P O Central University, Hyderabad, Telangana, India

Michael D. Röling, Mateusz Stoszko and Tokameh Mahmoudi
Department of Biochemistry, Erasmus Medical Centre, Rottrerdam, The Netherlands

Azam Bolhassani
Department of Hepatitis and AIDS, Pasteur Institute of Iran, Tehran, Iran

Miroslava Matuskova and Erika Durinikova
Cancer Research Institute, Slovak Academy of Sciences, Bratislava, Slovakia

Belén de la Hera and Elena Urcelay
Instituto de Investigación Sanitaria del Hospital Clínico San Carlos, IdISSC, Madrid, Spain

Hongxuan He
National Research Center for Wildlife Borne Diseases, Institute of Zoology, Chinese Academy of Sciences, Beijing, PR China

Alessandra Tenório Costa, Juliana Pereira Bravo, Rodrigo Kazuo Makiyama, Alessandra Vasconcellos Nunes and Ivan G. Maia
São Paulo State University "Júlio Mesquita Filho" Department of Genetics, Institute of Biosciences - UNESP, Botucatu Campus SP, Brazil

Index

www.ingramcontent.com/pod-product-compliance
Lightning Source LLC
LaVergne TN
LVHW080613210926
848485LV00003B/152
* 9 7 8 1 6 3 9 8 7 3 7 6 0 *